BATS
IN CAPTIVITY

Volume 4: Legislation and Public Education

BATS IN CAPTIVITY

Volume 4: Legislation and Public Education

edited by

Susan M. Barnard

Photo courtesy of Elaine Acker, Bat Conservation International

LOGOS
PRESS

BATS IN CAPTIVITY

Volume 4: Legislation and Public Education

Edited by Susan M. Barnard

Published in The United States of America
by
Logos Press, Washington, DC
WWW.LOGOS-PRESS.COM
INFO@LOGOS-PRESS.COM

First edition

Front cover photo: Rob Mies, Director of the Organization for Bat Conservation, showing a big brown bat (*Eptesicus fuscus*) to his daughter, Madison; photo courtesy of Stephen J. Getter, Wilderness Images.
Back cover photo: Greg Turner, Biologist with the Pennsylvania Game Commission (PGC), inspecting little brown myotis (*Myotis lucifugus*) for white-nose syndrome in Shindle Iron Mine, PA; photo courtesy of Joe Kosack, Wildlife Conservation Education Specialist, PGC.

10 9 8 7 6 5 4 3 2 1

ISBN-13
Hardcover: 978-1-934899-08-3
Softcover: 978-1-934899-09-0

Library of Congress Cataloging-in-Publication Data

Bats in captivity / edited by Susan M. Barnard.
v. cm.
Includes bibliographical references and index.
Contents: v. 1. Biological and medical aspects.
ISBN 978-1-934899-03-8 (softcover) -- ISBN 978-1-934899-02-1 (hardcover)
1. Captive bats. 2. Bats--Housing. I. Barnard, Susan M.
SF408.6.B38B385 2009
636.94--dc22
2009017988

In memory of

Arthur M. Greenhall

for his friendship.

Contents

Contributors

Suneth Attygalle
1020 Post Street, #201
San Francisco, CA 94109
U.S.A.

Ashley Averill
Memphis Zoo
2000 Prentiss Place
Memphis, TN 38112
U.S.A.

Nilima K. Badwaik
Department of Obstetrics & Gynecology
SUNY Downstate Medical Center
450 Clarkson Avenue, Box #24
Brooklyn, NY 11203
U.S.A.

Susan M. Barnard
106 Spooner Road
Hawthorne, FL 32640
U.S.A

Julie Batty
Memphis Zoo
2000 Prentiss Place
Memphis, TN 38112
U.S.A.

Fiona Colquhoun
Territory Wildlife Park
P.O. Box 771
Palmerston, NT 0831
Australia

Janet Cooper
Memphis Zoo
2000 Prentiss Place
Memphis, TN 38112
U.S.A.

Doug Dalton
Memphis Zoo
2000 Prentiss Place
Memphis, TN 38112
U.S.A.

Dianne Devison
2655 Forestream Trail
Pickering, ON L1X 2R4
Canada

Laura Seckbach Finn
Fly By Night, Inc.
P.O. Box 562
Osteen, FL 32764
U.S.A.

Thomas Gerald Finn
Fly By Night, Inc.
P.O. Box 562
Osteen, FL 32764
U.S.A.

Ron Haering
Office of Environment and Heritage
P.O. Box 1967
Hurstville, NSW 2220
Australia

Lisa Hailey
Memphis Zoo
2000 Prentiss Place
Memphis, TN 38112
U.S.A.

Stephen Jackson
Department of Primary Industries NSW
Locked Bag 21
Orange, NSW 2800
Australia

Mark Kiser
1740 Augustine Place
Tallahassee, FL 32301
U.S.A.

Selena Kiser
1740 Augustine Place
Tallahassee, FL 32301
U.S.A.

Kimberly Lengel
Philadelphia Zoo
3400 West Girard Avenue
Philadelphia, PA 19104
U.S.A.

Norma Lewis
3418 Tommy Lee Cook
Newnan, GA 30263
U.S.A.

Daniel Lunney
Office of Environment and Heritage
P.O. Box 1967
Hurstville, NSW 2220
Australia

Jennefer Maclean
Tolga Bat Rescue and Research, Inc.
P.O. Box 685
Atherton, QLD 4883
Australia

Mary Manson
Memphis Zoo
2000 Prentiss Place
Memphis, TN 38112
U.S.A.

Alison Matthews
Institute for Land, Water and Society
Charles Sturt University
P.O. Box 789
Albury, NSW 2640
Australia

A. J. Mitchell-Jones
29 Newark Avenue
Peterborough, Cambridgeshire PE1 4NH
United Kingdom

Patricia A. Morton
925 East Chicago Street
Whitewater, WI 53190
U.S.A.

Stephanie R. Ostrowski
Office of Public Health Preparedness and Response
Centers for Disease Control and Prevention
1600 Clifton Road
Atlanta, GA 30333
U.S.A.

Gregg Pittelkow
3230 McKinley Street NE
Minneapolis, MN 55418
U.S.A.

John J. Rasweiler, IV
Department of Obstetrics & Gynecology
SUNY Downstate Medical Center
450 Clarkson Avenue, Box #24
Brooklyn, NY 11203
U.S.A.

Chantal Routhier
Biodôme de Montréal
4777, avenue Pierre-De Coubertin
Montréal, QC H1V 1B3
Canada

Emily Schultz
Memphis Zoo
2000 Prentiss Place
Memphis, TN 38112
U.S.A.

Kathy Schwellenbach
7250 Guider Drive, #210
Woodbury, MN 55125
U.S.A.

John Seyjagat
National Aquarium in Baltimore
Pier#3, 501 East Pratt Street
Baltimore, MD 21202
U.S.A.

Sandi Shoemaker
Memphis Zoo
2000 Prentiss Place
Memphis, TN 38112
U.S.A.

Steve Templeton
Territory Wildlife Park
P.O. Box 771
Palmerston, NT 0831
Australia

Elizabeth Thunstrom
1388 Cambridge Drive
Coquitlam, BC V3J 2P7
Canada

Denise R. Tomlinson
1701 NW Eucalyptus Avenue
Arcadia, FL 34266
U.S.A.

Daniel van Dyk
Office of Environment and Heritage
P.O. Box 1967
Hurstville, NSW 2220
Australia

Steve Wing
Louisville Zoo
1100 Trevilian Way
Louisville, KY 40213
U.S.A.

Preface

Although it is unknown how long people have held bats in captivity, in his 1767 *Natural History of Selborne*, Gilbert White reported on a tame bat that took flies from the hand. Since then, both the popular and scientific literature have abounded with accounts of bats in captivity, increasing the pressure to improve all aspects of captive maintenance. Books written on the management of captive animals are generally written from the points of view and experiences of the authors. Because bats from around the world are being maintained in facilities outside their country of origin, I thought it would be in the best interest of these animals to produce a work that was cosmopolitan in ideas, thereby giving the reader choices in developing captive-management protocols. To this end, I solicited papers from knowledgeable bat workers around the world. The response was overwhelming; I received over 100 contributed papers. Therefore, I decided to divide the material into four volumes. This fourth volume contains useful information on the laws as they apply to maintaining and shipping bats and the successful exhibition of these animals. Authors were allowed broad latitude on material and style of presentation. Each contribution was later re-formatted for consistency and subjected to peer review.

There is some controversy surrounding the higher-level classification within the Chiroptera, including the traditional Megachiroptera and Microchiroptera suborders. Several recent molecular studies have suggested that the suborder Microchiroptera may not be monophyletic, and that six genera traditionally classified in the Microchiroptera (*Rhinolophus*, *Hipposideros*, *Megaderma*, *Macroderma*, *Craseonycteris*, and *Rhinopoma*) may actually be more closely related to the Old World fruit bats. Based on these molecular results, Teeling *et al.* (2005) proposed to reclassify bats into the suborders Yinpterochiroptera (the Pteropodidae plus the above six genera) and Yangochiroptera (the rest of the microchiropterans). However, these higher-level relationships have yet to be completely resolved, and therefore I have decided to follow the traditional classification system and use the two suborders, Megachiroptera and Microchiroptera, throughout this series.

Many photographs were taken when digital imaging was new, and some reproductions in this volume lack the quality of digital photos taken today. Unfortunately, the older images can not be replaced. I am of the opinion that the value of a poor image is better than no image at all. In such cases, where reproductions are of poor quality, I apologize.

This monumental task could not have been achieved without the assistance of many dedicated individuals, including the biologists who reviewed the many manuscripts appearing in this volume. I thank the photo and manuscript contributors for sharing their talents and expertise, exercising great patience, and being generous with their valuable time. I thank C. Dietrich Schaaf for his incredible endurance in editing the many manuscripts appearing in the four volumes of *Bats in Captivity*.

—Susan M. Barnard
Hawthorne, Florida

REFERENCES

Teeling, E. C., Springer, M. S., Madsen, O., Bates, P., O'Brien, S.J., and Murphy, W.J. 2005. A molecular phylogeny for bats illuminates biogeography and the fossil record. *Science*, 307:580–584.

White, G. 1767. *Natural History of Selborne*. Penguin Books, Ltd., London, England. 320 pp.

Section I

Bats and the Law

Bats and the Law

Susan M. Barnard

As the human population of the earth has increased, enormous numbers of plant and animal species are becoming endangered, and many of these are being driven to extinction. Today, more people are becoming aware of these losses to our natural heritage. Laws governing the conservation and utilization of the world's remaining wildlife populations and their habitats can help to protect them, providing that these laws are supported through public education and reduction in poverty.

Wildlife laws have long been in existence, but the primary focus was the preservation of animals for their utilitarian value to humans. For example, game animals were protected for the purpose of hunting and fishing. Wildlife law today has broadened to include the ecological, aesthetic, and moral benefits of protecting wildlife, and as a consequence, protective measures now encompass biodiversity (e.g, Conservation International [CI]; http://www.conservation.org/), wildlife habitat (e.g., The Nature Conservancy; http://www.nature.org/), and the welfare of animals (e.g., International Fund for Animal Welfare [IFAW]; http://www.ifaw.org/splash.php). However, individual nations can not realistically enforce their wildlife laws unless global wildlife markets are tightly controlled. This need has led to an increase in international agreements that include, but are not limited to, the Convention on International Trade in Endangered Species of Wild Flora and Fauna (CITES), Convention on the Conservation of Migratory Species of Wild Animals (CMS or Bonn Convention), World Heritage Convention, and Convention on Biological Diversity (CBD). In addition to the protection that each nation can provide for its native species through wildlife conservation efforts and monitoring of international trade, the World Trade Organization (WTO) Sanitary and Phytosanitary (SPS) Agreement oversees the protection of human, animal and plant health with bans on importing harmful exotic or invasive organisms.

Wildlife law is complicated and as diverse as the animals that range over the planet. Included in this section is a brief overview on wildlife law, as it applies to bats in Australia, Canada, Europe, and the United States of America. Also included are rules and regulations governing the transport of bats.

Chapter 1

Legislation

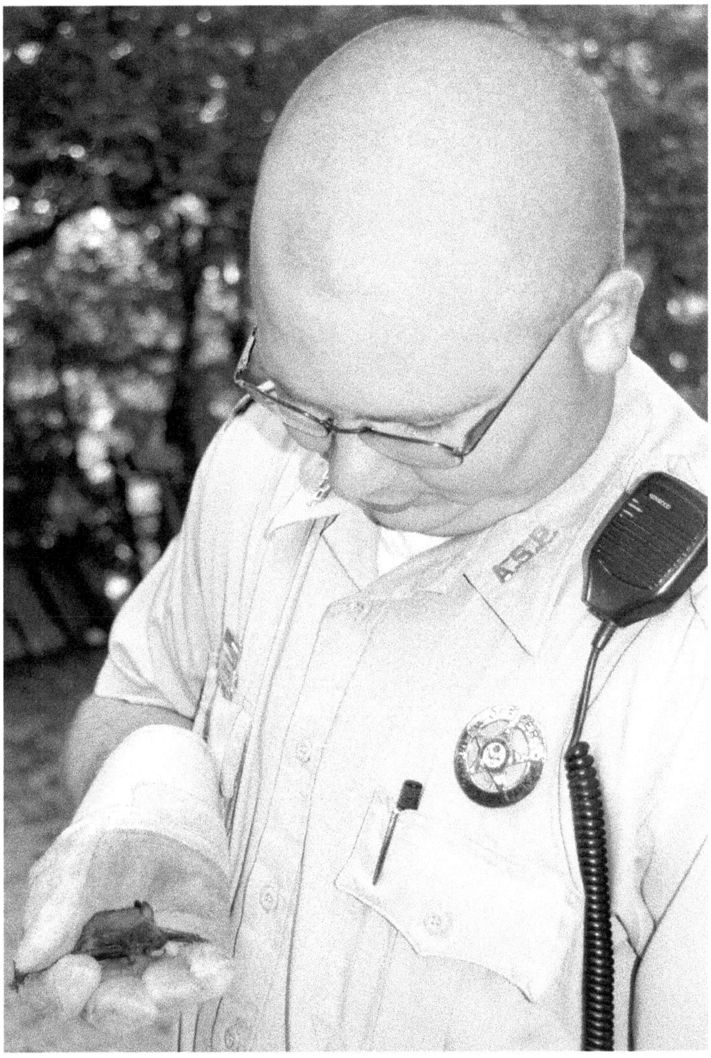

Photo courtesy of Devil's Den State Park

Australia

Daniel Lunney, Stephen Jackson, Alison Matthews, Ron Haering, and Daniel van Dyk

INTRODUCTION

The aim of this review is to describe the legal requirements for possessing, using, and conserving bats in Australia. The approach taken is to illustrate the legal status of captive bats by using one state, New South Wales (NSW), as an example. This is expanded to show the connection between NSW and the commonwealth government's legal role in bat protection and bat conservation. The paper deals primarily with the following pieces of legislation and how these are implemented: The commonwealth government's Environment Protection and Biodiversity Conservation Act 1999 (EPBC Act); and the NSW legislation: National Parks and Wildlife Act 1974 (NPW Act); Threatened Species Conservation Act 1995 (TSC Act); the Animal Research Act 1985, and the Exhibited Animals Protection Act 1986.

The use of Australian bats for research or education, their rehabilitation by wildlife rehabilitators, and their conservation and management are governed by different state and federal government laws. For foreign readers, there is a number of surprising features about Australian fauna legislation. Most laws are state laws and are administered by each state, and not by the commonwealth government. However, the Commonwealth, under the EPBC Act, protects and manages nationally and internationally recognized flora, fauna, ecological communities and heritage places, and influences decisions made by the states with respect to those species that are listed under this Act. For example, an action affecting significant assets may be a "controlled action" requiring EPBC Act approval. A recent striking example relates to the proposed removal of grey-headed flying foxes (*Pteropus poliocephalus*) from the Royal Botanic Gardens in the Sydney Central Business District (CBD). This species is listed as vulnerable under the EPBC Act. If it were not listed, then the Commonwealth would have no interest and the matter would be entirely handled by NSW under its legislation.

The role of the Commonwealth has been clearly outlined by Jackson (2011), where he lists the 11 bat species covered by the EPBC Act, as well as presenting examples where the Commonwealth has acted in the interests of bats. He cites the example of Booth vs. Bosworth, a legal case brought before the Federal Court of Australia, involving a lychee grower in Queensland and the spectacled flying fox (*Pteropus conspicillatus*). As Jackson pointed out, although the spectacled flying fox was not listed as a threatened species under the EPBC Act at the time of the subject activity, the Court found that the bat contributed to maintenance of the heritage values of the adjacent Wet Tropics World Heritage Area, an area protected under the provisions of the EPBC Act. Among the bats listed is the Christmas Island pipistrelle (*Pipistrellus murrayi*), which is now regarded as extinct, and that loss has shown that recovery actions need to be implemented in a timely manner, and that listing alone will not conserve bats (Lunney *et al.*, 2011a). The United Nations Year of the Bat 2011 (http://www.unep.org/Documents.Multilingual/Default.asp?DocumentID=647&ArticleID=6757&l=en) has put this matter of the need to conserve bats under an international spotlight. The decisions as to how the land, and what is on the land (including wildlife), is used and managed are primarily controlled by each state government's legislation. This division reflects the nature of the federal system of government in Australia, which has been in existence since Australia became a nation in 1901. The state laws are sufficiently similar for us to discuss bats and the law using NSW as our case study. Most bats (i.e., all Microchiroptera) have always enjoyed the protection afforded to most native mammals, along with such icon species as koalas (*Phascolarctos cinereus*) and kangaroos (*Macropus* spp.). However, a few species, namely all the large flying foxes, genus *Pteropus,* suborder Megachiroptera, have been unprotected until recently (1986 in NSW), and now some are officially recognized as threatened. The stand out example is the grey-headed flying fox, which is listed as vulnerable under both NSW legislation (TSC Act) and the EBPC Act. The management of this species has remained vexatious after its listing as threatened because it is still permitted to be shot in NSW, under license, when it damages orchards (Eby and Lunney, 2002).

The complexity of state and federal legislation can cause confusion when trying to make constructive efforts towards bat conservation. There may be a need to hold and study bats, or rehabilitate sick or injured bats, or exhibit bats to advance bat education. The laws in each circumstance are state laws and knowledge of them is fundamental to any understanding of what may be done with the native fauna.

The most recent synthesis of the world's diversity of bats (Wilson and Reeder, 2005) recognizes within the suborder Megachiroptera 186 species in 42 genera and one family, the Pteropodidae, and within the suborder Microchiroptera 17 families, 160 genera and 930 species. In Macdonald's (2009) encyclopedia of mammals, bats are noted as forming about one quarter of all living mammal species. For Australia, the proportions are the same, 1 in 4 of the native mammals, comprising 12 species of Megachiroptera (also called megabats) and 71 species of Microchiroptera (also called microbats) (compiled from Van Dyck and Strahan [2008], which lists 349 native mammal species). Another good guide to the diversity of Australian bats can be found in Churchill (2008). With nearly 83 described taxa, and a number of other species identified but not yet described, Australia has over 7% of the world's bat fauna and the State of New South Wales, with 37 species, has over 3% of the world's bats. This number is growing, with Parnaby (2009) distinguishing 4 species of long-eared bats (*Nyctophilus* spp.) within the taxon formally known as *N. timoriensis.* This increases the number of recognized Australian bats as being 86 species. Species lists for Australia are hard to compile, but state lists are even harder, largely because of the uncertainty of the boundaries of the distribution of many species. Of the 132 mammal species in NSW, 37 are bats. One extinct bat was confined to Lord Howe Island, off the coast of NSW, and 19 other species are recognized as threatened under the TSC Act. Thus 20 of 37 bat species (54%) are listed as threatened in NSW.

THE LEGAL STATUS OF BATS IN NEW SOUTH WALES

Bats are protected under the NSW National Parks and Wildlife Act 1974. Protected fauna are those not listed in Schedule 11 of this Act. Until 1986, this schedule included all three large species of flying fox; grey-headed flying fox, black flying fox (*Pteropus alecto*), and little red flying fox (*P. scapulatus*). In 1992, the black flying fox was included on the schedule of threatened species as vulnerable, but it was removed from the schedule in 2008. In 2001, the grey-headed flying fox was also listed as vulnerable. This is a dramatic legal change for a species that had been shot in the millions as an agricultural pest and whose roosts, locally known as camps, had been burned in efforts to force them to move elsewhere (Lunney and Moon, 1997).

Once a species is listed as protected, it automatically enjoys legal protection in the state in which it lives. To catch a bat for scientific, conservation or education purposes, a person needs a license from the Office of Environment and Heritage (OEH), the government organization that administers the NPW Act. The application for license renewal requires a record from the researcher of how many individuals of each species are to be caught and their fate (e.g., identified and released, or taken as a voucher specimen). If a researcher wants to import or export protected fauna from NSW an import/export license is required. Bats are exported from NSW by researchers working on Australia-wide distributions of species, mainly for taxonomic purposes and museum collections. Some bats are exported to overseas museums or other scientific institutions, and an export license is then also required from the commonwealth government. This export license is readily obtained if the overseas destination is a registered research institution. There are about 150 such registered locations in the United States, for example, which are mainly museums.

The biggest issue for bats and the law in NSW has been their status as threatened species, initially under the NSW Endangered Fauna (Interim Protection) Act 1991, and now under its replacement, the TSC Act. An interesting word used within this legislation is "harm." The TSC Act defines harm in relation to fauna as "hunt, shoot, poison, net, snare, spear, pursue, capture, trap, injure, or kill," but does not include harm by changing the habitat of an animal. However, the earlier legislation had been interpreted, in a significant case before the Land and Environment Court in September 1991, to include harm to an endangered animal's habitat.

There are 20 bat species listed on the TSC Act, and there are actions listed to assist their recovery. The link between the actions and the TSC Act is that the TSC Act seeks recovery of the species, with the ultimate aim of removing them from the schedules on the TSC Act because the species has recovered. Lunney *et al.* (2011b) examined the conservation of NSW bat species within the framework of the procedure for threatened species recovery, Priorities Action Statements (PAS), introduced in 2005 by the NSW government to replace the previous requirement to prepare recovery plans.

One of the merits of the PAS initiative is that it has produced a single statement of the parlous position of bats in NSW, the actions needed to reverse the threats, and a statement of where the priorities lie for action. The PAS process, and the 361 individual recovery actions that comprise the PAS for bats, are described and discussed within the context of current knowledge, so that bat biologists, managers, and policy makers can see the scale and emphases of the recovery process for bats (Lunney *et al.*, 2011b). Research actions comprise 163 (45%) of the 361 actions, and the 89 habitat management actions comprise 25% of all bat recovery actions. The primary message from the PAS process is that it has been possible to arrive at a set of actions to provide a basis for management, both at a funding level and a field level, for which research is an integral part. The PAS for bats represents the first official statement of recovery actions required for all bat species listed as threatened. It provides an indication of the scale and magnitude of threats confronting the bats in NSW. However, there is no legal obligation to act on the PAS statements; they are a guide only.

LAWS CONCERNING THE REHABILITATION OF BATS

Rehabilitation of native fauna is a major matter of community interest and thus of legal interest in NSW. Some species, such as koalas, have been so high profile that a specific policy was required (Lunney and Matthews, 1997). That policy is now being revised and updated. Rehabilitating bats has also become a major issue. In the 2008-2009 financial year (July to June), which is the formal period of record keeping, there were 1,930 flying foxes taken into care and 964 released back into the wild. There were also 629 microchiropterans taken in and 343 released. The figures for all fauna for that year were 70,852 taken into care, and 32,344 released. These figures come from the returns submitted by fauna rehabilitation groups. The magnitude of these figures shows that bats, particularly the grey-headed flying fox and its orphaned young, constitute a major group of fauna that is taken into care and rehabilitated. Most of the difference between those taken into care, and those released, is that these animals either died or were euthanized because they were deemed incapable of surviving in the wild. Bat rehabilitation is covered by OEH policy on the rehabilitation of protected fauna.

Under the NPW Act, OEH has a legislative responsibility for the protection and care of protected fauna (Section 92). The OEH may enter into arrangements for the carrying out of such works as the Director-General considers necessary for the protection and care of fauna (NPW Act s8(7)(b)). OEH therefore has a role in the licensing and oversight of organizations and individuals involved in the rehabilitation of injured, sick, and orphaned protected fauna and the release, or retention in captivity, of those animals. The Chief Executive Officer (CEO) of OEH (or delegate) may issue a general license under the NPW Act s.120 authorizing a person to obtain and hold any protected fauna for any specified purpose in NSW. The CEO of OEH may issue a license authorizing a person to liberate an animal anywhere within NSW (NPW Act s.127). Section 101(5)(c) of the NPW Act and selected provisions in the National Parks and Wildlife Regulation 2011 relate specifically to fauna rehabilitation. Sections 109 and 127 of the Act provide for the liberation of captive or captured animals.

OEH recognizes the important conservation and animal welfare role performed by wildlife rehabilitation groups. It is important that OEH and the rehabilitation groups maintain a partnership to achieve the optimum allocation of resources that may be available for the rehabilitation of protected fauna and to ensure compliance with the regulations is achieved.

REHABILITATION OF PROTECTED FAUNA POLICY

The Rehabilitation of Protected Fauna Policy (DECCW 2010) applies to the rescue, rehabilitation, and release of all injured, sick, and orphaned protected fauna in New

South Wales. Its objectives include ensuring that fauna rehabilitators are appropriately licensed and accountable for their activities and provide an appropriate level of animal care by operating in accordance with approved Codes of Practice and Guidelines; that they conduct these activities in an efficient and effective manner, and contribute to the maintenance of biodiversity through the successful return of temporarily disadvantaged animals to their natural habitat, and that they manage the risks associated with such actions.

The Code of Practice for Injured, Sick and Orphaned Protected Fauna (the Code) (OEH 2011a) sets standards for the care and housing of protected fauna that is incapable of fending for itself in its natural habitat. Compliance with the standards in this Code is a condition of fauna rehabilitation and release licenses (Section 120 of the NPW Act) issued by OEH. The Code does not apply to non-native fauna as their release into the wild has a detrimental impact on native fauna and is an offense under the NPW Act. The Code contains both standards and guidelines. Compliance with this Code does not remove the need to abide by the requirements of the Prevention of Cruelty to Animals Act 1979 and any other laws and regulations, such as the Local Government Act 1993. The Code was prepared by OEH in consultation with the NSW Wildlife Council, Taronga Conservation Society, and the Royal Society for the Prevention of Cruelty to Animals NSW (RSPCA). It is also endorsed by the NSW Animal Welfare Advisory Council.

The level of detail of the Code becomes apparent from the following examples of objectives in the Code:

> *Euthanasia*: To end an animal's life in situations where death is imminent; or recovery is impossible; or the likelihood of successful reintegration into the wild population is remote; or the animal poses an unacceptable health risk to wild animals.
>
> *Timing of release:* To ensure that fauna is released as soon as it is ready and at a time that minimizes stress and maximizes its chances of survival in its natural habitat.
>
> *Release site selection:* To ensure that the wild population and natural environment are not negatively impacted by the release. The welfare of the rehabilitated animal after release is a secondary objective.
>
> *Record keeping*: To maintain a database of fauna that has entered rehabilitation. Records of fauna admissions represent a vital resource for fauna rehabilitation groups, OEH and research institutions. They can be used to develop better treatments, educate rehabilitators, identify state-wide trends in fauna incidents, and identify threatening processes.

FLYING FOXES

A draft Code of Practice for Injured, Sick and Orphaned Flying Foxes has been prepared by OEH (2011b) and is awaiting ratification by the Australian Welfare Advisory Council (AWAC), which has indicated that it supports the implementation of the code. Thus, shortly there will be a separate code for flying foxes. The Code is intended for everyone authorized to rehabilitate and release flying foxes by the OEH, and has been developed to protect the welfare of animals being rehabilitated, and to help conserve the wild flying fox population. The Code contains both standards and guidelines for the care of flying foxes and is designed to be read in conjunction with the Code of Practice for Injured, Sick and Orphaned Protected Fauna.

Grey-headed flying foxes are listed as Vulnerable under the NSW TSC Act and the Commonwealth EPBC Act. Black flying foxes and little red flying foxes are not threatened, but are a protected species under the NSW NPW Act.

This Code was prepared by OEH in consultation with Tweed Valley Wildlife Carers, Northern Rivers Wildlife Carers, For Australian Wildlife Needing Aid (FAWNA), Wildlife ARC (Animal, Rescue, and Care Society), Native Animal Trust Fund, Sydney Metropolitan Wildlife Services, Wildlife Information Rescue and Education Service (WIRES), Ku-ring-gai Bat Conservation Society, Cabramatta Creek Flying-fox Committee, Native Animal Network Australia, Wildlife Rescue South Coast, individual flying fox rehabilitators and flying fox researchers. The Code is neither a complete manual on animal husbandry, nor a static document. It will be revised as necessary to take into account new knowledge of animal physiology and behavior; technological advances; developments in standards of animal welfare, and changing community attitudes and expectations about the humane treatment of animals. OEH will consult with license holders regarding potential changes to the Code, and give written notice when the Code is superseded.

PRIVATE FAUNA PARKS

The rehabilitation activities of licensed, private fauna parks are monitored and audited by OEH. OEH may carry out spot-checks or require parks to maintain and submit records of rehabilitation activities. Importantly, animals that are being rehabilitated are not permitted to be displayed within zoos and fauna parks to ensure there is no confusion with the animals kept on display, or for display. Animals on or for display are regulated by the Department of Primary Industries (DPI) NSW, which administers the Exhibited Animals Protection Act 1986.

A fauna park that can not demonstrate the continued availability of appropriate facilities and a commitment to legal and ethically responsible rehabilitation will not be permitted to hold protected fauna for rehabilitation. Operators of such parks may accept injured or displaced animals under the terms of Section 101 of the NPW Act, but must report their possession to OEH and then promptly pass them on to a licensed rehabilitator.

NUISANCE BATS

It is conceivable that some bats may be taken into captivity to remove them from an inappropriate place, such as a roof cavity. In NSW, pest control companies are licensed by OEH under the NPW Act to remove urban nuisance animals, particularly the common brushtail possum (*Trichosurus vulpecula*). Pest controllers are issued with a Section 120 (General) license; however, to date, no pest control company in NSW has been issued with a license to remove bats.

Individuals could also be licensed by OEH to remove problem bats. A Section 121 (Occupier's) license may be issued to an occupier or owner to capture non-threatened bats presenting a nuisance in homes and buildings. These licenses are issued from local offices, and because there is not a central register, the number of licenses that have been issued to remove bats, if any, is unknown.

BATS, RESEARCH ETHICS, AND THE LAW

In NSW, the Animal Research Act 1985 was promulgated to implement best practice in animal research in NSW. It came into force in 1991 following the completion of the Regulations in 1990 under that Act. Within this legislation, *animals* include all vertebrates, and thus include all bats. All researchers must obtain an Authority to conduct research from an accredited Animal Ethics Committee (AEC). Each committee must have a minimum of four members, one from each of four categories: a researcher, a veterinarian, a member of an animal welfare organization, and an independent (i.e., a person who is independent of having conducted animal research). Accreditation for each committee comes from the Animal Research Review Panel set up under the Act. Each research institution is to have its own AEC. Individual researchers who do not belong to institutions need to obtain their Authority from an AEC set up by NSW Agriculture, the state government department that administers the Act.

OEH has its own AEC. The principal author (DL) was a researcher on this committee for 15 years and has recorded this experience (Lunney, 2011). It was instructive to watch the progress of the implementation of this Act and interpretation of what constitutes research. Bats are regularly on application forms to the AEC because of their importance under the NSW TSC Act. This Act has ushered in more work under the NSW Animal Research Act 1985 because all developers need to comply with this legislation, and the Environmental Planning and Assessment Act 1979, in relation to threatened species.

The OEH AEC is presented with many research protocols; that is, detailed applications in relation to bats because of the need for widespread survey of forests, mines and tunnels, and accompanying ecological research. This has thrown up a number of issues that need resolution. The most pressing matter is knowing how to judge a protocol. 1) Is what has been presented in the protocol actually best practice? 2) Do other researchers agree? 3) If they do not agree, is there an acceptable range of actions associated with best practice? The welfare debate is a continuing issue. In NSW its sharp point is the nature of laws and their interpretation (Lunney, 1999, 2011; Lunney and Dickman, 2002). However, the intent of the Act is to pursue best practice, not close down research.

The Act relies on the nationally agreed and negotiated policy document, the *Care and Handling of Animals for Research and Teaching Purposes* (Australian Government 2004). Section 5 deals with Wildlife Studies. The principal author (DL) was a co-author of this section, but it covers all vertebrates, not just bats, so it lacks the specificity one requires to adequately care for an individual species of bat. Nevertheless, it does deal with typical wildlife research subjects such as catching, marking, release, temporary holding facilities, long-term holding facilities, and feeding and care of sick and injured animals. The remainder of the document on care and handling deals with permanent holding, as this matter was of uppermost importance when the Act was introduced and the early drafting of the National Health and Medical Research Council (NHMRC) care policy was underway. In NSW it is essential that bat researchers who submit a protocol to an AEC bear in mind the NHMRC policy in the preparation of their protocol. However, because almost all bat research in NSW is field survey, especially in assessing the fauna in new lands for possible reservation, many of the more difficult issues do not arise.

A major issue for bat researchers is marking study animals with bands. The commonwealth government has set up an Australian Bird and Bat Banding Scheme (ABBBS) to help researchers by supplying numbered bands for banding bats. The ABBBS also oversees the recruitment and training of bat banders. This is done through issue of Authorities that are recognized by state and territory nature conservation agencies. However, this scheme is by convention rather than law. The ABBBS Banding Authorities authorise the holder to use ABBBS bands, and may be used to confirm that the holder has a specified level of skill in capture, handling and banding bats. At present the ABBBS restricts new banding projects that involve certain species of microbats because the bands have been shown to cause injury to the bats.

Holding fauna is rarely a commonwealth matter, thus captive bats are of little interest to the Commonwealth. However, the matter of international import and export permits is exclusively a commonwealth issue. Any researcher, such as a bat taxonomist working in an overseas institution, needs an export permit from the Commonwealth. However, such bats are rarely live, so the question of holding and transporting captive bats does not arise. Commonwealth legislation is pertinent for bats that are proposed to be exported to overseas zoos.

EDUCATION

"Gollum" was the less-than-flattering but appropriate name given by wildlife rehabilitator, Helen George, to a grey-headed flying fox that became the first star of the Ku-ring-gai Bat Conservation Society's (KBCS) successful education program. The rise in the quality and reach of the KBCS education program was only matched by the rise in the legal status of grey-headed flying foxes in NSW. Gollum's first day on public display was in 1985 at the St. Ives show in suburban Sydney, and the grey-headed flying fox was removed from the schedule of unprotected species under the NPW Act in 1986. The education program included sophisticated teaching programs (e.g., Ford, 2002), yet the stars remain these magnificent bats themselves. They were held under a general license under s.120 of the NPW Act.

The emergence of the Australian Bat Lyssavirus (ABL) as an issue in 1996 has caused people to move away from education programs. As bat specialists, some authors of this chapter have been vaccinated with the rabies vaccine, but the death of a person from a bat bite, and its ramifications for grey-headed flying foxes (e.g., Field, 2002), was a blow to public education on the value of bats. The outbreak of the Hendra virus in mid 2011 has also increased public fear of flying foxes.

REGULATION OF THE EXHIBITION OF BATS IN NEW SOUTH WALES

There is a robust legal framework for licensing, inspecting, and regulating the display of bats in NSW. Bats are not widely displayed by zoos and fauna parks in NSW, or Australia generally, as only two species are held within the state by members of the Zoo and Aquarium Association (Hibbard and Wilkins, 2010). In 2011, these included about 20 grey-headed flying foxes held in two institutions and 9 ghost bats (*Macroderma gigas*) held in two institutions. Several small fauna parks also hold grey-headed flying foxes and one other park holds ghost bats. In the recent past, the Queensland blossom bat (*Syconycteris australis*) has also been displayed to the public within NSW.

In NSW, the display of animals by zoos, fauna parks, and mobile exhibitors is regulated by the Exhibited Animals Protection Act 1986 and the Exhibited Animals Protection Regulation 2010 that are administered by DPI NSW. In most cases, exhibitors who wish to hold bats already have an animal display establishment license as the owner of a zoo, fauna park, or mobile authority. In these cases, license owners need to obtain a permit before they can acquire any species of bat, as all species are listed as prescribed species in Schedule 2 of the Regulation. The requirement of a permit is due to the health precautions associated with the potentially lethal zoonotic disease, ABL. To acquire a permit for a species of bat, the applicant must demonstrate their expertise with the husbandry of the species, have a suitable exhibit and signage or other educational material, and demonstrate that they have been inoculated against ABL. When a permit has been obtained, the license holder can obtain the species of bat for which they have a permit by providing DPI NSW with an Application to Acquire or Dispose of Animals of a Prescribed Species (Form F) (http://www.hdpi.nsw.gov.au/agriculture/livestock/animal-welfare/exhibit).

If the species of bat requires a new enclosure, then an Application to Construct or Alter an Animal Display Establishment (Form G: see link above) must also be completed and sent to DPI NSW for assessment. The application must include diagrams of the enclosures that are to be constructed to show their size and the construction materials used. The application is assessed against the *Standards for Exhibiting Australian Mammals in New South Wales* (http://www.dpi.nsw.gov.au/__data/assets/pdf_file/0019/121537/australian-mammal-exhibition-standards.pdf), which has been developed to deal with the requirements for the exhibition of all Australian mammals, including bats, in NSW. These standards include the minimum enclosure sizes required for each genus of Australian bat.

In the event that someone wants to display bats to the public, but does not already have an established license for a zoo or fauna park, or an approval for a mobile display, they must first complete the appropriate form that requires them to provide plans of the enclosures they intend to build to hold the animals, and to demonstrate their skills and experience with the husbandry of the species they wish to display (for appropriate form, visit http://www.dpi.nsw.gov.au/agriculture/livestock/animal-welfare/exhibit). When DPI NSW is satisfied that all details of the application are satisfactorily addressed, permission is given to construct the animal display establishment. When this is completed, DPI

NSW is advised and an inspector personally checks to ensure the enclosures and other facilities are appropriate. Once this has been completed the exhibitor is issued with a license or an approval that allows the animals to be exhibited. The exhibitor can then acquire, or dispose of, bats by providing DPI NSW with an Application to Acquire or Dispose of Animals of a Prescribed Species (Form F: see link above). Licenses, permits and approvals need to be renewed each financial year. In addition, routine inspections are ideally undertaken every 24 months, or they can be more frequent or unannounced if the inspection is in response to a complaint.

The *Standards Recommended for the Care and Exhibition of Flying-Foxes* (Snell, 1994), developed by the Ku-ring-gai Bat Conservation Society Inc., has been utilized by the Department of Primary Industries NSW as the guide for exhibition of these large, fruit and blossom-eating bats. DPI NSW administers the display of animals in zoos and fauna parks under the Exhibited Animals Protection Act 1986. The aim in these standards for captive management is "to exhibit a contented healthy group of flying-foxes which will live a long time and function as a small colony with social interactions similar to wild flying-foxes and will need minimal replacements." The standards cover aspects of exhibition in captive groups, enclosure construction and furniture, diet, handling practices, treatment for sick and injured individuals, introduction of new animals, record keeping, and transport.

CONCLUSION

For bat specialists, Australia has an interesting and diverse bat fauna and an equally diverse set of laws to cover bats in captivity. In NSW, there are four laws relating to bats in captivity, and numerous peripheral laws. Further, it will be clear by now that it is the state laws within Australia's federal system of government that dictate land use and all that it implies, such as fauna and the keeping of that fauna, including bats, in captivity. Thus with four major laws in NSW, it follows that there are other major laws in the other states of Australia that govern bats in captivity (Appendix 1-1). The important point remains whether the matter of the laws relating to bats in captivity is well done in each jurisdiction across the world, and whether it is possible to examine the best example of laws and the relevant regulations and policies to produce even better laws and policies in every jurisdiction: Just as relevant is to revise laws and policies as new information becomes available about bats and the effectiveness of current laws.

ACKNOWLEDGMENTS

We thank Nancy Pallin and Jeff Hardy for making helpful comments on an early draft of the manuscript.

REFERENCES

Australian Government, National Health and Medical Research Council. 2004. *Australian Code of Practice for the Care and Use of Animals for Scientific Purposes* (7th ed). Department of Communications, Information Technology and the Arts, Canberra, ACT, Australia; http://www.nhmrc.gov.au/_files_nhmrc/publications/attachments/ea16.pdf.

Churchill, S. 2008. *Australian Bats* (2nd ed.). Jacana/New Holland, Crows Nest, NSW, Australia. 255 pp.

DECCW. 2010. *Rehabilitation of Protected Fauna Policy.* Department of Environment, Climate Change and Water NSW (DECCW); http://www.environment.nsw.gov.au/resources/wildlifelicences/100893PolicyFaunaRehab.pdf.

Eby, P. and Lunney, D. (eds.). 2002. *Managing the Grey-headed Flying-fox as a Threatened Species in NSW.* Royal Zoological Society of NSW, Sydney, NSW, Australia. 286 pp.

Field, H. 2002. The role of grey-headed flying-foxes in the ecology of Hendra virus, Menangle virus and Australian bat lyssavirus. In: *Managing the Grey-headed Flying-fox as a Threatened Species in NSW* (Eby, P. and Lunney, D., eds). Royal Zoological Society of NSW, Sydney, NSW, Australia. Pp. 139-141.

Ford, D. 2002. An historical perspective of changing community attitudes towards flying-foxes in Sydney. In *Managing the Grey-headed Flying-fox as a Threatened Species in NSW* (Eby, P. and Lunney, D.). Royal Zoological Society of NSW, Sydney, NSW, Australia. Pp. 146-159.

Hibbard, C. and Wilkins, R. (eds.). 2010. *Australasian Species Management Program: Regional Census and Plan* (20th ed.). Zoo and Aquarium Association, Sydney, NSW, Australia. 509 pp.

Jackson, D.G. 2011. The 2007 amendments to the EBPC Act 1999 and their relevance to bat conservation. In: *The Biology and Conservation of Australasian Bats* (Law, B., Eby, P., Lunney, D., and Lumsden, L., eds.). Royal Zoological Society of NSW, Mosman, NSW, Australia. Pp. 270-276.

Lunney, D. 1999. Is ethics opposed to science? A wildlife zoologist's viewpoint. In: *The Use of Wildlife for Research* (Mellor, D. and Monamy, V., eds.). Australian and New Zealand Council for the Care of Animals in Research and Teaching (ANZCCART), Glen Osmond, South Australia. Pp. 119-28.

Lunney, D. 2011. Ethics and Australian mammalogy: Reflections on 15 years (1991-2006) on an animal ethics committee. *Australian Mammalogy.* 17 pp.

Lunney, D. and Dickman, C.R. 2002. A zoological revolution: Utilising wildlife to conserve wildlife and landscapes. In: *A Zoological Revolution:Using Native Fauna to Assist in its Own Survival* (Lunney, D. and Dickman, C., eds.). Royal Zoological Society of NSW and the Australian Museum, Sydney, NSW, Australia. Pp. 166-74.

Lunney, D. and Moon, C. 1997. Flying foxes and their camps in the remnant rainforests of northeastern New South Wales. In: *Australia's Ever-Changing Forests III* (Dargavel, J., ed.). Centre for Resource and Environmental Studies (CRES), *Australian* National University (ANU), Canberra, ACT, Australia. Pp. 247-77.

Lunney, D. and Matthews, A. 1997. *Koala Care in New South Wales: Guidelines and Conditions.* Environmental Survey and Research Division, National Parks and Wildlife Service, Hurtsville, NSW, Australia; http://www.environment.nsw.gov.au/resources/wildlifelicences/KoalaRehabilitation1997.pdf.

Lunney, D., Law, B., Schulz, M., and Pennay, M. 2011a. Turning the spotlight on the conservation of Australian bats and the lessons drawn from the extinction of the Christmas Island pipistrelle. In: *The Biology and Conservation of Australasian Bats* (Law, B., Eby, P., Lunney, D., and Lumsden, L., eds.). Royal Zoological Society of NSW, Mosman, NSW, Australia. Pp. 485-498.

Lunney, D., Parnaby, H., Pennay, M., Haering, R., Law B., Eby, P., Schulz, M., and Turbill, C. 2011b. The Priorities Action Statement (PAS) for the threatened bats of New South Wales. In: *The Biology and Conservation of Australasian Bats* (Law, B., Eby, P., Lunney, D., and Lumsden, L., eds.). Royal Zoological Society of NSW, Mosman, NSW, Australia. Pp. 340-356.

Macdonald, D.W. (ed.). 2009. *The Encyclopedia of Mammals* (3rd ed.). Oxford University Press, Oxford, UK. 976 pp.

OEH. 2011a. *Code of Practice for Injured,Sick and Orphaned Protected Fauna.* Office of Environment and Heritage NSW, Sydney, NSW, Australia; http://www.environment.nsw.gov.au/wildlifelicences/RehabFaunaCode.htm.

OEH. 2011b. *Code of Practice for Injured, Sick and Orphaned Flying-foxes.* Office of Environment and Heritage NSW, Sydney, NSW, Australia. (in prep.).

Parnaby, H.E. 2009. A taxonomic review of Australian greater long-eared bats previously known as *Nyctophilus timoriensis* (Chiroptera: Vespertilionidae) and some associated taxa. *Australian Zoologist,* 35:39-81.

Snell, J. 1994. *Standards Recommended for the Care and Exhibition of Flying Foxes.* Ku-Ring-Gai Bat Colony Committee, Inc., Sydney, Australia. 22 pp.

Van Dyck, S. and Strahan, R. (eds). 2008. *The Mammals of Australia* (3rd ed.). New Holland, Chatswood, NSW, Australia. 888 pp.

Wilson, D.E. and Reeder, D.M. 2005. *Mammal Species of the World: A Taxonomic and Geographic Referenc* (3rd ed.). Harvard University Press, Baltimore, MD. 2,142 pp.

Canada

Elizabeth Thunstrom

INTRODUCTION

Canada comprises ten provinces and three territories that are regulated by federal agencies for the protection of endangered species, public health, research, import/export, and international trade. Unless wildlife is classified as at risk in the federal Species at Risk Act, it is primarily managed under provincial or territorial legislation. Several provinces place the emphasis on public health, and therefore list bats as disease vectors.

There are 19 species of bats currently known to occur in Canada. The distribution and diversity of bats within each province or territory is being substantially updated in the last decade with new surveys and technology (e.g., Jung *et al.*, 2006). The little brown myotis (*Myotis lucifugus*) is perhaps the most widely distributed species, occurring in each province and territory. The greatest number of bat species is found in British Columbia, which has 16 species. Climate likely plays a major role in bat distribution in Canada: The diversity and abundance of bats declines dramatically as one travels northward. Of the Canadian species, three (*Lasiurus borealis*, *L. cinereus*, and *Lasionycteris noctivagans*) migrate south of Canada during winter.

This paper is an overview of international, federal, provincial and territorial laws of Canada that oversee the keeping of bats in captivity for research, rehabilitation, and/or educational purposes.

FEDERAL REGULATORY AGENCIES

In Canada, the Minister of Environment is responsible for activities such as honoring the commitments in various international and national conventions, treaties and accords, including the 1998 National Accord for the Protection of Species at Risk, and the administration of national funding programs for species at risk, such as the Habitat Stewardship Program (HSP), and the Interdepartmental Recovery Fund (IRF). Through the 2003 Species at Risk Act, the Minister of Environment is empowered to enter into agreements with provinces and territories to assist them in developing recovery strategies for species under their management.

SPECIES AT RISK ACT

The purpose of the federal Species at Risk Act (SARA; http://laws-lois.justice.gc.ca/eng/acts/S-15.3/index.html) is to protect and recover endangered or threatened wild species and their habitats. It also manages species that are not yet threatened, but whose existence or habitat is in jeopardy. To ensure protection of species at risk, SARA contains prohibitions that make it an offence to kill, harm, capture, take, possess, collect, buy, sell or trade an individual of a species listed in Schedule 1. It is also an offence under SARA to damage or destroy critical habitat and residence of listed species.

SARA applies to all federal lands in Canada, and may apply to public lands owned or managed by the provinces and territories if it is deemed that a species listed under SARA is not being afforded adequate protection by provincial or territorial authorities. As of August 2011, two species of bats in British Columbia are currently listed as at risk in SARA; the pallid bat (*Antrozous pallidus*), which is listed as Threatened, and the spotted bat (*Euderma maculatum*), which is listed as a Species of Special Concern.

Species to be included under SARA are selected by the Committee on the Status of Endangered Wildlife in Canada (COSEWIC; (http://www.cosewic.gc.ca/eng/sct5/index_e.cfm). COSEWIC is a committee of academic, government and non-government wildlife experts that direct the assessments or classify species using the best available scientific, community, and aboriginal traditional knowledge. The committee determines the national status of wild species, sub-species, and nationally significant populations considered to be at risk, using ranking criteria.

The Minister of Environment must respond to COSEWIC's recommendation of status within 90 days. The federal cabinet then has 9 months to make a decision on the addition of the species to SARA, acknowledging that adding species to the list could have economic and social implications to Canadians.

THE CONVENTION ON INTERNATIONAL TRADE IN ENDANGERED SPECIES OF WILD FAUNA AND FLORA

The Wild Animal and Plant Protection and Regulation of International and Interprovincial Trade Act (WAPPRIITA) defines Canada's commitment to the principles of the Convention on International Trade in Endangered Species of Wild Fauna and Flora (CITES). CITES sets controls on international trade of plant and animal species that are or may be threatened due to commercial trade. Environment Canada is the lead agency responsible for implementing CITES on behalf of the Canadian government, and most other provinces and territories participate in the implementation of CITES within Canada. Within Environment Canada, the Canadian Wildlife Service administers CITES and interacts with the provinces, territories, and other federal agencies.

THE CANADIAN COUNCIL ON ANIMAL CARE

The Canadian Council on Animal Care (CCAC), also known as the French Conseil canadien de protection des animaux (CCPA) (http://www.ccac.ca/), is an autonomous and independent body that oversees the ethical use of animals in science in Canada. The CCAC is registered as a not-for-profit organization, and is financed primarily by the Canadian Institutes of Health Research (CIHR) and the Natural Sciences and Engineering Research Council of Canada (NSERC). The CCAC acts as a quasi-regulatory body and sets standards on animal use in science, which apply across Canada. Reviewed are guidelines for field study, animal collection, restraint, marking, moving, and holding, as well as medical and surgical procedures. Animal Welfare in Canada is also subject to Section 446 of the Criminal Code of Canada (http://www.petsinneedsociety.ca/pdf_files/Animal%20Cruelty%20&%20Criminal%20Code%20of%20Canada.pdf).

CANADIAN FOOD AND INSPECTION AGENCY

The Canadian Food and Inspection Agency (CFIA; http://www.inspection.gc.ca/english/anima/animae.shtml) plays an important role in minimizing and managing disease risk by protecting livestock, aquatic species, and wildlife from regulated diseases. The CFIA is responsible for the administration and enforcement of at least a dozen acts, including the Health of Animals Act (http://laws.justice.gc.ca/en/H-3.3). Among the many areas of regulation by the CFIA, rabies is a reportable disease under the Health of Animals Act and Regulations. For comprehensive information on rabies in Canada, the CFIA maintains a rabies website that contains annual reports on rabies listed by species and province (see http://www.inspection.gc.ca/english/anima/disemala/rabrag/statse.shtml). The website also provides comprehensive information on the disease and the CFIA rabies control program.

PROVINCIAL AND TERRITORIAL REGULATIONS

Bats are covered under the Wildlife Act of each province or territory. Bats held in captivity for research require special permits from the Ministries, outlining the full details of the project to be undertaken. Several provinces have zoo permits and resultant guidelines for care. Rehabilitation for release is permitted by the provinces, which specify the species that can be accepted.

TERRITORIES

Yukon

Yukon Department of Environment
10 Burns Road
Whitehorse, YT
Y1A 4Y9
Phone: 867-667-5652
Email: info@gov.yk.ca

The Yukon territory currently has 2 known species of bats, *Myotis lucifugus* and *Myotis septentrionalis*; however, there may be other species not yet recorded (Thomas Jung, pers. comm.). In the Yukon, bats are protected under the Yukon Wildlife Act (http://www.gov.yk.ca/legislation/acts/wildlife.pdf). *M. lucifugus* is common; whereas, *M. septentrionalis* is limited in its distribution in the Yukon and is considered a sensitive species.

The Yukon territory has no zoos, and there are currently no wildlife rehabilitation centers. The Yukon Wildlife Preserve is developing animal care facilities, and at a future date may be able to rehabilitate some species of native animals, including bats (Thomas Jung, pers. comm.). However, bats have not yet been presented to the preserve for care (Greg Meredith, pers. comm.).

To conduct research, a wildlife research permit is required under the Yukon Wildlife Act. For researchers from outside the Yukon, a permit is also required under the Scientists and Explorers Act.

The Yukon Wildlife Act also requires wildlife import/export permits for animals alive or dead.

NORTHWEST TERRITORIES (NWT)

Wildlife Division
Department of Environment and Natural Resources
Government of Northwest Territories
PO Box 1320
Yellowknife, NT
X1A 2L9
Phone: 867-920-8064
Email: wildlife@gov.nt.ca

The NWT has records for 6 species of bats, *Eptesicus fuscus, Lasiurus cinereus, Myotis evotis, M. lucifugus, M. septentrionalis,* and *M. volans*, and 1 species, *Lasiurus borealis*, whose presence has been recorded based on sonograms. Of the 6 known species, the 4 *Myotis* species are ranked "may be at risk" because of the threat of white-nose syndrome (http://nwtspeciesatrisk.ca/pdf/NWT_Species_2011_2015.pdf).

The NWT has no bat rehabilitation facilities or zoos, but under Section 53 of the NWT Wildlife Act (http://www.justice.gov.nt.ca/pdf/ACTS/Wildlife.pdf) is information covering captive wildlife. Under the Act, permits are required to conduct research (Section 98) and to import/export wildlife (Section 59, [1, 2]). Anyone applying for a wildlife research permit, where the study involves the handling of vertebrates other than fish, must also submit an application to handle wildlife to the Chair of the NWT-Wildlife Care Committee (WCC). The Committee provides to the authority issuing wildlife research permits a recommendation on whether or not an application to handle wildlife should be approved. For more information on the NWT-WCC, see http://www.enr.gov.nt.ca/_live/pages/wpPages/Wildlife_Care_Committee.aspx.

NUNAVUT (NU)

Department of Environment
P O Box 1000, Stn. 1300
Iqualuit, NU
XOA OHO
Phone: 867-975-7700
Email: info@gov.nu.ca

Nunavut has 2 known species of bats, *Lasiurus cinereus* and *Myotis lucifugus*, that are covered under the Nunavut Wildlife Act (http://www.canlii.org/en/nu/laws/stat/snu-2003-c-26/latest/snu-2003-c-26.html). Van Zyll de Jong (1985) also places *L. borealis* at Coral Harbor, South Hampton Is. in Hudson Bay.

There are no rehabilitation facilities or zoos. Research conducted in Nunavut requires a permit and approval of the Nunavut Wildlife Management Board (Sec 152-2). Under the Nunavut Wildlife Act, permits are required to import and export wildlife (see Section 105-6).

PROVINCES (WEST TO EAST)

For a summary of zoo regulations in each of Canada's provinces, see Appendix 1-2.

BRITISH COLUMBIA (BC)

BC Ministry of Forests, Lands and Natural Resource Operations
Fish and Wildlife Branch
PO Box 9339, STN. Prov. Gov't.
Victoria V8W 9M1
Phone: 250-387-5669
Website: http://www.gov.bc.ca/env/contacts.html

British Columbia has the highest diversity of bat species in Canada; 16 species have been recorded, including *Antrozous pallidus, Corynorhinus townsendii, Eptesicus fuscus, Euderma maculatum, Lasionycteris noctivagans, Lasiurus blossevillii, L. cinereus, Myotis californicus, M. ciliolabrum, M. evotis, M. keenii, M. lucifugus, M. septentrionalis, M. thysanodes, M. volans,* and *M. yumanensis.* Of these, *A. pallidus and E. maculatum* are considered Species at Risk under SARA, Schedule 1. Nevertheless, all bats are protected under Schedule A of the BC Wildlife Act (http://www.bclaws.ca/EPLibraries/bclaws_new/document/ID/freeside/00_96488_01).

Of the licensed rehabilitation centers in BC, six accept bats for care. The BC Wildlife Act prohibits the possession of native wildlife without a permit, and each center has its own specific requirements for handling bats in care. Handlers are required to have pre-exposure rabies immunization. For additional information on rehabilitation, see the website of the Wildlife Rehabilitators' Network of BC (http://www.wrnbc.org/).

Under the BC Wildlife Act, permits are required to conduct wildlife research. A detailed plan of the proposed research is required, which has to be approved by the originating scientific or educational agency, or institution prior to any project moving forward, and CCAC guidelines must be followed.

Since 1 April 2010, commercial zoos are required to meet the accreditation standards of the Canadian Association of Zoos and Aquaria (CAZA). The 2008 Environmental Management Act Amendment (Species and Public Protection; http://www.leg.bc.ca/38th4th/1st_read/gov29-1.htm) amended the provincial Wildlife Act (see 7 Section 1 [1] of the Wildlife Act, R.S.B.C. 1996, c. 488) to cover regulatory gaps for managing alien (exotic) species, as well as protecting the public and native wildlife. Under the new controlled alien species regulation (http://www.env.gov.bc.ca/fw/wildlifeactreview/cas/species.html), exotic bats are not specifically listed, but individual municipalities can prohibit the possession of them as pets. Currently, there is a display of fruit bats at the Vancouver Aquarium. They are kept in an enclosed habitat under CAZA regulations.

Import/export permits are required under the BC Wildlife Act for cross-border or international movement of animals.

If handling bats is required during their removal from a structure, a permit is required. After bats have left a structure for winter hibernation, property owners are advised to make necessary repairs to prevent them from re-entering in spring (Purnima

Govindarajulu, pers. comm.). Bats are not included under Schedule C of the BC Wildlife Act that allows the killing of pest wildlife.

ALBERTA

Alberta Sustainable Resource Development
Government of Alberta
9920 – 108 Street
Edmonton, Alberta T5K 2M4
Canada
Telephone: 780-427-9503
Website: http://srd.alberta.ca/FishWildlife/WildSpecies/Mammals/Bats/

Alberta has 9 known species of bats (*Eptesicus fuscus, Lasionycteris noctivagans, Lasiurus borealis, L. cinereus, Myotis ciliolabrum, M. evotis, M. lucifugus, M. septentrionalis,* and *M. volans*), all of which are designated as Wildlife under the Alberta Wildlife Act (http://www.qp.alberta.ca/documents/Acts/w10.pdf).

There are eight licensed rehabilitation centers in Alberta. Currently, the rehabilitation of bats is not allowed because they are considered rabies vectors. For additional information on rehabilitation, see the website of the Alberta Wildlife Rehabilitators' Association (http://www.albertawildliferehab.org/).

Under the Alberta Wildlife Act, permits are required to conduct scientific research using wildlife, including bats, whether captive or free-living. The researcher must submit a research permit application to the provincial Wildlife Animal Care Committee for animal care review and then forward it to local Fish and Wildlife Managers, who determine whether or not the permit is issued. Research must be conducted under CCAC guidelines and an annual permit renewal is required. Research permit guidelines and applications are available on the Fish and Wildlife web pages http://srd.alberta.ca/FishWildlife/.

Permits are required by zoos for the public display and education of native and exotic wildlife species. Although there is no specific legislation for the maintenance and care of bats, an appointed advisory committee reviews development plans and animal care protocols for non-accredited zoos.

Movement of live wildlife into or out of Alberta is regulated through import or export permits. In the case of import, bats can only be brought in by those possessing a current research or zoo permit.

SASKATCHEWAN (SK)

Ministry of the Environment
Fish and Wildlife Branch
3200 Albert Street
Regina, SK S4S 5W6
Phone: 306-787-6218
Website: http://www.environment.gov.sk.ca/wildlife

Bats are covered under the Saskatchewan Wildlife Act, 1998 (http://www.qp.gov.sk.ca/documents/English/Statutes/Statutes/W13-12.pdf). There are 6 known species of bats in the province (*Eptesicus fuscus, Lasionycteris noctivagans, Lasiurus borealis, L. cinereus, Myotis lucifugus, M. septentrionalis*), and currently no species is listed as threatened or endangered.

Bats may be rehabilitated under a special provincial permit based upon city bylaws. Currently, there are two wildlife centers that take in bats. Pre-exposure rabies vaccinations are required by anyone handling bats. For additional information on rehabilitation of bats, see the Wildlife Rehabilitation Society of Saskatchewan website (http://wrsos.org/contact-us).

A permit is required to conduct research, which includes approval under the Veterinarians Act (http://www.qp.gov.sk.ca/documents/English/Statutes/Statutes/V5-1.pdf) and the appointed Animal Care Committee.

A permit is required to hold bats in captivity. Permits are issued only to approved zoos and rehabilitation centers; no private individuals may keep bats in captivity.

Import/export permits are required to transport wildlife across provincial or international borders.

There are no special regulations for removing bat colonies from buildings.

Manitoba (MB)

Manitoba Conservation
Wildlife and Ecosystem Protection Branch
Box 24, 200 Saulteaux Crescent
Winnipeg, MB R3J 2W3
Phone: 204-945-7775
Website: http://www.manitoba.ca/conservation/wildlife

All 6 species of bats found in Manitoba (*Eptesicus fuscus, Lasionycteris noctivagans, Lasiurus borealis, L. cinereus, Myotis lucifugus,* and *M. septentrionalis*) are listed as Protected Species by the Designation of Wild Animals Regulation under the Manitoba Wildlife Act (http://www.gov.mb.ca/conservation/wildlife/legislation/wildlife_act.html).

Permits are required to possess and rehabilitate native wildlife, including bats. Applicants must meet the husbandry conditions set forth by the government. The Wildlife Haven Rehabilitation Centre of the Manitoba Wildlife Rehabilitation Organization and the Prairie Wildlife Rescue Centre, rehabilitate bats under Manitoba conservation permits. If there is a suspicion that an animal might have rabies, or if a person has been bitten, the animal is required to be euthanized and tested. For additional information on rehabilitation in Manitoba, see Wild Haven's website at http://www.wildlifehaven.ca/ and the Prairie Wildlife Rehabilitation Centre's website at http://www.pwildlife.ca/.

Wildlife Scientific Permits (provincial) are required to study any wild animals; also approval from either a University Animal Care Committee or the Manitoba Provincial Animal Care Committee is required. Studies conducted in a provincial or national park, or a protected area require additional permits.

There are two zoos in the province of Manitoba. Permits are required to maintain wildlife for scientific or educational display purposes. Facility and husbandry standards are regulated by conditions attached to the permit. Zoos have adopted the CAZA requirements for health, welfare, and transportation of animals, and the Assiniboine Park Zoo has also applied for AZA accreditation. Zoos undergo annual inspections, which are required in Manitoba (see Appendix 1-2).

Permits are required to import/export wildlife into the province of Manitoba. A health certificate is required from CFIA to import/export non-native species. If the non-native species is classified as endangered, a CITES permit is also required.

Because all bat species are protected in Manitoba, a permit is required to move bats off a person's property.

Ontario (ON)

Ontario Ministry of Natural Resources
Main Office
300 Water Street
PO Box 7000
Peterborough, ON K9J 8M5
Phone: 705-755-2000
Toll free: 800-667-1940
Website: http://www.mnr.gov.on.ca/en/

Ontario has 8 known species of bats: *Eptesicus fuscus, Lasionycteris noctivagans, Lasiurus borealis, L. cinereus, Myotis leibii, M. lucifugus, M. septentrionalis,* and *Perimyotis subflavus.* All are covered under the Fish and Wildlife Conservation Act (FWCA) (http://www.e-laws.gov.on.ca/html/statutes/english/elaws_statutes_97f41_e.htm) as Specially Protected Mammals.

A FWCA permit is required to maintain wildlife in captivity. To become a licensed wildlife custodian, authorization is required from the local Ontario Ministry of Natural Resources (OMNR) Office. All new applicants must pass the Ontario wildlife rehabilitation exam. If a potential rehabilitator wishes to rehabilitate rabies vector species, they must also pass the Ontario rabies vector species course and exam. For additional information on rehabilitation, see the Ontario Wildlife Rehabilitation and Education Network (OWREN) (http://www.owren-online.org).

Animals used in research are regulated by the Animals for Research Act (http://www.e-laws.gov.on.ca/html/statutes/english/elaws_statutes_90a22_e.htm). All other animals in captivity are covered by the Ontario SPCA (OSPCA) Act, 2009 (http://www.ontariospca.ca/press_releases/2009/09_sept22.html).

Currently, zoos are covered by the OSPCA Act. A zoo license is required only for the possession of native wildlife scheduled as Game or Specially Protected wildlife. There are no licensing requirements for other species of wildlife in zoos. Seven wildlife facilities maintaining captive animals are accredited by CAZA. The OSPCA Act allows for inspections (see http://ontariospca.ca/1-bill50.shtml).

Ontario provincial permits for importation and exportation cover furbearers and some game animals, but not all wildlife (Pamela Gomer, pers. comm.).

A person or their agent can remove bats from their own property and buildings without a permit. The Ontario Fish and Wildlife Conservation Act, Subsection 31(5), requires that the person not cause unnecessary suffering to bats. Property owners are encouraged by the OMNR to prevent conflicts, use harassment techniques to encourage bats to leave on their own, and use lethal methods as a last resort (see http://www.mnr.gov.on.ca/en/Business/FW/2ColumnSubPage/290050.html). To address disease concerns, the Act also states that all wildlife must be released in close proximity to where it was captured.

Québec

Ressources Naturelles et Faune (MRNF)
880, Chemin Sainte-Foy, 2è étage
Québec G1S 4X4
Phone: 418-627-8694
Website: http://www.mrnf.gouv.qc.ca/

The province of Québec has 5 non-migratory species of bats (*Eptesicus fuscus, Myotis leibii, M. lucifugus, M. septentrionalis*, and *Perimyotis subflavus*) and 3 migratory species (*Lasionycteris noctivagans, Lasiurus borealis*, and *L. cinereus*). Hibernacula have protection under the Ministère des Ressources Naturelles et de la Faune du Québec.

A license is required to operate rehabilitation centers. Centers are permitted by individual districts.

Research projects must follow guidelines of the CCAC. A provincial license is required.

Licenses are required by the Québec Wildlife Act, 2011 (http://www.canlii.org/en/qc/laws/stat/rsq-c-c-61.1/latest/) to keep bats in zoos. Licenses cover the general care of native and non-native wildlife. Inspections are required for license holders. For more information on keeping bats in captivity, see http://www2.publicationsduquebec.gouv.qc.ca/dynamicSearch/telecharge.php?type=3&file=/C_61_1/C61_1R5_A.HTM.

In general, the importation and exportation of animals varies with species and by permission of CITES.

Article 67 of the Act respecting the conservation and development of wildlife (http://www2.publicationsduquebec.gouv.qc.ca/dynamicSearch/telecharge.php?type=2&file=/C_61_1/C61_1_A.html; C-61.10) states that, “Before killing or injuring an animal that has caused damage to your property, efforts must be made to scare the animals away or otherwise prevent them from doing damage.”

New Brunswick (NB)

Department of Natural Resources
Fish and Wildlife Branch
PO Box 6000,
Fredericton, NB E3B 5H1
Phone: 506-453-2440 or 506-453-3826
Website: http://www.gnb.ca/0078/wildlife-e.asp

There are 7 known species of bats in New Brunswick: *Eptesicus fuscus, Lasionycteris noctivagans, Lasiurus borealis, L. cinereus, Myotis lucifugus, M. septentrionalis*, and *Perimyotis subflavus*. Sensitive species include *E. fuscus, M. lucifugus, M. septentrionalis, and P. subflavus* (New Brunswick Natural Resources). The status of *L. noctivagans* and the two lasiurines is undetermined. For the status of bats in New Brunswick, visit http://www1.gnb.ca/0078/WildlifeStatus/search-e.asp).

Under the New Brunswick Fish and Wildlife Act (http://www.gnb.ca/0062/PDF-acts/f-14-1.pdf), permits are required to keep wildlife in captivity. Only one rehabilitation center is provincially approved to accept wildlife, the Atlantic Wildlife Institute (http://www.atlanticwildlife.ca).

Wildlife research in New Brunswick requires a permit from the department of Natural Resources, Fish and Wildlife Branch (Section 90[1][d]).

There are five zoos in the province of New Brunswick that have Ministry permits to maintain both native and exotic wildlife in captivity. These zoos have adopted CAZA requirements for animal welfare, veterinary care, safety and security, and accommodation requirements. There is no requirement for regular inspection of the facilities.

To import/export wildlife, an authorization permit is required from the Ministry. The importation of exotic wildlife also requires a permit under the New Brunswick Fish and Wildlife Act (Section 38.1[1][a] and 90.1[1][a]).

NOVA SCOTIA (NS)

Department of Natural Resources
Wildlife Division
136 Exhibition Street
Kentville, NS B4N 4E5
Phone: 902-679-6091
Website: http://www.gov.ns.ca/natr/wildlife/

There are 6 known bat species in Nova Scotia (*Lasionycteris noctivagans, Lasiurus borealis, L. cinereus, Myotis lucifugus, M. septentrionalis*, and *Perimyotis subflavus*), which have full legal protection under the Nova Scotia Wildlife Act (http://www.gov.ns.ca/natr/wildlife/laws/captivewildlife.asp).

Permits are required to maintain wildlife in captivity for research, rehabilitation, and education (zoos) (http://www.gov.ns.ca/natr/wildlife/biodiversity/permits_approvals.asp). The Director of Wildlife determines which wildlife species may be imported into Nova Scotia and the rules for keeping them. Although there are two licensed wildlife facilities in the province, due to human health concerns, these facilities may not accept bats for care.

There are five licensed education facilities (zoos). One facility is not open to the public. Permits cover both indigenous and exotic species. There are some general requirements for animal welfare, but no veterinary, safety, or housing requirements. Inspections are annual, but sometimes they occur monthly.

Both native and non-native wildlife may be imported into Nova Scotia. Species that are listed as "not dangerous" to the Nova Scotia eco-system are exempt.

Bats that are considered a nuisance or of concern in buildings can not be killed without a permit, which would only be granted if other attempts to discourage the animals are ineffective. A Nuisance Wildlife Operator's License is required and a Conservation Officer must be satisfied that scaring techniques did not work.

PRINCE EDWARD ISLAND (P.E.I.)

Department of Environment, Energy and Forestry
Forests Fish and Wildlife Division
183 Upton Road
P.O. Box 2000
Charlottetown, P.E.I. C1A 7N
Website: http://www.gov.pe.ca/eef/_

Prince Edward Island is 45% forested and is farmed extensively. It is arguable that prairie ecosystems have been as affected on a similar scale.

There are 3 species of bats found in P.E.I.: *Lasiurus cinereus*, a rare migrant, and *Myotis lucifugus* and *M. septentrionalis*. These bats are not protected under the P.E.I. Wildlife Conservation Act (WCA) (http://www.gov.pe.ca/law/statutes/pdf/w-04_1.pdf), or its regulations. Their status has not been evaluated regarding risk of extirpation.

A Department permit is required for scientific or educational purposes, or species defined as wildlife, including exotic bats, but not for native bats, which are not considered wildlife (see WCA, Section 10[c]).

The Atlantic Veterinary College (AVC) at the University of Prince Edward Island is the only licensed wildlife rehabilitation facility in the province. The only animals AVC handles are primarily species of concern and raptorial birds (Dibblee, 2011).

Research projects must comply with CCAC guidelines. Note: The AVC is currently conducting necropsies of birds and bats collected under wind turbines. In addition, surveillance is underway for the possible appearance of white-nose syndrome in bats.

P.E.I. has no zoos. Exotic wildlife held in captivity must be registered and meet regulatory standards.

Import/export permits are required (see Section 10[b] by the WCA).

Removal of nuisance wildlife is allowed only under permit, and only if deterrence methods have been unsuccessful.

NEWFOUNDLAND AND LABRADOR (NL)

Environment and Conservation
Wildlife Division
P O Box 8700
4th Floor, West Block
Confederation Building
St. John's, NL A1B 4J6
Phone: 709-729-2664 Toll free: 800-563-6181
Website: http://www.env.gov.nl.ca/env/wildlife/

The Island of Newfoundland has 3 known species of bats (*Lasiurus cinereus, Myotis lucifugus*, and *M. septentrionalis*; Labrador, on the mainland, has one, *M. lucifugus. L. cinereus* is a rare visitor to the Island of Newfoundland and does not breed in the province (Bruce Rodrigues, pers. comm.).

Under the NL Wildlife Act (http://www.assembly.nl.ca/legislation/sr/statutes/w08.htm) a permit is required to hold bats in captivity. To date, the Ministry has had no requests to keep bats in captivity. Injured wildlife must be reported to a Regional Conservation Officer who in turn would contact the Salmonier Nature Park regarding any rehabilitation initiative. The Salmonier Nature Park (http://www.env.gov.nl.ca/env/snp/index.html) is the only rehabilitation facility in the province. It was created under Section 5 of the NL Wildlife Act. It was developed as a center for environmental education, wildlife rehabilitation, research, and environmental monitoring.

A permit is necessary for the import/export of wildlife (see CNLR 1156/96 of the NL Wildlife Act).

There are no specific protections or requirements for the removal of bats from buildings. Wildlife species considered by the Minister to be a nuisance or a menace may be taken, by shooting or trapping, with the written permission of the Minister or his or her agent (see Section 88 of the NL Wildlife Act).

ACKNOWLEDGMENTS

I thank the following people for their helpful comments and assistance with the manuscript: Donald F. McAlpine, New Brunswick Museum; Robert M. Barclay, University of Calgary; R. Mark Brigham, University of Regina; M. Brock Fenton, University of Western Ontario; Hugh Broders, St. Mary's University; Melissa Matlow, World Society for the Protection of Animals; Rob Laidlaw and Jennifer Aziz, Zoocheck Canada, Inc.; Janna Mariano-Groza, Alberta Wildlife Rehabilitators' Association; Jenith Dack, Assiniboine Park Zoo; Thomas Jung, Yukon Fish and Wildlife; Joanna Wilson and Suzanne Carrière, Northwest Territory Environment and Natural Resources; Steve Pinksen and Elissa McKinnon, Nunavut Department of Environment; Purnima Govindarajulu, British Columbia Ministry of Environment; Brent Gurd, British Columbia Ministry of Natural Resource Operations; Margo Pybus, Wildlife Diseases, Alberta Fish and Wildlife Division; Marvin Hlady, Saskatchewan Ministry of Environment, Fish and Wildlife Branch; Bill Watkins, Jack Dubois and James R. Duncan, Manitoba Wildlife and Ecosystem Protection Branch, Manitoba Conservation; Tamara Gomer, Ontario Department of Natural Resources; Jacques Jutras and Nathalie Desrosiers, Québec Ressources Naturelles et Faune; Bob Petrie and Pamela Mills, Nova Scotia Department of Natural Resources; Rosemary Curley, Prince Edward Island Department of Environment, Energy and Forestry; Bruce Rodrigues, Newfoundland and Labrador Department of Environment and Conservation.

LITERATURE CITED/BACKGROUND READING

Broders, H.G., Quinn, G.M., and Forbes, G.J. 2003. Species status of spatial and temporal patterns of action of bats in south western Nova Scotia, Canada. *Northeastern Naturalist*, 10:383-398.

Dibblee, R. 2011. Leave wildlife alone: P.E.I. biologist. CBS News; http://www.cbc.ca/news/canada/prince-edward-island/story/2011/06/05/pei-rescue-animals.html, accessed July 2011.
Forbes, G.J., McAlpine, D.F., and Scott, F.W. 2010. Mammals of the Atlantic maritime ecozone. In: *Assessment of Species Diversity in the Atlantic Maritime Ecozone* (McAlpine, D.F. and Smith, I.M., eds.). National Research Council Press, Ottawa, Ontario, Canada. Pp. 693-718.
Government of Alberta Sustainable Resource Development; http://srd.alberta.ca/AboutUs/UsingThisSite/Search.aspx?q=bats&btnG=Search&client=goa_default&sitesearch=www.srd.alberta.ca&proxystylesheet=goa_default&site=pub_goa&as_ft=i&as_filetype=All%20file%20formats, accessed June 2011.
Government of Newfoundland and Labrador Department of Environment and Conservation; http://www.env.gov.nl.ca/env/snp/programs/education/animal_facts/mammals/#lm3, accessed June 2011.
Henderson, L. E., Farrow, L.J., and Broders, H.G. 2009. Summer distribution and status of bats of Prince Edward Island, Canada. *Northeastern Naturalist,* 16:131-140.
Hinterland Who's Who; http://www.hww.ca/hww2.asp?id=63, accessed June 2011.
Jung, T.S., Slough, B.G., Nagorsen, D.W., Dewey, T.A., and Powell, T. 2006. First records of the northern long-eared bat, *Myotis septentrionalis*, in the Yukon territory. Canadian Field-Naturalist, 120:39-42.
Nagorsen, D.W. and Brigham, R.M. 1993. *Bats of British Columbia*. Royal British Columbia Museum Reference Handbook, BC, Canada. 164 pp.
New Brunswick Natural Resources Species and Status List. 2005. http://www.gnb.ca/0078/Wildlife-e.asp, accessed July 2011.
Ontario Woodlot Association Newsletter. 2008. S&W Report; http://www.ont-woodlot-assoc.org/sw_bats.html.
Slough, B.G. and Jung, T.S. 2007. Diversity and distribution of the terrestrial mammals of the Yukon: A review. *Canadian Field-Naturalist*, 121:119-127.
van Zyll de Jong. 1985. *Handbook of Canadian Mammals 2: Bats*. National Museum of Natural Sciences, National Museums of Canada, Ottawa, ON. 212 pp.

Europe

Anthony J. Mitchell-Jones

INTRODUCTION

Europe consists of approximately 50 separate states, depending on the geographic definition used, ranging in size from Vatican City (0.4 km^2) to Russia (approx. 4,000,000 km^2 for the European part), most of which have some species protection legislation. Twenty-seven European countries are members of the European Union (EU) and are legally obliged to implement EU Directives in their own national legislation. This ensures some commonality of legislation across the Union, although there remains room for national interpretation. In addition, many European countries are signatories to multilateral environmental agreements (conventions) and thus voluntarily commit themselves to certain obligations. The Eurobats Agreement (see below) includes an area beyond Europe that could share some bat populations, so for this account a number of Middle Eastern countries are included.

The country accounts cover the great majority of European states, although to varying levels of detail. In many federal states, at least some nature conservation functions, often including the issuing of permits, are delegated to regional authorities. It has not generally been possible to list all these, so researchers wishing to work on bats in these countries are advised to contact the federal authority in the first instance. The Eurobats website (http://www.eurobats.org/) is a useful source of information that is updated regularly.

THE EU HABITATS DIRECTIVE

The Habitats Directive (Council Directive 92/43/EEC on the Conservation of Natural Habitats and of Wild Fauna and Flora 1992) is the main EU instrument for protecting habitats and species. The aim of this Directive is to conserve fauna, flora, and natural and semi-natural habitats of importance in the EU, including through designation of the Natura 2000 network of protected sites. The Directive also requires Member States to strictly protect species listed on Annex IV, which includes all Microchiroptera, in their natural range. The capture and killing of bats, their deliberate disturbance (particularly during breeding, rearing, hibernation or migration), keeping, transporting, selling or exchange are all prohibited, as is damage or destruction of breeding sites or resting places.

THE BERN CONVENTION

The Bern Convention (Convention on the Conservation of European Wildlife and Natural Habitats, 1979) aims to maintain populations of wild flora and fauna with particular emphasis on endangered and vulnerable species, including migratory species. All European bat species are listed in Appendix II (strictly protected) except for *Pipistrellus pipistrellus,* which is listed in Appendix III (protected). Parties agree to implement appropriate protection in their national legislation.

THE BONN CONVENTION

The Bonn Convention (Convention on the Conservation of Migratory Species of Wild Animals, 1979) recognizes that endangered migratory species can only be properly protected if they receive protection over their entire migratory range. The Convention operates primarily through a series of regional agreements covering particular species in particular areas. The Agreement on the Conservation of Populations of European Bats (Eurobats - http://www.eurobats.org/), which came into force in 1994, aims to protect all 45 species of bats identified in Europe through legislation, education, conservation measures, and international co-operation with Agreement members, Range States and with those who have not yet joined. Thirty-three European states are currently (2011) Parties to the Agreement and agree to implement certain protection measures for bats. In addition, a recent resolution introduced guidelines on ethics for research and field work practices that should be adopted by all Parties (see: http://www.eurobats.org/documents/pdf/MoP6/record_MoP6/MoP6.Record.Annex8-Res.6.5-Ethics.pdf).

COUNTRY ACCOUNTS

ALBANIA (EUROBATS, BERN)

Wildlife is protected by Law Nr.10 006, dated 23.10.2008. This gives basic protection to all species and greater protection to species, including all bats, listed in the national Red Data Book, or listed under the Bern Convention. The law also prohibits the keeping of wild animals, except by authorized rescue centers. Permits are issued by the Ministry of Environment, Forestry and Water Management (http://www.moe.gov.al/).

AUSTRIA (EU, BERN, BONN)

Nature conservation legislation is under the authority of the nine individual federal states (Bundesländer), which all have different laws. International laws and agreements have to be incorporated into federal state legislation. The current laws can be found at http://www.ris.bka.gv.at/. All species of bats and their roosts are protected in all the federal states. It is forbidden to intentionally disturb, catch, take from the wild, transport, injure or kill bats, as well as to possess, acquire, or pass on bats or parts of them. Damaging bat roosts or resting places is also forbidden. A special permit is necessary to keep bats for rehabilitation. The competent authorities for this permit vary between federal states (see below). There is also a federal law on animal welfare, stating that keeping wild animals with special maintenance requirements, like bats, is subject to registration, and minimum requirements have to be met. The following are the competent authorities for permit applications in each Bundesland:

Burgenland: local Bezirkshauptmannschaft or Magistrat (the local district administrative authority); if more than one district is involved, the Amt der Burgenländischen Landesregierung (office of the provincial government), Abteilung 5: Hauptreferat für Natur- und Umweltschutz.

Carinthia: local Bezirkshauptmannschaft or Magistrat for each district.

Lower Austria: Amt der Niederösterreichischen Landesregierung, RU5 - Abteilung Naturschutz.

Salzburg: local Bezirkshauptmannschaft or Magistrat; if more than one district is involved or within protected areas, the Amt der Salzburger Landesregierung, Abteilung 13: Naturschutz.

Styria: local Bezirkshauptmannschaft or Magistrat, each one if the permit is needed for the whole federal state.

Tyrol: local Bezirkshauptmannschaft or Magistrat; if more than one district is involved, the Amt der Tiroler Landesregierung, Abteilung Umweltschutz.

Upper Austria: outside of protected areas: local Bezirkshauptmannschaft or Magistrat; if more than one district is involved, one Bezirkshauptmannschaft will handle the application process for all of them. Within protected areas: Amt der Oberösterreichischen Landesregierung, Abteilung Naturschutz.

Vienna: Magistrat der Stadt Wien, Magistratsabteilung 22 - Wiener Umweltschutzabteilung and MA 60 - Veterinäramt.

Vorarlberg: local Bezirkshauptmannschaft or Magistrat, each one if the permit is needed for the whole federal state.

BELARUS (BONN)

The Natural Environment Protection Law, the latest revision that was adopted in 2002, is the basic domestic legal document regulating nature protection activities in Belarus. There are also the Law on Hunting and Hunting Management (1997), which prohibits the capture of bats, and the Provision on Zoological Collection (1997), which provides rules on the keeping of wildlife, including bats. The statutory Red Book of Belarus 2004 includes 6 of the 16 bat species currently found in the country; these are *Myotis dasycneme* (EN), *M. nattereri* (NT), *M. brandtii* (VU), *Barbastella barbastellus* (EN), *Nyctalus leisleri* (VU) and *Eptesicus nilssonii* (VU). The National Academy of Sciences (http://nasb.gov.by/eng/index.php/) is the responsible body for bat conservation and management.

BELGIUM (EU, EUROBATS, BERN)

Each region of Belgium has a legal framework in place to ensure an adequate protection of bats and their roosting and hibernation sites. These regional laws fall under the umbrella of the federal law on the Conservation of Nature Act of 12 July 1973.

In the Walloon Region, the Natura 2000 Decree of 6 December 2001 on the conservation of Natura 2000 sites and species protection of fauna and flora, implements the EU Habitats Directive. All bats are strictly protected (Appendix IIa Decree Natura 2000). The Service public de Wallonie (SPW) Direction générale de l'Agriculture, des Ressources naturelles et de l'Environnement (DGARNE) http://agriculture.wallonie.be is responsible for the issuing of permits for capturing or keeping bats.

In the Flemish region, the Decree of 21 October 1997 on nature conservation and the natural environment (the Nature Decree) and the Decree of the Flemish Government of 15 May 2009 on species protection and species management (BS: 13/08/2009) implement the Habitats Directive. The Flemish Agency for Nature and Forests (ANB) (http://www.natuurenbos.be/) is responsible for the issuing of permits or capturing or keeping bats.

BOSNIA AND HERZEGOVINA (BERN)

Bats are protected by the Nature Protection Law at entity level (Federation of Bosnia and Herzegovina and Republic of Srpska). There is no single list of protected species. Some species are protected by the Law referring to the species listed in the EU Habitats Directive. It is forbidden to keep or deliberately kill in the wild all protected species, including bats; to deliberately disturb them, especially during breeding, hibernation or migration; to damage or destroy their nests and habitats; to keep, transport, sell or exchange species from the wild; perform actions that may cause disappearance or reduction of populations. Competent entity Ministries grant licenses for exemption from protection for scientific research and educational purposes.

BULGARIA (EU, EUROBATS, BERN)

Under the Biodiversity Act (SG 77/09.08.2002 and subsequent Regulations) all bats are strictly protected (Annex 3) and financial penalties apply. Regulation No. 8 (SG 4/16.01.2004) requires everyone who is engaged in bat studies and has direct contact with them to possess a valid permit issued by the Ministry of Environment and Water (http://www.moew.government.bg/index_e.html). The permits apply for a specific term and are issued for an individual or institution and a specifically defined area. It is illegal to keep bats in captivity without a permit.

Rescue centers are the only organizations allowed to tend and rehabilitate (but not to breed) injured bats and return them into the wild once recovered. The work of these centers is regulated by the Biodiversity Act and specialized Regulations. The Wildlife Rescue Centre of the Green Balkans NGO (http://www.greenbalkans-wrbc.org/index.php?language=en_EN) is the only one in the country that has the necessary knowledge and professional experience to care for injured bats.

Around 40% of the important bat caves in Bulgaria are included in the protected territories defined by the Protected Areas Act (SG 133/11.11.1998), which include national parks, protected areas, and natural landmarks. Some of these caves have the status of natural landmarks and some are Natura 2000 sites. In the majority of the cases, the protected area also includes adjacent land.

The National Biodiversity Monitoring System lists 9 species of bats (*Miniopterus schreibersii, Myotis bechsteinii, M. blythii, M. capaccinii, M. emarginatus, M. myotis, Rhinolophus blasii, R. euryale* and *R. ferrumequinum*) as priority species for winter and summer monitoring. In 2011, seventy-three underground bat roosts were considered to be highly important on national and European levels.

CROATIA (EUROBATS, BERN)

All bat species are legally protected by the Nature Protection Act as strictly protected species listed in the Ordinance on Proclamation of Wild Taxa as Protected and Strictly Protected. Deliberate taking from the wild, capturing, killing, disturbing in nurseries, hibernacula or migratory roosts, deterioration and/or destruction of roosts, or trading specimens either dead or alive are forbidden. For such actions, fines are proscribed according to the Nature Protection Act. Out of 35 bat species recorded in Croatia, 17 are listed in the Red List of Endangered Plants and Animals of the Republic of Croatia. The Ministry of Culture (http://www.min-kulture.hr/) may issue permits for bat research with nature protection conditions that should be fulfilled. Any kind of action on bats or their habitat is also under the jurisdiction of the same Ministry. Injured or exhausted bats may be transferred to two out of five authorized rescue centers. Exhausted or injured bats may be rehabilitated by any member of the public providing he/she has skills and can provide appropriate conditions. All such cases should be reported and approved by the Ministry of Culture.

CYPRUS (EU, BERN)

All bats and their roosts are protected by the Law on Protection and Management of Nature and Wild Life (No. 153(I)/2003). For all protected species, including bats, it is prohibited to disturb, kill, hunt, capture, keep or stuff them, or to damage or destroy their reproductive, resting, or hibernation sites. It is also prohibited to acquire, transport, import, export, exchange or sell them, dead or alive, in whole or in part.

Bats can be kept for scientific research by national and international research groups under a permit issued by the Department of Environment (http://www.moa.gov.cy/moa/agriculture.nsf/environment_en/environment_en), which is the national authority for issuing research permits.

CZECH REPUBLIC (EU, EUROBATS, BERN)

All species of bats are protected by the Nature and Landscape Act No. 114/1992 (updated in 2004). All species are also listed as critically endangered or endangered by novel regulation 175/2006 of regulation 395/1992. Important hibernacula are designated as, or included in, Particularly Protected Areas (under Act No. 114/1992). Capturing or handling bats or interference with bat roosts requires a permit. Exceptions (permits) to the Act are granted by competent authorities (Ministry of Environment‘s Agency for Nature and Landscape Protection [http://www.nature.cz], Ministry of Agriculture, and regional authorities). Offenses against the law are penalized by the Czech Environmental Inspectorate (http://www.cizp.cz/lang/l2).

DENMARK (EU, EUROBATS, BERN)

All bats have been protected by law since 1931 (Game Act of 1931). Since the Game Act of 1967 and the Hunting and Game Act of 1993, all mammals and birds without an open season are fully protected and must not be hunted or killed. This has been further strengthened by Statutory Order No. 901 of 11 July 2007. Furthermore it is forbidden to ring or mark bats without a license. The licensing authority is the Danish Forest and Nature Agency (http://www.naturstyrelsen.dk/English/English/).

The habitats of bats are protected by the Nature Act of 1992. This act prohibits altering the state of natural lakes of more than 100 m^2 and of watercourses. Moors, bogs, marshes, swamps, coastal meadows, humid permanent grasslands and uncultivated, dry meadows are covered by the same prohibition when the habitat types extend to more than 2,500 m^2 either separately, jointly, or in connection with the aforementioned lakes.

ESTONIA (EU, EUROBATS, BERN)

All bat species have been legally protected in Estonia since 1958 and are now protected by the Nature Conservation Act 2004 (as amended), which implements the requirements of the Habitats Directive. Bats are in Category II of protection, so it is illegal to kill, injure, or disturb them during the breeding, hibernation, or migration seasons. Permits to capture and ring bats are available from the Environmental Board (http://www.keskkonnaamet.ee/eng), which also controls the possession of live or dead bats.

FINLAND (EU, EUROBATS, BERN)

All bats are protected by the Nature Conservation Act from killing, capturing, deliberate harming, or disturbing. It is forbidden to destroy or damage the breeding or resting sites of bats. Permission for derogations from these provisions, for reasons mentioned in the Habitats Directive, can be given by the regional Centres for Economic Development, Transport and the Environment.

According to the Nature Conservation Act, a sick or injured animal of a protected species should be helped. This is actually an obligation concerning any animals based on the Animal Welfare Act. Temporary possession of the animal is allowed for transportation and first aid. There are no clear restrictions on who can take care of sick animals, but when the animal needs surgery or administration of veterinary medicines, advice should be sought from a veterinarian. Otherwise the Animal Welfare Act regulations should be followed. Licenses to possess bats in captivity for scientific purposes are issued by the regional Centres for Economic Development, Transport and the Environment. Advice on legal and licensing matters can be sought from these regional centers (http://www.ely-keskus.fi/fi/Sivut/default.aspx) and the Ministry of Environment (http://www.ymparisto.fi). Licenses for animal experiments (e.g., blood sampling) are issued by the Animal Experiment Board in Finland (http://www.intermin.fi/lh/etela/hankkeet/ellapro/home.nsf/pages/indexeng).

The Åland Islands have their own nature conservation legislation. Under Section 14 of the Nature Conservation Act of the Åland Islands, 1998, all mammals are protected except game species. According to the Animal Welfare Act (1998:95), sick or injured wild animals should be helped. Landskapsregereringen (The State Department of Åland) may issue licenses to possess bats in captivity for scientific purposes (http://www.regeringen.ax/).

FRANCE (EU, EUROBATS, BERN)

All bats are protected by law (Code de l'environnement). A special regulation (April 2007) fully protects the animals and their habitats as required by the Habitats Directive. A further regulation from the Code forbids the keeping of bats in captivity without two special administrative authorizations, one relating to the quality of the establishment, and the other relating to the competence of the keeper. In effect, this means that bats are only kept in rescue centers under veterinary control. Licenses can also be issued, on the advice of the National Council for Nature Protection, to allow the capture of bats. For threatened species, licenses are issued by the Ministry of Ecology (http://www.developpement-durable.gouv.fr/), for others, Préfets of Departements in the French regions.

GEORGIA (EUROBATS, BERN)

The protection and use of animal resources is regulated by the Law on Fauna (1996), shortly to be updated. Any taking of animals, including bats, from the wild is prohibited unless authorized by the Ministry of Environmental Protection and Natural Resources. Exceptions are allowed only when an animal needs emergency veterinary care, with an obligation to return the animal to the wild afterwards. The revised law permits the killing of any animal that attacks a person or causes damage to property. The latter case allows for different interpretations when defining the damage caused.

The Georgian Criminal Code includes articles stipulating punishment for the destruction of protected species' habitats (Articles 287-289), punishable with imprisonment for 3-5 years. Illegal taking of animals from the wild entails a fine for the offense and compensation for damage. The amounts of the fine and compensation are currently being revised.

Twenty-eight species of bats occur in Georgia. The following 4 bat species are included in the Red List of Georgia (2006): *Rhinolophus mehelyi, R. euryale, Myotis bechsteinii* and *Barbastella barbastellus.*

GERMANY (EU, EUROBATS, BERN)

Bats are protected by the federal Nature Protection Law (Bundesnaturschutzgesetz) and the respective legislation of the German federal states (Länder). There are two levels of protection in Germany, with some species receiving stronger protection than others. All German bat species receive the stronger protection. It is an offense to catch, kill or injure bats, or to take them from the wild. It is forbidden to damage or destroy bat roosts, or to disturb bats while they are in their roosts (including taking photographs). It is illegal to keep or trade bats. Exceptions (e.g., for scientific purposes or for injured bats) are possible. The Ministries in charge of nature conservation (i.e., Minister for Environment, Nature, Conservation, and Nuclear Safety) for each of the 16 federal states (Länder) are responsible for issuing the necessary permits.

GREECE (EU, BERN)

The legislation on the protection of biodiversity was updated and consolidated in Law 3937/2010 (Conservation of Biodiversity and other provisions). This law aims to secure a favorable conservation status for biodiversity, especially habitats and species classified as Important, Rare, or Threatened. The law implements the EU Habitats and Species Directive and introduces a National Red List in compliance with the criteria of the International Union for Conservation of Nature (IUCN). The latter will be updated every 10 years. For listed species, deliberate killing, hunting, trapping, possessing or trading are prohibited. Research on endemic and other protected species and habitat types is under the jurisdiction of the Ministry of the Environment, Energy and Climate Change (http://www.ypeka.gr), which can issue any necessary permits.

A new version of the Red Data Book of Threatened Animals of Greece was published in 2009 by the Hellenic Zoological Society and WWF Greece. According to this version, among the 35 bat species reported from Greece, 2 species (*Barbastella barbastellus* and

Eptesicus bottae) were classified as Endangered, 6 species (*Rhinolophus mehelyi, Myotis daubentonii, Nyctalus lasiopterus, Pipistrellus hanaki, Plecotus auritus* and *P. macrobullaris*) as Vulnerable, and 8 species as Near Threatened. Eleven species were classified as Data Deficient due to lack of knowledge and recent changes in their systematic status (e.g., genus *Plecotus*).

HUNGARY (EU, EUROBATS, BERN)

All bats have been protected by law since 1901, with amendments added in 1954, 1974 and 1975. Bats are currently protected by the Nature Conservation Law Decree of 1982 and subsequent amendments. There is also a new Act on the Conservation of Nature of 1996.

It is prohibited to capture, kill, keep, or disturb bats in any way. Their roosts are also protected. A total of 2,924 caves are protected of which 124 are strictly protected.

The law does not allow keeping any protected animal (dead or alive) without license. The National Inspectorate for Environment, Nature and Water (http://www.orszagoszoldhatosag.gov.hu/) is the licensing authority for strictly protected animals and County Departments for protected animals.

There is no restriction on who can possess sick bats, but surgery or the administration of veterinary medicines is controlled by legislation that restricts this to qualified veterinary surgeons. Strictly protected bat species are *Barbastella barbastellus, Miniopterus schreibersii, Myotis bechsteinii, M. dasycneme, M. emarginatus, Nyctalus lasiopterus, Rhinolophus euryale*, and *R. ferrumequinum*.

IRELAND (EU, EUROBATS, BERN)

All bats and their breeding places are protected under the Wildlife Act (1976). The Wildlife (Amendment) Act of 2000 provides protection for all resting places, not just maternity roosts. This Act also introduced the requirement to possess a license when filming or photographing bats at their roosts. Bats are also protected by the European Communities (Natural Habitats) Regulations of 1997 (Statutory Instrument 94 of 1997), which enacts the Habitats Directive.

It is not an offense to capture an injured bat for the purposes of tending it for later release, or to humanely kill such a bat if it can not be rehabilitated. Licenses can be issued to allow the long-term possession of injured bats deemed unlikely to survive unaided in the wild. Licenses can also be issued for the capture and tagging of bats for scientific study. License applications are assessed by the National Parks and Wildlife Service of the Department of the Environment, Heritage and Local Government (http://www.npws.ie/).

ISRAEL (BONN)

The Wildlife Protection Law 1955 protects wildlife species and includes some animal welfare provisions such as a clause banning hunting and catching wildlife with traps, snares, nets, and poisons. The Law and its regulations incorporate the CITES (Convention on International Trade in Endangered Species) treaty into Israeli legislation. All species of bats except the Egyptian fruit bat (*Rousettus aegyptiacus*) are protected by this law.

Bats may not be captured, moved or killed, and their roosts may not be destroyed or disturbed. However, the Israel Nature and Parks Authority (http://www.parks.org.il/ParksENG/) may issue licenses to capture or move bats. Scientific licenses for banding bats or catching for identification and research are also issued by this Authority.

ITALY (EU, EUROBATS, BERN)

The first Italian law protecting animals was the Royal decree of 5 June 1939, No. 1016, forbidding killing or capturing bats of all species. The main Italian law regarding wildlife and hunting (L 11 February 1992, No. 157) also protects bats. The Habitats and Species Directive is implemented by Presidential Decree (DPR) 357/97. All species of bats are listed in Annex IV of the Directive, which gives them full protection; thirteen species are also listed in Annex II, thus requiring the designation of Natura 2000 sites.

Bats can be kept in captivity for scientific purposes provided this activity is authorized by the Environment Ministry and the Regional authorities. According to current procedures, all scientific activities concerning bat capture and handling must be authorized by the Ministry; proposals should be submitted both to the Ministry and the Higher Institute for Environmental Research (ISPRA) (http://www.isprambiente.gov.it/site/it-IT/), which acts as the Ministry's scientific advisor. A separate permit should also be issued by the relevant regional authorities according to local laws.

Injured bats may be kept in captivity by officially recognized rescue centers only. They must be recognized by regional authorities and meet well-defined requirements in terms of animal welfare and public health.

The relevant national authority issuing permits is the Nature Protection Directorate (Direzione per la Protezione della Natura) of the Environment Ministry. Regional authorities have their own offices (environmental sections) issuing local permits.

JORDAN (BONN)

All wild animals, including bats, are protected under Law 44 (2002). In 2008, Law No. 43 listed 4 species (*Rousettus aegyptiacus, Hypsugo* (*Pipistrellus*) *ariel, Myotis capaccinii* and *M. nattereri*) in Appendix 3.

Two non-governmental organizations (Humane Center for Animal Welfare and the Society for the Protection of Animals World Wide) provide veterinary services for injured animals, including all forms of wild animals. Very few cases involving bats have been reported over the past five years. Bats can be kept for scientific research under a permit issued by the Royal Society for the Conservation of Nature (RSCN). Bat research under laboratory conditions is limited; however, field research is regulated by the RSCN for national and international research groups. The RSCN (http://www.rscn.org.jo/) is the national authority for issuing research permits and transport of specimens abroad for scientific studies.

LATVIA (EU, EUROBATS, BERN)

Bats were legally protected in April 1957 and are now protected by The Law on Conservation of Species and Habitats. This came into force on 5 April 2000. It is illegal to capture or kill bats, deliberately disturb bats, particularly during reproduction, hibernation or migration, destroy roosts, or to keep bats in captivity.

All 15 species of bats found in Latvia are listed as Specially Protected Species. Permits are needed to take bats from the wild or keep them in captivity for scientific purposes, although there is no specific limitation on tending injured bats. The Nature Protection Agency (http://www.daba.gov.lv/public/eng/) is the responsible institution for issuing permits related to protected species.

LEBANON

All wildlife, including bats, is protected by the hunting law with Decree No. 580 approved on 25/4/2004. The wildlife center, Educational Centre for Wildlife Conservation, is the only center in Lebanon that cares for and rehabilitates Lebanese wildlife, although bats are rarely brought to the center. In general, bats are in a safer situation compared to other countries, except in very few accessible caves where fire made by people is the cause of disturbance. Keeping bats for scientific purposes is not covered by regulations. However, at present, bats are not kept for scientific research or display in Lebanon. The authorities responsible for issuing permits are the Lebanese National Council for Scientific Research, Ministry of Environment, and Ministry of Agriculture.

LITHUANIA (EU, EUROBATS, BERN)

All bats are protected by the Law on Wild Fauna 1997 (as amended) and the Law on Protected Plant, Animal and Fungi Species 2009. Together, these implement the requirements of the Habitats Directive. The Nature Protection Department of the Ministry of Environment (http://www.am.lt/VI/index.php) is the authority responsible for issuing permits to capture or keep bats.

LUXEMBOURG (EU, EUROBATS, BERN)

All bats and their roosts were protected by Grand-Ducal Regulations in November 1972. This was reinforced by the Nature Protection Legislation of 11 August 1982 and subsequently 19 January 2004, and their respective Grand-Ducal Regulations of 8 April 1986 and 9 January 2009. For all protected species, including bats, it is prohibited to disturb, kill, hunt, capture, keep or stuff them, or to damage or destroy their reproductive, resting and hibernation sites. Moreover, it is prohibited to acquire, transport, import, export, exchange or sell them, be it dead or alive, in whole or in part.

Licenses can be issued by the Department of Environment (http://www.environnement.public.lu/) of the Ministry of Sustainable Development and Infrastructure, for scientific reasons, for reasons of public safety or other public interest, for translocation or reintroduction purposes, or for keeping a limited number in captivity.

MACEDONIA (EUROBATS, BERN)

Bats are protected by the Nature Protection Law of Macedonia, which implements the requirements of the Bern and Bon Conventions. There is no single list of protected species or national Red List. It is forbidden to keep or deliberately kill in the wild all protected species, (but also non-protected if there is not a hunting ground), including all species of bats; also, it is forbidden to deliberately disturb them, especially during breeding, hibernation or migration; to damage or destroy their reproduction and rearing places or habitats; to keep, transport, sell or exchange species from the wild; or to perform actions that may cause the disappearance or destruction of populations. The Ministry of Environment and Physical Planning (http://www.moepp.gov.mk/) grants licenses for scientific research and educational purposes; in the case of protected areas, a similar license must also be issued by the relevant management authority.

MALTA (EU, EUROBATS, BERN)

The Flora, Fauna and Natural Habitats Protection Regulations, 2006 (Legal Notice 311 of 2006, as amended) prohibits, via Regulation 25, the pursuit, capture or killing of any bat, as well the deliberate destruction, keeping, transport, or sale of all bat species (being covered under Schedule V[a] to the Regulations). Deliberate disturbance of any species, particularly during periods of breeding, rearing, hibernation or migration, as well as the destruction or deterioration of breeding sites or resting places is also prohibited. These regulations also require the designation of Special Areas of Conservation (SACs) for certain bat species, such as *Rhinolophus hipposideros* and *Myotis punicus.*

The Malta Environment and Planning Authority (MEPA) (http://www.mepa.org.mt/), as the competent authority for wildlife legislation, issues a number of permits every year, allowing research and monitoring of bat populations. The results obtained from this research allow further analysis of the status of bats in the Maltese Islands, and direct conservation efforts. Permits can be issued for *bona fide* studies and would normally be accompanied by a number of conditions.

MOLDOVA (EUROBATS, BERN)

Eighteen of the 21 bat species known to occur in Moldova are protected by the Law on Animal Kingdom. The other 3 species have only recently been recorded and so are not yet protected. Six of these species are included in the second edition of the Red Book of Moldova (Law on Red Book of Moldova No. 325, 25.12.2005). Bats are also protected by the Law on State Protected Natural Areas. The authority for issuing permits is the Biodiversity and Natural Resources Department, Ministry of Environment (http://www.mediu.gov.md/).

MONACO (EUROBATS, BERN)

Bats are not currently protected. An environmental law, Code de l'Environnement, that will implement the requirements of the Bern Convention is being prepared.

MONTENEGRO (EUROBATS, BERN)

The main wildlife law is The Law on Nature Protection, which protects all bats. This forbids the harassment, capture, and injuring of protected wild animals, reducing their populations or destroying or damaging their habitat or changing their living conditions. It is forbidden to catch, keep, or kill Strictly Protected wild animal species, disturb then, especially during breeding, rearing, migration or hibernation, damage or destroy their nests, eggs or breeding, or resting areas.

Exception can be made by the administrative authorities, provided it does not threaten the survival of the species concerned. Some Strictly Protected and Protected species can be used for scientific purposes and for exhibition, with the permission of administrative authorities. The Law on Nature Protection states that specimens can be kept in captivity only with the permission of the competent authority. Anyone who finds dead, sick or injured specimens of Strictly Protected and Protected wild animals should immediately notify the competent authority, which may allow keeping the sick or injured animals in captivity for treatment and recovery. The competent authority is the Environment Protection Agency of Montenegro (http://www.epa.org.me/index.php?lang=en).

THE NETHERLANDS (EU, EUROBATS, BERN)

On 1 April 2002, the Flora and Fauna Act came into force, replacing several earlier pieces of legislation such as the section of the Nature Conservation Act 1973 that protected bats. The Flora and Fauna Act implements the species protection requirements of the Habitats Directive and thus protects all species of bats.

Under the Flora and Fauna Act it is forbidden to catch or kill bats, or to attempt to do so. It is also forbidden to possess live or dead bats, or parts of bats, whether preserved or not, or to offer bats for sale. It is forbidden to intentionally disturb bats. It is also forbidden to damage, destroy or disturb maternity roosts, resting places, or other regularly used roosts of bats. In the case of roosts, intentional or non-intentional destruction, damage or disturbance is forbidden. Licenses are required for research or any other handling or disturbing of bats. The keeping of bats by the public is not allowed unless a license is issued to rehabilitate diseased or injured bats.

The Ministry of Economic Affairs, Agriculture and Innovation (http://english.minlnv.nl) is the competent authority to issue licenses in accordance with this Act.

NORWAY (EUROBATS, BERN)

All bats are fully protected in Norway. The Nature Diversity Act of 19 June 2009, and its amendments of 17 September 2010, provides species management and protection formerly provided by the Wildlife Act of 1981. The new Act imposes a stronger protection regime for bats (and other wildlife) by explicitly requiring permission from the authorities for potentially damaging activities (Article 6). For example, roosts are now given a higher degree of protection than previously, even at times when they are not occupied.

Article 15 (Protection and Management Principles) of the Nature Diversity Act states that unnecessary harm and suffering caused to animals occurring in the wild and their nests, lairs and burrows must be avoided.

According to Article 7 of the Wildlife Act of 1981, wildlife may not be kept in captivity unless permission is obtained from the competent authority, the Directorate for Nature Management (http://www.dirnat.no/). Only one such license for bats has ever been granted, and that was to the Norwegian Zoological Society's bat rehabilitation center.

POLAND (EU, EUROBATS, BERN)

On October 2008, the Act on Nature Conservation was amended to implement the requirements of the EU Habitats Directive, and thus protect all bats from being killed, injured and their roosts from damage or destruction. Regulations under the Law on Experiments on Animals, 2005, allow the capture of bats under license from the Ministry of Environment (http://www.mos.gov.pl/), but only recognized institutions can carry out the full range of scientific research.

PORTUGAL (EU, EUROBATS, BERN)

Portuguese bats have been protected since 1967 by the Hunting Law. National and international legislation protects all species from being killed, injured, disturbed or captured, and their roosts from being damaged or destroyed. The law allows for sick or injured bats to be kept for rehabilitation. Only official rehabilitation and rescue centers and their staff are allowed to possess bats after being licensed by the wildlife authority. Surgery or the administration of veterinary medicines is controlled by legislation that restricts this to qualified veterinary surgeons. Licenses can also be issued to possess bats in captivity for scientific purposes, although this is rarely done. The wildlife licensing authority, which can advise on legal and licensing matters, is the Institute for Conservation of Nature and Biodiversity (http://portal.icnb.pt/ICNPortal/vEN2007/).

ROMANIA (EU, EUROBATS, BERN)

All bats are protected by law. Emergency Government Ordinance (GO) 57/2007 regarding the regime of natural protected areas, the conservation of natural habitats, wild flora and fauna, through Annex 3, protects many species, including bats. Annex 4 also protects all microchiroptera, declaring them Strictly Protected species in Romania. The two Annexes are the implementation of the EU Habitat Directive in Romanian law. General protection measures are also included in Emergency Government Ordinance 195/2005, confirmed with modifications and extensions by Law 265/2006.

Paragraphs 31-40 of GO 57/2007 recognize the need to monitor bat populations and the possibility of adding new species to the Natura 2000 list if necessary. Paragraph 33 forbids collecting, capturing, hunting, killing or injuring bats, and also the disturbance of bats during the reproduction or hibernation period, or the damage, destruction or disturbance of bat roosts. It is forbidden to keep dead or live bats in captivity. Trading or transporting of bats is also forbidden.

Paragraphs 41-49 protect important bat habitats including underground habitats.

Enforcement of the legal framework is led by the National Environment Guard, which is a subordinate body of the Ministry of Environment and Forests.

State institutions (museums, research institutions, and universities) can keep bats for scientific purpose provided they have a permit from the Romanian Academy (http://www.acad.ro/def2002eng.htm) and the Ministry of Environment and Forests (http://www.mmediu.ro/index_en.html). In the case of research in a specific habitat or territory, a permit is needed from the administrator of the territory, which can be a local authority, environmental protection agency, national (natural) park administration, or protected area administration.

RUSSIAN FEDERATION

The Russian Federal Law on Fauna (1995) is the main act that prohibits the capture, collection, storage, purchase, selling or transport of any animal species, including their products, parts or derivatives, without adequate permission, or be in violation of conditions imposed by a permit, or in breach of any other established procedure. Under the Federal Law, On the Protection of Environment, all species included in federal or regional Red Data Books are strictly protected. Russia criminalized both actions and omissions that might lead to death, reduction in number, or disturbance to the species listed in the Red Data Book of the Russian Federation. Penalties include confiscation of the captured items and equipment, as well as substantial fines. Destruction of critical habitats for species included in the Russian Red Data Book that results in the loss of their populations is punishable under Article 259 of the RF Criminal Code with imprisonment for up to 3 years. Russian procedures for assessing Red Data Book category only partially comply with the IUCN Procedures and are rather original in many aspects (Regulation On Keeping Up The Red Data Book Of The Russian Federation 1997). Protected species are allocated to one of the following categories:

- I endangered (the threat of extinction is very high);
- II declining population;
- III rare (stable or slowly increasing population);
- IV not numerous, poorly studied, sporadically distributed, uncertain status;
- V restored population (due to conservation measures), not liable to economic use.

The Red Data Book lists the major factors threatening the species and recommends protective measures. Federal or regional authorities should take this into account if they establish a new nature reserve, or approve an industrial project that can affect any of the listed species. There are about 50 species of bats known from the territory of the Russian Federation. Seven are listed in the Red Data Book of Russia: *Miniopterus schreibersii* (I), *Myotis blythii* (II), *M. emarginatus* (II), *Nyctalus lasiopterus* (III), *Rhinolophus mehelyi* (II), *R. ferrumequinum* (III), and *R. hipposideros* (III). Many others are listed in the regional registers, and thus are legally protected.

A special permit from the federal environmental regulator Rosprirodnadzor (http://rpn.gov.ru/) is needed for any kind of interference with species listed in the federal Red Data Book. Thus a permit is also needed for keeping species in captivity for scientific purposes. As a rule, the legislation of Russian regions also require a permit from regional environmental ministries or institutions for keeping or affecting bat species included in regional Red Data Lists or unlisted.

SERBIA (BERN, BONN)

The Law on Environmental Protection (Official Gazette of Republic of Serbia No. 2004), the Law on Nature Protection (Official Gazette of Republic of Serbia No. 36/09, 88/10), and the Rulebook on Proclamation and Protection of Strictly Protected and Protected Wild Species of Plants, Animals and Fungi (Official Gazette of Republic of Serbia No.

5/10) all cover the protection of species and give protection to all bats. The Ministry of Environment and Spatial Planning, Directorate of Environmental Protection is responsible for issuing permits to allow otherwise illegal acts.

SLOVAKIA (EU, EUROBATS, BERN)

The first legal protection for bats and other wildlife in former Czechoslovakia was the Act on the State Protection of Nature (1955), which protected all bats. Since 2002, all Slovak bats (28 species) have been protected by the Act of NC SR No. 543/2002 on Nature and Landscape Protection. This act was prepared in line with nature protection legislation of the European Union. The disturbance of bats, killing, wounding, trapping, catching or removing is prohibited. Damage and destruction of the habitats and roosts of bats are also prohibited. This prohibition does not apply in cases of immediate threat or damage to human life or health, or damage to property, or if covered in special regulations. Practical issues concerning this act are implemented by the Decree of Ministry of Environment SR No. 24/2003. This Ministry (http://www.minzp.sk/en/) is also responsible for issuing permits.

All caves (that could be used as roosts by bats) are protected as natural monuments or natural reserves (the highest degree of protection). Two EU Natura 2000 sites were selected especially for bats, Dubnícke bane (mine hibernacula) and Bradlo (hibernacula and nursery roost in mine).

SLOVENIA (EU, EUROBATS, BERN)

All species of bats in Slovenia are included in Rules on the Inclusion of Endangered Plant and Animal Species in the Red List and are therefore protected by the Decree on Protected Wild Animal Species, as well as by the Nature Conservation Act. This legislation creates offenses for taking from the wild, disturbing, trading, breeding, acquisitioning of specimens (dead or alive), keeping in captivity, or introducing into or repopulating the wild. The Environmental Agency of the Republic of Slovenia (http://www.arso.gov.si/en/), a body of the Ministry for Environment and Spatial Planning, may issue permits for catching, collecting, keeping in captivity, moving, researching or exporting bats, where it is necessary for expert or scientific research or for other justifiable reasons, with consideration of preserving healthy populations in the natural environment. Injured or exhausted bats may be kept for rehabilitation by any member of the public for up to 30 days, although the authorities should be notified within 3 days; after that period, bats should be taken to an official animal rehabilitation center. All bat carcasses and other remains should be transferred to the Slovenian Museum of Natural History.

SPAIN (EU, BONN, BERN)

All fauna is protected by law (Ley 42/2007, de 13 de diciembre, del Patrimonio Natural y de la Biodiversidad). However, most endangered species are covered by a special regulation (Real Decreto 139/2011, de 4 de febrero; http://www.boe.es/boe/dias/2011/02/23/pdfs/BOE-A-2011-3582.pdf). This regulation includes all 31 Spanish bat species as Specially Protected Species that need action to protect them (Listado de Especies Silvestres en

Régimen de Protección Especial). Thirteen of those species are listed in the Red List of Endangered Spanish Species (Catálogo Español de Especies Amenazadas), but just *Myotis capaccinii* is considered Endangered with a conservation program required within 3 years. The other 12 species are listed as Vulnerable and a program should be launched within 5 years.

This national legal framework has been adapted by each of Spain's 17 autonomous communities. The departments in charge of nature conservation for each of the communities, plus Ceuta and Melilla, are responsible for issuing any permits needed to capture or keep bats. The autonomous communities are not able to have less restrictive regulations than the national requirements, but they are able to develop their own red lists, which must include at least species present in the Red List of Endangered Spanish Species.

The autonomous communities are responsible for creating sites of special natural interest (lugares de interés comunitario), as required by the Habitats Directive, to protect ecosystems and roosts of cave dwelling bats and forest bats. These sites could be included in the Natura 2000 network. The communities also have protected areas, such as natural parks and micro-reserves, where bats can be monitored.

SWEDEN (EU, EUROBATS, BERN)

Bats receive protection under the Hunting legislation and the Species Protection Ordinance, which enacts the EU Habitats Directive. Both state that all bats are protected. Bats may not be captured, moved or killed, and their roosts may not be destroyed or disturbed. However, a competent authority, usually the County Administration, may issue licenses to handle or move bats. To handle injured, sick or orphaned bats also requires a license. Scientific licenses for banding bats or catching them for identification are issued by the Swedish Environmental Protection Agency (http://www.naturvardsverket.se/fladdermus).

SWITZERLAND (BERN, BONN)

At the federal level, all bats are protected by the federal law on Nature and Heritage Protection (Bundesgesetz über den Natur- und Heimatschutz [NHG]) and Regulations that arise from it. In line with the requirements of the Bern Convention, bats are protected from killing, injuring or capturing, and their roosts are protected from damage or destruction.

Individual Cantons are responsible for implementing the legislation and can issue permits to allow the study or possession of bats. Injured bats may be kept in captivity until they can be released, but permission is needed from the veterinary authorities of the Canton. A national project manages bat conservation at the local level, with bat conservation officers in most Cantons (see: http://www.fledermausschutz.ch/).

TURKEY (BERN)

Bats are legally protected by the Ministry of Environment and Forestry under the category of 'Wild animals under protection by the Ministry of Environment and Forestry.' This also involves a prohibition of hunting bat species, among other species, during the hunting season. Any type of research to be undertaken on bats in the wild or in captivity requires a permit from the General Directorate of Nature Conservation and National Parks (http://www.milliparklar.gov.tr/DKMP/HomePage.aspx?sflang=tr). The same is true for any bat research that is to be done in captivity, which also requires a permit from the Ethics Committee of the university in which the research is being conducted.

UKRAINE (EUROBATS, BERN)

Bats are protected by several laws including the Law on Environment Protection (1991), the Law on the Animal Kingdom (2001) (amended in 2009 and 2010) and the Law on the Red Data Book of Ukraine (2002), which enhanced the conservation status of listed species, including all bats. The Ministry of Environment and Nuclear Safety has prepared legislation on penalties relating to the illegal capturing or killing of animals, including bats.

A more recent law should provide improvements for the protection of bats. This is the All-Ukrainian Program of Establishment of the National Ecological Network for the period 2001-2015. There is also a law, On Ecological Network (2004), which additionally provides a basis for protection of bat habitats.

The taking of protected species from the wild is only allowed under special permits that are granted by the competent authority, the Ministry of Environment and Natural Resources of Ukraine (http://www.menr.gov.ua/), after assessment by the National Commission on Red Data Book. Bats can not be hunted, killed, captured, exported or imported without such permits. The use of protected species for scientific or breeding purposes is allowed, but only by state scientific programs that have been endorsed by the Government. Bats' breeding and resting sites are also protected. According to the Law of Ukraine, On the Protection of Animals Against Cruelty (2006), the taking of injured animals from the wild without a permit is allowed only for the purposes of providing them with assistance.

UNITED KINGDOM (EU, EUROBATS, BERN)

In all four countries of the UK, all bat species are protected from killing, injuring, disturbing or capturing, and their roosts are protected from damage or destruction, although the legislation varies slightly. In England and Wales, the most important relevant law is the Conservation of Habitats and Species Regulations 2010; in Scotland it is the Conservation (Natural Habitats &c.) Regulations 1994 and in Northern Ireland, the Conservation (Natural Habitats &c.) Regulations (Northern Ireland) 1995. In all cases, the law allows for sick or injured bats to be kept for rehabilitation or, if unavoidable, killed; although, long-term captives or dead specimens must be held under license. There is no restriction on who can possess sick bats, but surgery or the administration of veterinary medicines is controlled by legislation that restricts this to qualified veterinary surgeons. Licenses can be issued to possess bats in captivity for scientific purposes, although this is rarely done. The wildlife licensing authorities, who can advise on legal and licensing matters, are Natural England (http://www.naturalengland.org.uk/), Scottish Natural Heritage (http://www.snh.gov.uk/), the Countryside Council for Wales (http://www.ccw.gov.uk/Splash.aspx), and the Northern Ireland Environment Agency (http://www.doeni.gov.uk/niea/).

The UK also has a number of Crown Dependencies, which have their own wildlife legislation. For the Isle of Man, the Wildlife Act 1990 gives full protection to all species of bats occurring on the island. The Act is administered by the Department of Environment Food and Agriculture (http://www.gov.im/daff/Wildlife/), which is responsible for issuing licenses. Jersey, largest of the Channel Islands (http://www.gov.je/Environment/), has the Conservation of Wildlife (Jersey) Law 2000, which gives protection to listed species including all bats; the Bailiwick of Guernsey, which includes Alderney, Sark and Herm, has no specific wildlife legislation, but like Jersey, is included in the UK's membership of Eurobats.

ACKNOWLEDGMENTS

Many colleagues across Europe and the Middle East very generously contributed to this chapter by drafting paragraphs for their country. Any errors introduced during editing are entirely my responsibility. Grateful thanks to: Anna Nele Herdina (AT), Jasminko Mulaomerović (BH), Boyan Petrov (BG), Hans-Peter Stutz (CH), Daniela Hamidovic (HR), Elena Erotokritou (CY), Helena Jahelkova (CZ), Matti Masing (EE), Michel Perret (FR), Ioseb Natradze (GE), Oliver Schall (DE), Eeva-Maria Kyheröinen (FI), Panagiotis Georgiakakis (GR), Zoltan Bihari (HU), Ferdia Marnell (IE), Carmi Korine (IL), Danilo Russo (IT), Zuhair Amr (JO), Mounir Abi-Said (LB), Laurent Schley (LU), Gunārs Pētersons (LV), Branko Micevski (MK), Lisa Schembri Gambin (MT), Azra Haveric (ME), Peter Lina (NL), Per Ole Syvertsen & Øystein Størkersen (NO), Luisa Rodrigues (PT), Abigel Szodoray-Paradi (RO), Suren Gazaryan (RU), Martin Cel'uch (SK), Primož Presetnik (SI), Carles Flaquer & Ricardo Gómez Calmaestra, Marie Nedinge (SE), Raşit Bilgin (TR), Volodymyr Domashlinets (UA).

United States Of America

John Seyjagat

INTRODUCTION

In the United States, bats are maintained under captive conditions primarily for research, rehabilitation, zoological display, and education. Bat workers must have a basic knowledge of state and federal wildlife laws and the agencies overseeing them. This paper is an overview of agencies that oversee wildlife regulations applying to bats.

UNITED STATES FISH AND WILDLIFE SERVICE

The United States Fish and Wildlife Service (FWS) is a federal government agency within the United States Department of the Interior (Fig. 1-1). By working with partners, the FWS is dedicated to the management of fish, wildlife, and habitats for the continuing benefit of the American people.

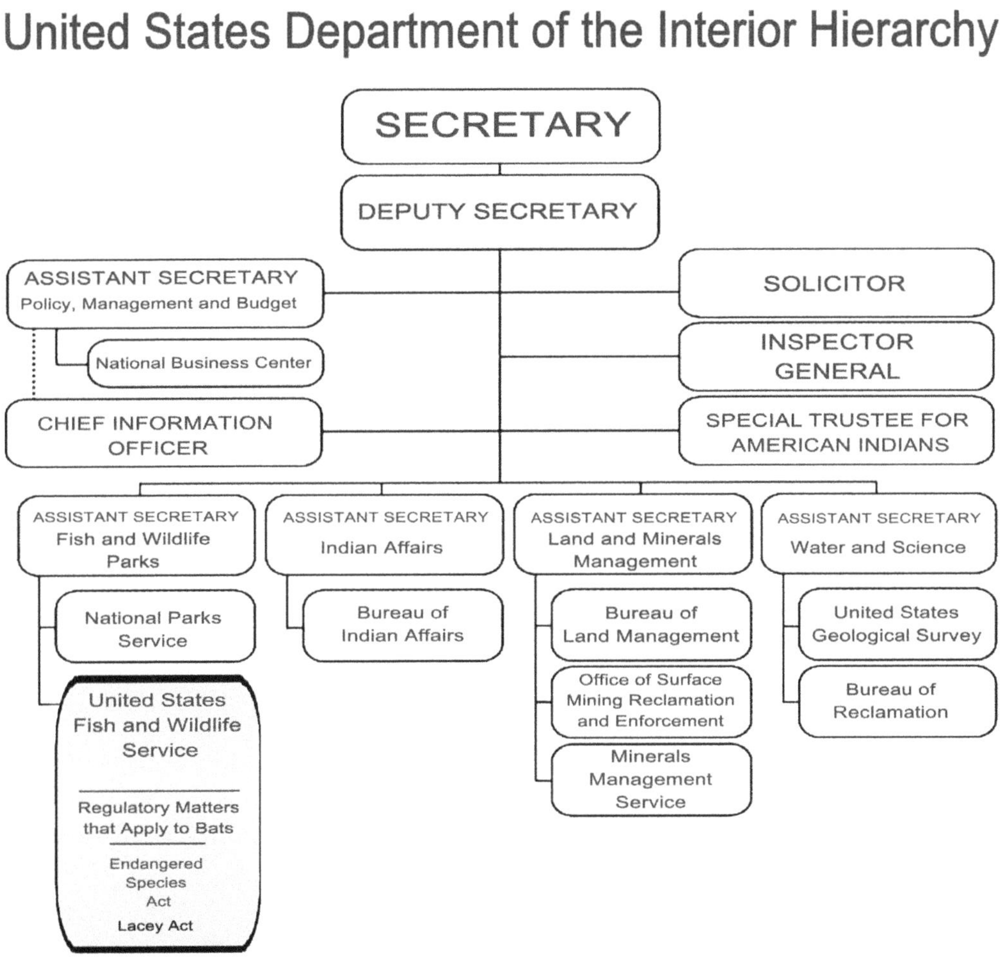

Fig. 1-1. Hierarchy of the United States Department of the Interior with respect to regulatory matters that apply to bats. Illustration modified by S. M. Barnard from Wikimedia Commons.

The agency regulates the importation of all wildlife entering the United States. To import bats, a permit is required from the country of origin. If the species to be imported is endangered, a CITES permit (Convention on International Trade in Endangered Species of Wild Fauna and Flora) is also required. Endangered species of wildlife may not be sold or traded between institutions unless accompanied by a Captive Bred Wildlife (CBW) Certificate (see http://www.fws.gov/forms/3-200-41.pdf). To obtain appropriate permits issued by the FWS, visit http://www.fws.gov/permits/overview/overview.html, or contact the agency at:

United States Fish and Wildlife Service
Office of Management Authority
4401 N. Fairfax Drive, Room 212
Arlington, VA 22203-1610
Phone: 800-358-2104 or 703-358-2104
Fax: 703-358-2281

Endangered Species Act

The Endangered Species Act (ESA) (http://www.nmfs.noaa.gov/pr/pdfs/laws/esa.pdf) is administered jointly by the FWS, which has jurisdiction over terrestrial and freshwater species, and NOAA Fisheries, a bureau of the Department of Commerce, which is responsible for marine and anadromous species. The Act provides broad protection for species of fish, other wildlife, and plants that are listed as threatened or endangered in the United States or elsewhere. Provisions are made for listing species, as well as for recovery plans and the designation of critical habitat for listed species. The Act outlines procedures for federal agencies to follow when taking actions that may jeopardize listed species, and contains exceptions and exemptions. The Endangered Species Act is also the enabling legislation in the United States for CITES. Criminal and civil penalties are provided for violations of the Act and the Convention.

The ESA provides two levels of protection. Listed as Endangered, and provided the most significant protection, are any species of animals in danger of extinction. Listed as Threatened, and protected by somewhat less restrictive regulations, are animals vulnerable to endangerment in the near future. With limited exceptions, the Act prohibits the using, taking, possessing, selling, or advertising for sale or trade of listed species in the United States and abroad, unless authorized by permit.

A species is listed as Threatened or Endangered for one or more of the following reasons:

1) current or threatened destruction, modification, or curtailment of its habitat or range;
2) over utilization for commercial, recreational, scientific, or educational purposes;
3) decline due to disease or predation;
4) inadequacy of existing regulatory mechanisms, and
5) other natural or man-made factors affecting its continued existence.

Once a listing is confirmed, critical habitat must be designated. Critical habitat is defined as 'the geographic area that is essential to, or requires special protection for, the conservation of a species.' The Act also requires the Secretary of the Interior to develop and implement recovery plans for listed species. Recovery teams formulate recovery criteria and site-specific management plans.

Convention on International Trade in Endangered Species of Wild Fauna and Flora

For purposes of the Convention on International Trade in Endangered Species of Wild Fauna and Flora (CITES), the Secretary of the Interior is designated as the Management Authority and the Scientific Authority and the functions of both Authorities are carried out by the Fish and Wildlife Service. CITES recognizes that unrestricted commercial exploitation is a major threat to species' survival and establishes worldwide controls over trade in certain species of endangered and threatened wildlife and plants. CITES categorizes animals and plants in one of three Appendices as follows:

> Appendix I includes species for which it is determined that any commercial trade is detrimental to the survival of the species. Therefore, no commercial trade is allowed in Appendix-I species. Non-commercial trade in such species is allowed if it does not jeopardize the species' survival in the wild. Permits are required for the exportation and importation of Appendix-I species.
>
> Appendix II includes species for which it has been determined that commercial trade may be detrimental to the survival of the species if that trade is not strictly controlled. Trade in these species is regulated through the use of export permits.
>
> Appendix III includes species listed by a range country that requires the assistance of other parties to ensure that exports of their native species are legal. Permits are used to control and monitor trade in native species. Any CITES party may place a native species in Appendix III.

For additional information on CITES, visit http://www.cites.org/.

Lacey Act

Under the Lacey Act, it is unlawful for any person to import, export, transport, sell, receive, acquire, or purchase fruit bats or their offspring of the genus *Pteropus* without an Injurious Wildlife Permit (http://www.fws.gov/forms/3-200-42.pdf). Bats taken in violation of the Act are subject to forfeiture as well as all vessels, vehicles, aircraft, and other equipment used to aid in the illegal act. Furthermore, such bats may be promptly exported or destroyed at the expense of the importer or consignee. The Lacey Act also applies to any territory of the United States, the District of Columbia, the Commonwealth of Puerto Rico, or any possession of the United States.

Pteropus species held under an Injurious Wildlife Permit require special facilities. These facilities must be permanent, escape-proof, and consist of double enclosed containment such as an enclosure within a building, or cage within another cage. The double-containment concept can also be met with a single enclosure having double-meshed doors, walls, and ceiling. When mesh is used to double-wall an enclosure, it should be separated by at least two inches to minimize the risk of a bat getting its foot trapped between the two walls (Fascione, 1995).

The Injurious Wildlife Permit must be placed on the outside of the enclosure at all times. Injurious wildlife and their progenies may not be removed from the designated enclosure or institution without an Injurious Wildlife Transportation Permit.

On rare occasions, a bat covered by the Lacey Act may be used in outreach educational programs by obtaining a special permit from the FWS Office of Management Authority. It must be shown that the bat in question is permanently marked and identifiable, medically certified non-flighted, and can be transported to and from programs in a double-enclosed, escape-proof carrier.

For additional information on the Lacey Act, see http://www.animallaw.info/articles/ovuslaceyact.htm.

CENTERS FOR DISEASE CONTROL AND PREVENTION

The Centers for Disease Control and Prevention (CDC) considers exotic bats etiological agents and vectors of human disease. The CDC's Foreign Quarantine Regulations (42 CRF - Part 71; http://law.justia.com/us/cfr/title42/42cfr71_main_02.html) prohibit the importation of etiological agents, and distribution after importation, unless accompanied by a permit issued by the CDC (http://www.cdc.gov/od/eaipp/forms/Permit_to_Import_or_Transport_Live_Bats1-17.pdf). Application for a permit to import or transport exotic live bats are managed by the Etiologic Agent Import Permit Program. Completed application forms may be returned by FAX (404-718-2093) or by mail to:

Centers for Disease Control and Prevention
Etiologic Agent Import Permit Program
1600 Clifton Road, N.E. Mailstop A-46
Atlanta, GA 30333
http://www.cdc.gov/od/eaipp/

Bats intended for transfer between institutions should meet the veterinary requirements of the receiving institution. All transfers must be accompanied by:

1) A standard Certificate of Veterinary Inspection (CVI) signed by a licensed veterinarian (Fig. 1-2).
2) FWS Injurious Wildlife Permit when shipping bats in the genus *Pteropus* (http://www.fws.gov/forms/3-200-42.pdf).
3) United States Department of Agriculture/Animal and Plant Health Inspection Service (USDA/APHIS) permit showing Record of Acquisition, Disposition or Transport of Animals (Form 7020; http://www.aphis.usda.gov/animal_welfare/downloads/forms/aph7020.pdf).
4) Animal Data Transfer Form from the American Association of Zoo Keepers (http://www.aazk.org/pdf/2007_AAZK_ADT_Final_Revision1.pdf)

To import into and transfer bats within the United States, the CDC requires that the responsible parties be:

(1) knowledgeable and skilled in the handling of the bats;
(2) directly responsible for pre- and post-quarantine and animal welfare
(3) sufficiently trained in zoonotic disease control to supervise import quarantine isolation and testing (requires an attending licensed veterinarian);
(4) able to institute and enforce appropriate preventive occupational health measures;
(5) able to recognize and diagnose diseases that are of potential human or veterinary public health significance, and
(6) responsible for prompt reporting to the CDC of any disease conditions recognized.

Moreover, all imported bats must undergo a quarantine period of at least 6 months for rabies. All imported bats (and transferred bats not isolated and tested prior to shipment) must be tested, examined, and treated for diseases and parasites during the quarantine period. Employees who work with bats at such facilities should receive pre-exposure rabies vaccination. Bats should be housed in a manner that precludes direct contact with the public.

AMERICAN ASSOCIATION of ZOO VETERINARIANS
STANDARD CERTIFICATE OF VETERINARY INSPECTION
Nº 87085

OWNER .... CONSIGNEE .... DATE Issued ....
ADDRESS .... ADDRESS .... Mode of Transport ☐ Land ☐ Air ☐ Sea
NAME OF AGENT ....
PHONE .... PHONE ....

ANIMAL I.D. Tatoo Band Tag Etc.	NAME Common Scientific	PHYSICAL DESCRIPTION Age Sex Weight Other	HISTORY

HISTORY
Date of Vaccination including name of product used
Date of last deworming including name of product used
Housed with .... other animals
Recent Health problems with similar or adjacent species
Given name and dosage of *all* drugs given prior to or during transport (including antibiotics, restraint or immobilizing drugs)
Tuberculin Used and Dosage .... Site of injection ....
☐ Mammalian old tuberculin
☐ Bovine PPD results: ☐ Pos ☐ Neg ☐ Suspicious
Other (specify)
EIA results for Equine ☐ Positive ☐ Negative
Brucellosis Test for Ruminants Test Dilution 1-50 1-100 1-200 Results

Method of Examination ☐ Visual ☐ Physical

Permit Obtained If Required Yes ☐ No ☐

"I certify, as an accredited veterinarian, that the above described animals have been inspected by me and that they are not showing signs of infectious, contagious, and/or communicable disease, (except where noted). The vaccinations and results of tests are as indicated on the certificate. To the best of my knowledge, the animals listed on this certificate meet the state of destination and federal interstate requirements. No further warranty is made or implied."

White Copy – State Veterinarian Office
Green Copy – State Veterinarian Office
Yellow Copy – Issuing Veterinarian
Pink Copy – Accompany Shipment
Orange Copy – Owner

SIGNATURE .... DVM AAZV Member ☐
ADDRESS .... Non Member ☐
APPROVED BY ....
State Veterinarian

Camden Printing 7279 Rev 02/06
OWNER/AGENT STATEMENT: The animals in this shipment are as certified to and listed on this certificate.

Fig. 1-2. Example of a standard Certificate of Veterinary Inspection (CVI).

ANIMAL WELFARE ACT

Congress passed the Animal Welfare Act (AWA) in 1966 (7 USC, 2131 - 2159; http://awic.nal.usda.gov/nal_display/index.php?info_center=3&tax_level=3&tax_subject=182&topic_id=1118&level3_id=6735&level4_id=0&level5_id=0&placement_default=0), and since then the Act has been amended six times (1970, 1976, 1985, 1990, 2002, and 2007). The AWA requires that minimum standards of care and treatment be provided by businesses (a.k.a. regulated businesses) that buy or sell animals, exhibit them to the public, transport them commercially, and/or use them in experiments or teaching, and that such businesses be licensed or registered by the USDA. Businesses such as retail pet stores, hobby breeders, public animal shelters, private animal shelters, and boarding kennels are not covered by the AWA.

The AWA is of limited purpose and scope. It is not a broadly stated anti-cruelty law and does not deal with all species of animals. Laboratory rats and mice are not covered under the Animal Welfare Act and therefore not regulated by the USDA. No ectothermic vertebrates (e.g., fish or frogs) or invertebrates (e.g., fruit flies or worms) are covered by the Animal Welfare Act, nor are animals used for food. Although the AWA mandates standards of care for animals in captivity with regard to their housing, handling, sanitation, nutrition, water, veterinary services, and protections from extreme weather, these animal care standards are minimal and do not adequately protect the regulated animals from mistreatment, neglect or improper handling.

STATE REGULATORY AGENCIES

Bat species found within the continental United States are all of the suborder Microchiroptera and are, in most states, protected, native, nongame wildlife species. However, each state differs in its laws concerning bats. Some states design laws in the interest of public health, addressing bats as vectors of disease, rather than as mammals needing protection (NPWRC, 2006). Some states have laws protecting bats in caves; but not from inappropriate removals from man-made structures. Although the names of the state regulatory agencies vary (Appendix 1-3), they are responsible for overseeing regulations on the capture, containment, and holding of native bat species.

Persons wishing to collect or be in possession of native bat species for any period of time must obtain a scientific collecting permit from state authorities. Persons intending to use bats listed on the Endangered Species List must also obtain an Endangered Species Permit from the FWS (see *United States Fish and Wildlife Service* above). The possession of bats, both native and exotic, is regulated by the appropriate state authority (Appendix 1-3) and state public health veterinarians (Appendix 1-4).

STATE PUBLIC HEALTH VETERINARIANS

State Public Health Veterinarians (SPHVs) (Appendix 1-4) are the local and state professionals who regularly consult with physicians, emergency rooms, legislators, local officials, schools, health departments, and the general public on preventing exposures and controlling diseases that humans can get from animals and animal products. Many SPHVs have singular state authority on rabies exposures.

The SPHV works for his or her state's health department and is usually housed in the epidemiology, toxicology, or environmental health division. Public health departments in each state have instituted their own guidelines and regulations for general public health and safety when dealing with bats. Certain states may also have regulations initiated at the county level. To obtain information on county and state public health regulations concerning bats, see Appendix 1-4.

OCCUPATIONAL SAFETY AND HEALTH ADMINISTRATION

The Occupational Safety and Health Administration (OSHA) is part of the United States Department of Labor. With the Occupational Safety and Health Act of 1970 (http://www.osha.gov/pls/oshaweb/owasrch.search_form?p_doc_type=oshact), Congress created OSHA to ensure safe and healthful working conditions for working men and women by setting and enforcing standards and by providing training, outreach, education and assistance. The Act mandates that all employees must receive advance training on potentially hazardous agents in the work place. Although there is no specific OSHA regulation for those working with bats, the following information may be helpful:

Public Law 91-596 of 29 December 1970:

- 29 CFR 1904.7 – General Recording Criteria (http://www.osha.gov/pls/oshaweb/owadisp.show_document?p_table=STANDARDS&p_id=9638);
- 29 CFR 1910:132 – Personal Protective Equipment (http://www.osha.gov/pls/oshaweb/owastand.display_standard_group?p_toc_level=1&p_part_number=1910);
- 29 CFR 1910:134 – Respiratory Protection (http://www.osha.gov/pls/oshaweb/owadisp.show_document?p_table=STANDARDS&p_id=12716);
- 29 CFR 1910:1200 – Hazard Communication (http://www.osha.gov/pls/oshaweb/owadisp.show_document?p_table=STANDARDS&p_id=10099).

For questions concerning OSHA regulations, contact:

U.S. Department of Labor
Occupational Safety & Health Administration
200 Constitution Avenue
Washington, D.C. 20210
800-321-6742 (Toll Free United States)

REFERENCES

Fascione, N. 1995. *Fruit Bat Husbandry Manual*. Lubee Foundation, Inc. and the American Zoo and Aquarium Association's Bat Taxon Advisory Group, Gainesville, FL. 39 pp.

NPWRC (Northern Prairie Wildlife Research Center). 2006. House Bat Management: Discussion and Recommendations. http://www.npwrc.usgs.gov/resource/mammals/housebat/discuss.htm, accessed December, 2010.

ACKNOWLEDGMENTS

I thank Valerie Lounsbury for reviewing the manuscript.

Appendix 1-1. Australian commonwealth and state agencies governing bats

State	State Agency and Link	Mailing Address	Phone
Australian Capital Territory	Department of the Environment, Climate Change, Energy and Water http://www.tams.act.gov.au/live/environment/licensing_of_plants_and_animals	Licensing and Investigations, Territory and Municipal Services Level 4, Macarthur House 12 Wattle Street, Lyneham ACT 2602 GPO Box 158 Canberra City, ACT 2601 Email: tams.lc@act.gov.au	(02) 6207 6376
Victoria	**A.** Department of Sustainability and Environment **B.** Wildlife licenses are issued through the Wildlife and Game Licensing Service	**A.** Environmental Research Coordinator Locked Bag 3000 Box Hill, VIC 3128 Email: environmental.research@dse.vic.gov.au. **B.** PO Box 500 East Melbourne, VIC 8002 Email: customer.service@dse.vic.gov.au	**A.** (03) 9296 4651 **B.** DSE Customer Service Center Ph: 136 186.
New South Wales	Office of Environment and Heritage http://www.environment.nsw.gov.au/wildlifelicences/	Wildlife Licensing and Management Unit Parks and Wildlife Division NPWS, PO Box 1967 Hurstville, NSW 1481 Email: wildlife.licensing@environment.nsw.gov.au	(02) 9585 6406
South Australia	Department of Environment and Natural Resources http://www.environment.sa.gov.au/Plants_Animals/Permits_licences	Research Permits Science Resource Center Department of Environment and Natural Resources GPO Box 1047 Adelaide, SA 4661 Email: DENRresearchpermits@sa.gov.au	Fauna permits (+61 8) 8124 4972
Queensland	Department of Environment and Resource Management http://www.derm.qld.gov.au/services_resources/item_details.php?item_id=203593	Permit and License Management Department of Environment and Resource Management GPO Box 2454 Brisbane, QLD 4001 Email: palm@derm.qld.gov.auLevel	1300 130 372

Appendix 1-1 continued

State	State Agency and Link	Mailing Address	Phone
Tasmania	Department of Primary Industries, Parks, Water and Environment http://www.dpiw.tas.gov.au/inter.nsf/Topics/LVAE-55535W?open	Biodiversity Conservation Branch Department of Primary Industries, Parks, Water and Environment 134 Macquarie Street Hobart, TAS 7000 Email: Wildlife.Reception@dpipwe.tas.gov.au	(03) 6233 6556
Northern Territory	Natural Resources, Environment, The Arts and Sport http://www.nt.gov.au/nreta/wildlife/permits/	Parks and Wildlife Service Permits Office Ground Floor, Goyder Center Chung Wah Terrace PO Box 496 Palmerston, NT 0830 Email: pwpermits.nretas@nt.gov.au	(+61 8) 8999 4795 or (+61 8) 8999 4814
Western Australia	Department of Environment and Conservation (DEC) http://www.dec.wa.gov.au/content/view/864/1992/	Norm Press Admin Officer Fauna DEC Nature Protection Branch Locked Bag 104 Bentley Delivery Centre, WA 6983 Email: norm.press@dec.wa.gov.au	(+61 8) 9334 0434
Commonwealth	Department of Sustainability, Environment, Water, Population and Communities http://www.environment.gov.au/epbc/permits/index.html	Department of Sustainability, Environment, Water, Population and Communities GPO Box 787 Canberra, ACT 2601	(+61 2) 6274 1111

Appendix 1-2. Provincial zoo regulations summary (prepared by the World Society for the Protection of Animals, September 2011)

Jurisdiction	Does a permit cover native wildlife?	Does a permit cover exotic wildlife?	Is liability insurance required?	Are there animal welfare requirements?	Are there veterinary care requirements?	Are there safety and security requirements?	Are there accommodation requirements?	Are there Inspections?	What are the permit fees?
Newfoundland and Labrador 1 reptile zoo. 1 provincially-run zoo in a park. Strong guidelines for evaluating applications.	Yes	Yes	No	Yes	Yes Facilities are required to practice preventative medicine. Each animal must receive a general examination every 6 months.	Yes	Yes 80 different species are listed and assigned a specific category of accom-modation.	Inspections are random, using a checklist to ensure conformity with guidelines.	No fee
Alberta 13 permitted facilities.	Yes	Yes	Yes As of 1 April 2006, new standards require minimum coverage of $2 million.	Yes As of 1 April 2006, animal behavior husbandry standards and general animal care standards were implemented.	Yes As of 1 April 2006, new animal health care standards were implemented.	Yes	Yes In addition to general exhibit design standards, as of 1 April 2006, more specific standards for mammals, which were developed by the AZA, are now used to evaluate permit applications/ renewals.	Yes An appointed advisory committee reviews zoo development plans and animal care protocols for non-accredited zoos.	$100 per year
New Brunswick 5 known zoos.	Yes	Yes	No	Yes Adopted CAZA requirements.	Yes Adopted CAZA requirements.	Yes Adopted CAZA requirements.	Yes Adopted CAZA requirements.	No	No fee

Appendix 1-2 continued

Jurisdiction	Does a permit cover native wildlife?	Does a permit cover exotic wildlife?	Is liability insurance required?	Are there animal welfare requirements?	Are there veterinary care requirements?	Are there safety and security requirements?	Are there accommodation requirements?	Are there Inspections?	What are the permit fees?
Saskatchewan Province recognizes 15 zoos.	Yes	Yes	Must maintain adequate liability insurance.	Yes Some general welfare conditions.	Yes Must keep records of tests for diseases.	Yes Specific standards for some species.	Yes Specific standards for some species.	No policy for regular inspections, but conservation officers have the authority to inspect facilities.	No fee
Québec 12 zoos listed on government website.	Yes	Yes	Unknown	General requirements, but not for specific species.	Yes	Yes General requirements; nothing specific.	Yes General requirements, but not for specific species.	Yes	$219.90 per year
P.E.I. Has no operating zoos and has a policy to discourage their establishment.	Yes	Yes	Must have sufficient liability insurance.	Yes No specific standards.	Yes No specific standards.	Yes No specific standards.	No	Officers may, at all reasonable times, enter and search the premises (with a warrant in the case of a private dwelling; without a warrant in other cases).	$250 per 5-year term
Manitoba 2 zoos. Permits to maintain wildlife are required.	Yes	Only members of the family Cervidae and wild boar – species are listed under exotic wildlife regulations.	No	Yes Adopted CAZA requirements. There are also welfare conditions that must be met during transport.	Yes Adopted CAZA requirements.	Yes Adopted CAZA requirements.	Yes Adopted CAZA requirements.	Yes Annual inspections by a veterinarian from, or recommended by, the Manitoba government.	$100 per year

Appendix 1-2 continued

Jurisdiction	Does a permit cover native wildlife?	Does a permit cover exotic wildlife?	Is liability insurance required?	Are there animal welfare requirements?	Are there veterinary care requirements?	Are there safety and security requirements?	Are there accommodation requirements?	Are there Inspections?	What are the permit fees?
B.C. 6 known zoos open to the public.	Yes	No	Must maintain minimum coverage for $1 million.	Yes	No	Yes	Yes	Unknown	$130 per year
Ontario 45 known captive wildlife facilities, but number fluctuates (only 7 are accredited by CAZA).	Yes A permit is required for keeping native wildlife scheduled as "game" or "specially protected" wildlife.	No	No	Some general conditions are attached to the permit, but only apply to some native species. Standards are vague and unenforceable.	Some general conditions are attached to the permit, but only apply to some native species. Standards are vague and unenforceable.	No	Some general conditions are attached to the permit, but only apply to some native species. Standards are vague and unenforceable.	A conservation officer has the authority to enter and inspect a facility where licenses are issued, but there are no regular zoo inspections. The decision of when to inspect is left to the discretion of each district office.	$100 per year
Nova Scotia 5 zoos (1 not open to the public).	Yes	Yes	No	Yes Some general conditions.	No	No	No	Yes Inspections are typically once a year, but some are monthly.	No fee

Appendix 1-3. United States state agencies governing nongame wildlife

State	State Agency and Link	Mailing Address	Phone
AL	Department of Conservation and Natural Resources (http://www.dcnr.state.al.us/)	64 North Union Street Montgomery, AL 36130	334-242-3469
AK	Department of Fish and Game (http://www.dfg.ca.gov/)	1255 West 8th Street Juneau, AK 99811	907-465-4100
AZ	Game and Fish (http://www.azgfd.gov/)	5000 W. Carefree Highway Phoenix, AZ 85086	602-942-3000
AR	Game and Fish Commission (http://www.agfc.com/)	2 Natural Resources Drive Little Rock, AR 72205	501-223-6300
CA	Department of Fish and Game (http://www.dfg.ca.gov/)	1416 9th Street (12th Floor) Sacramento, CA 95814	916-445-0411
CO	Division of Wildlife (http://wildlife.state.co.us/)	6060 Broadway Denver, CO 80216	303-297-1192
CT	Department of Environmental Protection (http://www.ct.gov/dep/site/default.asp)	79 Elm Street Hartford, CT 06106	860-424-3000
DE	Division of Fish and Wildlife (http://www.fw.delaware.gov/Hunting/Pages/Wildlife.aspx)	3002 Bayside Drive Dover, DE 19901	302-739-4782
FL	Fish and Wildlife Conservation Commission (http://www.myfwc.com/)	Farris Bryant Building 620 S. Meridian Street Tallahassee, FL 32399	850-488-4676
GA	Department of Natural Resources (http://www.georgiawildlife.com/)	2070 U.S. Highway 278 SE Social Circle, GA 30279	770-918-6400
HI	Department of Land and Natural Resources Division of Forestry and Wildlife (http://hawaii.gov/dlnr/dofaw)	1151 Punchbowl Street Room 325 Honolulu, HI 96813	808-587-0066
ID	Fish and Game (http://fishandgame.idaho.gov/cms/wildlife/nongame/)	600 S. Walnut Boise, ID 83712	208-334-3700
IL	Department of Natural Resources (http://www.dnr.state.il.us/home.htm)	One Natural Resources Way Springfield, IL 62702	217-785-8687

Appendix 1-3 continued

State	State Agency and Link	Mailing Address	Phone
IN	Division of Fish and Wildlife (http://www.in.gov/dnr/fishwild/2356.htm)	402 West Washington Street Indianapolis, IN 46204	317-232-4200
IA	Department of Natural Resources (http://www.iowadnr.gov/)	502 East 9th Street Des Moines, IA 50319	515-281-5918
KS	Department of Wildlife and Parks (http://www.kdwp.state.ks.us/Other-Services)	512 SE 25th Avenue Pratt, KS 67124	620-672-5911
KY	Department of Fish and Wildlife Resources (http://fw.ky.gov/)	1 Sportsman's Lane Frankfort, KY 40601	800-858-1549
LA	Department of Wildlife and Fisheries (http://www.wlf.louisiana.gov/)	2000 Quail Drive Baton Rouge, LA 70808	225-765-2800
ME	Department of Inland Fisheries and Wildlife (http://www.maine.gov/ifw/wildlife/)	41 State House Station Augusta, ME 04333	207-287-8000
MD	Department of Natural Resources (http://www.dnr.state.md.us/wildlife/)	Wildlife & Heritage Service Headquarters Tawes State Office Building E-1 580 Taylor Avenue Annapolis, MD 21401	410-260-8540
MA	Division of Fisheries and Wildlife (http://www.mass.gov/dfwele/dfw/)	1 Rabbit Hill Road Westborough, MA 01581	508-389-6300
MI	Department of Natural Resources and Environment (http://www.michigan.gov/dnr/)	Mason Building Lansing MI 48909	517-373-2352
MN	Department of Natural Resources (http://www.dnr.state.mn.us/index.html)	500 Lafayette Road St. Paul, MN 55155	651-296-6157
MS	Department of Wildlife, Fisheries, and Parks (http://home.mdwfp.com/)	1505 Eastover Drive Jackson, MS 39211	601-432-2400
MO	Department of Conservation (http://mdc.mo.gov/discover-nature)	Conservation Headquarters 2901 W. Truman Boulevard Jefferson City, MO 65102	573-522-4115
MT	Fish, Wildlife, and Parks (http://fwp.mt.gov/wildthings/tande/default.html)	1420 East Sixth Avenue Helena, MT 59620	406-444-2535

Appendix 1-3 continued

State	State Agency and Link	Mailing Address	Phone
NE	Game and Parks Commission (http://outdoornebraska.ne.gov/admin/commission/divisions.asp)	2200 N. 33rd Street Lincoln, NE 68503	402-471-0641
NV	Department of Wildlife (http://www.ndow.org/about/facts/)	1100 Valley Road Reno, NV 89512	775-688-1500
NH	Fish and Game Department (http://www.wildlife.state.nh.us/Wildlife/nongame_and_endangered_wildlife.htm)	11 Hazen Drive Concord, NH 03301	603-271-2461
NJ	Department of Environmental Protection (http://www.state.nj.us/dep/)	501 East State Street 3rd Floor Trenton, NJ 08625	609-292-3541
NM	Department of Game and Fish (http://www.wildlife.state.nm.us/index.htm)	One Wildlife Way Santa Fe, NM 87507	505-476-8000
NY	Department of Environmental Conservation (http://www.dec.ny.gov/)	625 Broadway Albany, NY 12233	518-402-8013
NC	Wildlife Resources Commission (http://www.ncwildlife.org/Wildlife_Species_Con/index.htm)	1722 Mail Service Center Raleigh NC 27699	919-707-0050
ND	Game and Fish Department (http://gf.nd.gov/)	100 N. Bismarck Expressway Bismarck, ND 58501	701-328-6300
OH	Department of Natural Resource (http://www.dnr.state.oh.us/)	2045 Morse Road Building D Columbus OH 43229	614-265-6565
OK	Department of Wildlife Conservation (http://www.wildlifedepartment.com/)	1801 N Lincoln Oklahoma City, OK 73105	405-521-3851
OR	Department of Fish and Wildlife (http://www.dfw.state.or.us/wildlife/)	3406 Cherry Avenue N.E. Salem, OR 97303	503-947-6000
PA	Department of Conservation and Natural Resources (http://www.dcnr.state.pa.us/)	Rachel Carson State Office Building 400 Market Street Harrisburg, PA 17105	717-787-2869

Appendix 1-3 continued

State	State Agency and Link	Mailing Address	Phone
RI	Department of Environmental Management Division of Fish and Wildlife (http://www.dem.ri.gov/programs/bnatres/fishwild/)	3 Fort Wetherill Road Jamestown, RI 02835	401-423-1920
SC	Department of Natural Resources (http://www.dnr.sc.gov/)	Rembert C. Dennis Building 1000 Assembly Street Columbia, SC 29201	803-734-3886
SD	Department of Game, Fish, and Parks (http://gfp.sd.gov/)	523 East Capitol Avenue Pierre, SD 57501	605-773-3387
TN	Wildlife Resources Agency (http://www.state.tn.us/twra/nongame.html)	Ellington Agricultural Center 440 Hogan Road Nashville, TN 37220	615-781-6500
TX	Parks and Wildlife Department (http://www.tpwd.state.tx.us/)	4200 Smith School Road Austin, TX 78744	512-389-4800
UT	Department of Natural Resources (http://naturalresources.utah.gov/)	1594 W. North Temple Salt Lake City, UT 84114	801-538-7200
VT	Fish and Wildlife Department (http://www.vtfishandwildlife.com/)	10 South 103 South Main Street Waterbury, VT 05671	802-241-3700
VA	Department of Game and Inland Fisheries (http://www.dgif.virginia.gov/)	4010 West Broad Street Richmond, VA 23230	804-367-1000
WA	Department of Natural Resources (http://www.dnr.wa.gov/Pages/default.aspx)	1111 Washington Street SE Olympia, WA 98504	360-902-1000
WV	Division of Natural Resources (http://www.wvdnr.gov/)	Building 74 324 Fourth Avenue South Charleston, WV 25303	304-558-2754
WI	Department of Natural Resources (http://dnr.wi.gov/)	101 S. Webster Street Madison, WI 53707	608-266-2621
WY	Game and Fish Department (http://gf.state.wy.us/)	5400 Bishop Boulevard Cheyenne, WY 82006	307-777-4600

Appendix 1-4. Contact information for United States public health veterinarians

State	Public Health Department and Link	Mailing Address	Phone/FAX
AL	Department of Public Health, Epidemiology (http://www.adph.org/epi/index.asp?ID=3385)	201 Monroe Street, RSA Tower Suite 1480 Montgomery, AL 36104	334-206-5969 334-206-3842
AK	Division of Public Health Section of Epidemiology (http://www.epi.hss.state.ak.us/default.jsp)	3601 C Street, Suite 540 Anchorage, AK 99503	907-269-8002 907-562-7802
AZ	Department of Health Services (http://www.azdhs.gov/phs/edc/)	150 N 18th Avenue, Suite 140 Phoenix, AZ 85007	602-364-4562 602-364-3198
AR	Department of Health and Human Services (http://www.healthy.arkansas.gov/programsServices/ epidemiology/ Pages/default.aspx)	P.O. Box 1437, Slot H-32 Little Rock, AR 72205	501-280-4136 501-661-2428
CA	Department of Health Services Veterinary Public Health Section (http://www.cdph.ca.gov/programs/vphs/ pages/default.aspx)	PO Box 997413, MS 7308 Sacramento, CA 95899	916-552-9740 916-552-9725
CO	Department of Public Health & Environment Communicable Disease Epidemiology Program (http://www.cdphe.state.co.us/dc/Epidemiology/ dc_guide.html)	4300 Cherry Creek Drive South Denver, CO 80246	303-692-2700 303-782-0338
CT	Department of Public Health (http://www.ct.gov/dph/cwp/view.asp?a=3136&q=396178)	410 Capitol Avenue MS #11 EPI PO Box 340308 Hartford, CT 06134	860-509-7994 860-509-7910
DE	Department of Agriculture (http://dda.delaware.gov/)	2320 S. DuPont Highway Dover, DE 19901	302-698-4560 302-697-4451
FL	Department of Health (http://www.doh.state.fl.us/Planning_eval/about/ index.html)	4052 Bald Cypress Way Bin A08 Tallahassee, FL 32399	850-245-4732 850-922-8473
GA	Division of Public Health (http://health.state.ga.us/epi/index.asp)	2 Peachtree Street, N.W. Suite 14-234 Atlanta, GA 30303	404-657-2629
HI	State Department of Health (http://hawaii.gov/health/)	1250 Punchbowl Street Honolulu, HI 96813	808-586-4400 808-586-4444

Appendix 1-4 continued

State	Public Health Department and Link	Mailing Address	Phone/FAX
ID	Department of Health and Welfare Office of Epidemiology, Food Protection and Immunization (http://www.healthandwelfare.idaho.gov/)	450 W State Street 4th Floor PO Box 83720 Boise, ID 83720	208-334-5939 208-332-7307
IL	Department of Public Health Division of Infectious Disease (http://www.idph.state.il.us/about/ohp.htm)	525 W Jefferson Street Springfield, IL 62761	217-785-7165 217-557-4049
IN	State Department of Health Epidemiology Resource Center (http://www.in.gov/isdh/24769.htm)	2 North Meridian Street Indianapolis, IN 46204	317-233-7272 317-771-2132
IA	Department of Public Health (http://www.idph.state.ia.us/)	Lucas State Office Building Des Moines, IA 50319	515-281-4933 *No fax listed*
KS	Department of Health and Environment (http://www.kdheks.gov/)	1000 SW Jackson Street Suite 210 Topeka, KS 66612	785-296-2501 785-925-1123
KY	Department of Public Health Division of Epidemiology and Health Planning (http://chfs.ky.gov/dph/epi/default.htm)	275 E Main Street HS2GWC Frankfort, KY 40621	502-564-3418 ext. 3569 502-564-9626
LA	Office of Public Health Infectious Disease Epidemiology Section (http://www.dhh.louisiana.gov/offices/?id=249)	2323 Saint Charles Avenue New Orleans, LA 70130	504-219-4593 504-219-4522
ME	Department of Agriculture Division of Animal Health and Industry (http://www.maine.gov/agriculture/ahi/)	28 State House Station Augusta, ME 04333	207-287-7615 207-624-5044
MD	Department of Health and Mental Hygiene Center for Zoonotic and Vector-Borne Diseases (http://www.edcp.org/vet_med/)	201 W. Preston Street Room 317 Baltimore, MD 21201	410-767-6703 410-383-2420
MA	Department of Public Health Hinton State Laboratory Institute (http://www.mass.gov/dph/bls)	305 South Street, 5th Floor Jamaica Plain, MA 02130	617-983-6800 617-983-6840

Appendix 1-4 continued

State	Public Health Department and Link	Mailing Address	Phone/FAX
MI	Department of Community Health Communicable Disease Division Bureau of Epidemiology (http://www.michigan.gov/mdch/0,1607,7-132-2945_5104---,00.html)	Capitol View Building 201 Townsend Street 5th Floor Lansing, MI 48913	517-335-8165 517-335-8263
MN	Department of Health Acute Disease Investigation and Control Section (http://www.health.state.mn.us/divs/idepc/adic.html)	625 Robert Street North St. Paul, MN 55155	651-201-5107 651-201-5082
MS	Board of Animal Health State Department of Health (http://www.mbah.state.ms.us/disease_programs/rabies/ ms_rabies_laws.htm)	45 Sunny Meadows Drive Hattiesburg, MS 39402	601-953-3800 601-359-1177
MO	Department of Health & Senior Services, SDCEE (http://www.dhss.mo.gov/)	930 Wildwood Drive Jefferson City, MO 65109	573-751-6114 573-526-0235
MT	Department of Public Health & Human Services (http://www.dphhs.mt.gov/)	111 North Sanders Street SRS Building Helena, MT 59601	406-444-0936 *No fax listed*
NE	Department of Health and Human Services Public Health Support (http://www.dhhs.ne.gov/Public_Health/)	301 Centennial Mall S PO Box 95007 Lincoln, NE 68509	402-471-1374 402-471-3601
NV	State Health Division Office of Epidemiology (http://health.nv.gov/Epidemiology.htm)	4150 Technology Way Suite 211 Carson City, NV 89706	775-684-5911 775-684-5999
NH	Department of Health & Human Services Bureau of Disease Control (http://www.dhhs.state.nh.us/)	29 Hazen Drive Concord, NH 03301	603-271-4820 603-271-0545
NJ	Department of Health and Senior Services Infectious & Zoonotic Disease Program (http://www.state.nj.us/health/cd/izdp/index.shtml)	PO Box 369 Trenton, NJ 08625	609-826-4872 609-826-4874
NM	Department of Health Epidemiology and Response Division (http://www.health.state.nm.us/erd/index.shtml)	1190 St. Francis Drive Room N-1350 Santa Fe, NM 87505	505-827-0006 505-827-0013

Appendix 1-4 continued

State	Public Health Department and Link	Mailing Address	Phone/FAX
NY	Department of Health Bureau of Communicable Disease Control (http://www.health.state.ny.us/diseases/communicable/)	Corning Tower, Room 651 Empire State Plaza Albany, NY 12237	518-473-4439 518-408-1745
NC	Department of Health and Human Services Division of Public Health (http://publichealth.nc.gov/)	1902 Mail Service Center Raleigh, NC 27612	919-733-0391 919-870-4807
ND	Department of Agriculture State Board of Animal Health (http://www.agdepartment.com/Programs/Livestock/ BOAH/BOAH.html)	600 E. Boulevard Avenue Dept. 602 Bismarck, ND 58505-0020	701-328-2354 701-328-4567
OH	Department of Health Zoonotic Disease Program (http://www.odh.ohio.gov/odhprograms/dis/zoonoses/ zoodis1.aspx)	900 Freeway Drive Building 8 Columbus, OH 43229	614-466-0283 614-644-1057
OK	State Department of Health Office of the State Epidemiologist (http://www.ok.gov/health/)	1000 NE Tenth Street Oklahoma City, OK 73117	405-271-4060 405-271-6680
OR	Department of Human Services (http://www.oregon.gov/DHS/)	800 NE Oregon Street Suite 772 Portland, OR 97232	971-673-1111 971-673-1100
PA	Department of Health Division of Infectious Disease Epidemiology (http://www.portal.state.pa.us/portal/server.pt/community/laboratories/14158/rabies/556798)	Health and Welfare Building Room 933 7th and Forster Streets Harrisburg, PA 17120	717-346-4524 717-772-6975
RI	Department of Environmental Management Division of Agriculture (http://www.dem.ri.gov/programs/bnatres/agricult/ index.htm)	235 Promenade Street Room 370 Providence, RI 02908	401-222-2781 ext. 4503 401-222-6047
SC	Department of Health and Environmental Control Division of Acute Epidemiology (http://www.scdhec.gov/health/disease/acute/index.htm)	1751 Calhoun Street Columbia, SC 29211	803-898-4042 803-898-0897
SD	Animal Industry Board (http://aib.sd.gov/)	411 South Fort Street Pierre, South Dakota 57501	605-773-3321 605-773-5459

Appendix 1-4 continued

State	Public Health Department and Link	Mailing Address	Phone/FAX
TN	Department of Health Communicable and Environmental Disease Services (http://health.state.tn.us/ceds/index.htm)	Cordell Hull Building, 1st Floor 425 5th Avenue N Nashville, TN 37243	615-741-5948 615-741-3857
TX	Department of State Health Services Zoonoses Control Branch (http://www.dshs.state.tx.us/idcu/health/zoonosis/)	PO Box 149347, MC 1956 Austin, TX 78714	512-458-7111 ext. 6628 512-458-7454
UT	Department of Agriculture and Food (http://www.ag.utah.gov/)	P.O. Box 146500 Salt Lake City, UT 84114	801-538-7162 801-538-7169
VT	Department of Health Division of Health Surveillance (http://healthvermont.gov/admin/hs.aspx)	108 Cherry Street Burlington, VT 05402	802-863-7240 802-865-7701
VA	Department of Health (http://www.vdh.state.va.us/)	109 Governor Street Madison Building, 4th Floor Richmond, VA 23219	804-864-8113 804-864-8131
WA	Department of Health Zoonotic & Vector-borne Diseases Program Office of Environmental Health, Safety & Toxicology (http://www.doh.wa.gov/ehp/ts/zoo.htm)	Town Center Building 3 243 Israel Road Tumwater, WA 98501	360-236-3369 360-236-2261
WV	Department of Health and Human Resources Infectious Disease Epidemiology (http://www.wvidep.org/)	350 Capitol Street Room 125 Charleston, WV 25301	304-558-5358 304-558-6335
WI	Division of Public Health Bureau of communicable Disease (http://www.dhs.wisconsin.gov/aboutdhs/dph/bcder.htm)	1 W Wilson Street, Room 318 Madison, WI 53703	608-266-2154 608-261-4976
WY	Wyoming Department of Health (http://wdh.state.wy.us/)	6101 Yellowstone Road Suite 510 Cheyenne, WY 82003	307-421-8591 307-777-5402

Chapter 2

Shipping

Photo by Susan M. Barnard

Transporting Bats Within Australia

Jennefer Maclean

INTRODUCTION

Bats vary enormously in size, temperament, dietary preferences, microclimate requirements, and the degree to which they can withstand handling and transporting. Many aspects of transportation such as whether bats are shipped by air or road, container design, duration of travel, and the competence of handlers can have a significant impact on their welfare. The stressors that bats experience during a trip may include handling, separation from their social group, confinement in an unfamiliar transport container, loading and unloading, movement and vibrations, attempts to maintain their balance, unfamiliar sights, sounds and smells, and lack of food and water. The primary goal when shipping bats is to reduce the impact of such stressors.

PREPARING MICROCHIROPTERANS FOR TRANSPORT

Microchiropterans in Australia are transported primarily by researchers from trapping sites to field camps or to laboratories for study, and by wildlife rehabilitators from rescue sites to care facilities.

Some bats, particularly the cave-dwelling horseshoe bats (*Rhinolophus* spp.) and leaf-nosed bats (*Hipposideros* spp.), require humid environments and are delicate to handle. They are not held easily in captivity, and even when placed in the softest of cotton bags, they tend to rub skin from their forearms and legs. Keeping these species in captivity or transporting them should be done only by trained personnel. The Australasian Bat Society (ABS) (http://ausbats.org.au/) should be contacted for helpful advice. ABS is currently developing a handling and transport guide for Australian bats in conjunction with the Australian Wildlife Health Network.

Other microchiropteran species are more robust and can be transported with relative ease when the basic guidelines below are followed:

1. Use cloth bags to transport microbats, and place only one bat in a bag to prevent conflicts and over-heating. Bags must be tied securely so bats do not escape. Turn bags inside out and remove loose threads that may entangle bats before putting animals in them.
2. Bags containing bats can be placed in polystyrene containers. Polystyrene containers protect bats from injuries and temperature extremes. When temperatures exceed the range of comfort for a particular species, hot or cold packs can be placed in the lid or base of the container. Rods can be inserted through holes drilled on both sides of the container at the top, from which cloth bags containing the bats are hung. When bats are transported by automobile, containers should be placed on the back seat because the force of deployed air bags can damage them.

3. Airflow is impeded while bats are held in cloth bags, and in hot weather poor airflow can magnify the effects of heat. Never expose bats to direct sunlight. Maintain them in an air-conditioned environment whenever possible. Researchers working with bats at field camps can maintain them in a Coolgardie safe (http://www.aussiethings.biz/coolgardie_safe.html) until they are released. A Coolgardie safe designed for bats (Fig. 2-1) is a vertical stack of shelves with a cotton cover on all sides. Cloth bags containing bats can be hung from each shelf. Water drawn from a pan on top of the safe wets the cotton cover, and bats are cooled by evaporation. Bags holding bats in a Coolgardie safe should never be allowed to become wet because air can not flow readily through wet cloth, and the animals may suffocate.

Fig. 2-1. When air conditioning is unavailable, bats can be kept cool in a Coolgardie safe. Photo courtesy of Michael Pennay.

Rescued crevice-dwelling bats are usually transported by wildlife rehabilitators in tied cloth pouches or in cloth that is folded and placed inside small plastic containers (e.g., Exo Terra® Faunariums). Such containers can also be lined with cloth, insect screen or other mesh material (see Fig. 13-32, Volume 2) to allow non-crevice-dwellers to hang from the container sides. The size of a plastic container is determined by the bat species, number of bats, and distance that they must be transported. Whether or not a heating or cooling source is used depends on the ambient temperature and general condition of the bat(s). When rescued bats are cold, many caregivers put them in pouches (see Fig. 13-29, Volume 2) that they wear under a shirt. Caution should be used, however. Even though the danger of infection by Australian Bat Lyssavirus (ABL) carried by microbats is extremely low, bat rehabilitators who come into close physical contact with their charges must be vaccinated against ABL, especially when caring for the yellow-bellied pouched bat (*Saccolaimus flaviventris*).

PREPARING MEGACHIROPTERANS FOR TRANSPORT

The vast majority of bats transported in Australia are orphaned flying foxes. Healthy juvenile and adult flying foxes are usually transported in standard airline pet carriers (Fig. 2-2; see also Appendix 2-2d,e), but other types of cages are used as well. Transport containers must have a mesh-covered ceiling (Fig. 2-3) from which megabats can hang. The container must also be of sufficient height to allow bats to hang freely without their heads touching the floor. Any mesh used to cover ceilings, walls, and doors of transport containers must not allow the wings of flying foxes to pass through and get caught. Preferable mesh size is no larger than 1 in. by 1 in. (25 mm by 25 mm).

As is the case with all wild animals, the stress of transport can be reduced by preventing bats from seeing what is outside a transport container. A sheet draped over the container works well, but if it is too heavy it can impede airflow. Transport containers must never be over-crowded, and bats known to fight or hang onto each other should not be placed in the same container.

The weight and volume of transport containers must be considered. For the safety of personnel and bats alike, containers must be lifted with ease when loading them in and out of vehicles. Some rehabilitators use containers constructed of shade cloth built on a very light timber frame (Fig. 2-4), which can be customized to fit through the door or loading hatch of any vehicle. This can be ideal for short trips in very hot climates because the shade cloth allows good airflow. This type of container also eliminates the need for a cover. Bats weighing less than 400 g can hang directly on the shade cloth ceiling. A rigid mesh ceiling should be provided for heavier animals.

Fig. 2-2. Standard pet carriers used for transporting flying foxes. The size of the carrier is determined by the size of the bat. Photo by the author.

Bats rescued from barbed wire fence or entangling netting, or those rescued with other kinds of physical trauma, are best transported wrapped in a towel with their feet free and lying down. Wounded animals feel more secure if they can hold on to something with their feet and/or thumbs (Fig. 2-5), so a basket of sufficient depth should be used to contain them. Similar containers (Fig. 2-6) are used for bats with tick paralysis, an illness caused by a neurotoxin produced in the salivary glands of ticks (*Ixodes holocyclus*). Although bats suffering from this disease can not hang, they are more secure if they can hold on with their thumbs.

Fig. 2-3. Standard pet carrier with wire mesh attached to the ceiling (arrow). Photo by the author.

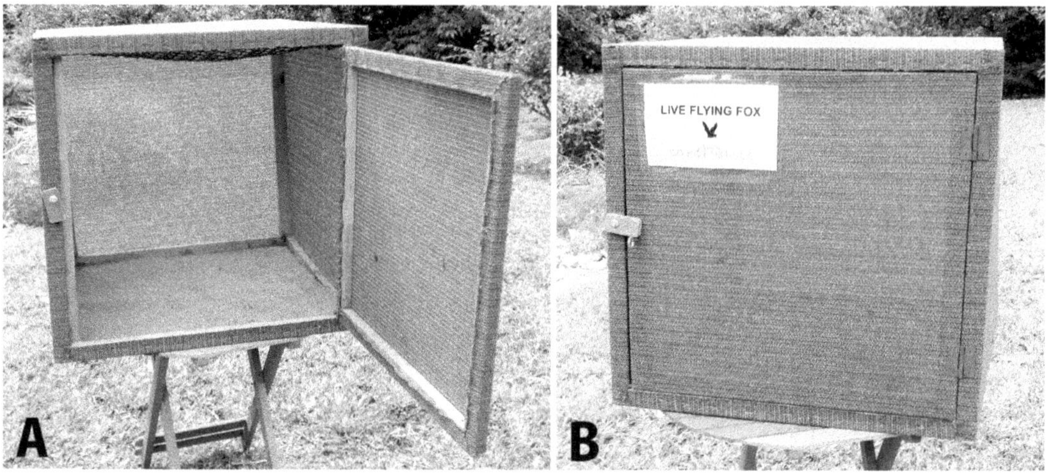

Fig. 2-4. Shade cloth transport container. A) container open showing interior and lightweight construction; B) container closed, latched, and labeled. Photo by the author.

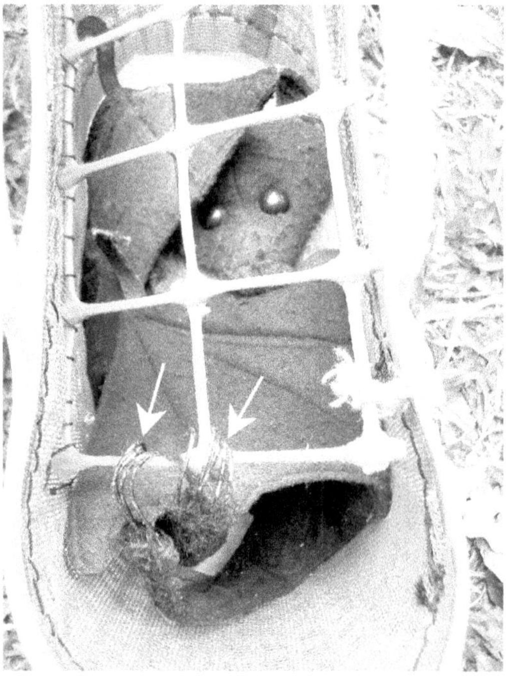

Fig. 2-5. Flying fox holding onto mesh with feet (arrows) for security. Photo by the author.

Fig. 2-6. A) flying fox with tick paralysis being rescued from forest floor. B) fly screen cover prevents flies from laying eggs on rescued flying foxes with tick paralysis. C) the type of basket shown here must be placed in a container to keep them from falling over. Photo A courtesy of Ashleigh Johnson; photos B and C by the author.

Whether shipped by road or air, very young flying foxes up to 2 weeks of age should be transported wrapped in cloth (Fig. 2-7a). In many situations, a heat source may have to be placed with the young animals because they are unable to thermoregulate.

Some megachiropterans, such as the tube-nosed fruit bat (*Nyctimene robinsonii*) and blossom bats (*Sychonycteris australis* and *Macroglossus minimus*), are extremely nervous and do not accept handling well. Although they are not presented for care very often, they are best transported and maintained in soft mesh containers such as the Exo Terra® Flexarium (Fig. 2-8) because their skin is delicate and damaged easily. They usually require only short-distance transport from the point of rescue to the place of care, and back again. Like all bats, these animals have individual temperaments and each requires individual assessment. Although most specimens can be transported in the same flexarium that houses them during care, others are better transported if they are placed in small containers that prevent too much movement.

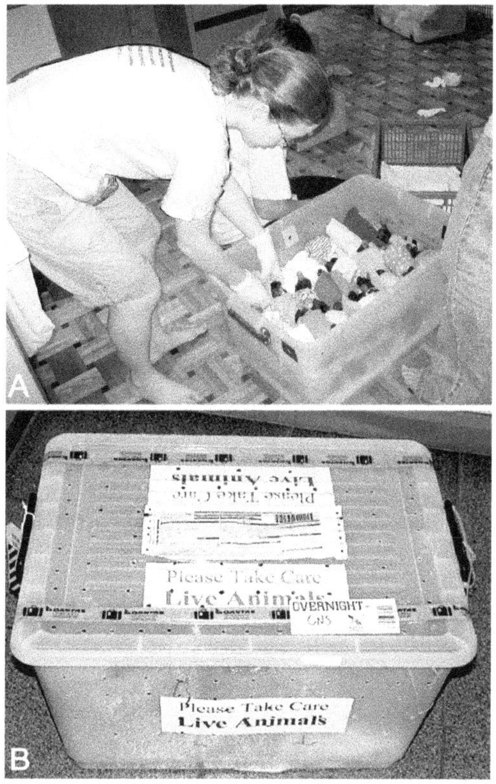

Fig. 2-7. Tick paralysis occurs during the flying fox reproductive season on the Atherton Tableland, leaving more orphaned babies than can be cared for by local wildlife rehabilitators, who may have to ship orphans elsewhere in Australia for care. A) orphans being prepared for air transport; B) shipping container with orphaned flying foxes. Photos by the author.

Fig. 2-8. Exo Terra® Flexarium with a *Nyctimene* bat. Photo courtesy of Ashleigh Johnson.

METHODS OF TRANSPORT

The most common method of transporting bats in Australia is by road, although bats needing long-distance travel are sometimes shipped by air.

AIR TRANSPORT

The International Air Transport Association (IATA) has Live Animal Regulations (LAR) that are the international standard for transporting animals by commercial airlines (see Schwellenbach *et al.*, this chapter). The objective of IATA LAR is to ensure that all animals are transported safely, humanely, and in a uniform manner, which requires shippers to meet specific regulations and container requirements. Container Requirement 77 (see Fig. 2-12) for bats and flying lemurs is not detailed enough for all Australian bats such as baby flying foxes (see also http://www.daff.gov.au/animal-plant-health/welfare/nccaw/guidelines/transport/air).

Air transport is preferred when long distance and timely delivery are factors, and when the budget for transportation is not restricted. For example, bats may need to be shipped by air to rehabilitators at some distance if local personnel can not care for them.

Adult bats travel well in stackable airline pet carriers (see Figs. 2-2 and 2-3). Adults also travel well in non-crushable containers divided into separate compartments (Fig. 2-9; see also Appendix 2-2), as do young bats conditioned to handling in captivity. Such

Fig. 2-9. Two- to three-month-old juvenile flying foxes in transport containers to be returned to the state where they were rescued for release into the wild. Photo by the author.

containers have been used for young bats by Tolga Bat Hospital in Atherton, Queensland when as many as 400-500 orphans needed to be transported for care because tick paralysis killed their mothers.

It is usually not necessary or advisable to provide food or water during air transport. Adequate food and water must be provided up to 3 hours before a flight for adult bats. This ensures that most individuals have emptied their bowels and bladders 2 hours before flight time, which is the interval required by airlines for check-in.

When bats are transported by air, qualified personnel who have been vaccinated against ABS should be on stand-by at all airports where there are stop-overs, so that the animals can be fed and watered in case of delays.

Transporting bats during extremely hot weather should be avoided. Young flying foxes weighing less than 250 g require monitoring and water as soon as possible on arrival.

Commercial airlines sometime prefer that customers use one of several carriers that handle animals (e.g., Dogtainers Pet Transport). However, most of these companies specialize in domestic animals and typically have little or no experience transporting native wildlife.

ROAD TRANSPORT

Air-conditioning is essential when bats are transported by road during hot weather, and they should never be left in a parked vehicle during rest stops unless the air conditioner is operating and a qualified person is with the vehicle. Animals left inside a vehicle may die of heatstroke and suffocation in just a few minutes. Uncarpeted metal vehicle floors can become very hot. Newspapers placed under bat containers help to insulate them from the hot floor, and containers must be checked regularly during a journey. An uninterrupted trip is preferable to one broken up by rest stops, especially if bats must be unloaded and re-loaded.

Security devices like seat belts prevent containers from toppling during transport. Rubber mats can also be placed under containers to prevent them from sliding, and to insulate them from the heat of uncarpeted metal floors.

It is important to observe bats inside their containers during transit. This may be difficult while one is driving and when containers are covered with sheets to protect bats from outside disturbances. In such cases, newspapers can be placed on the floor inside containers. If bats fall during transport, they can be heard rustling the papers.

Anyone transporting bats in a vehicle should be able to recognize when they may be in discomfort, pain, or distress. Transporters should know how to alleviate discomfort, and determine when veterinary attention may be necessary.

Bats should never be placed in the trunks (boots) of transport vehicles. Trunks do not provide appropriate ventilation, and they can become extremely hot. Moreover, bats should not be placed in the beds of trucks, or in any situation where they are exposed to the elements, harsh winds, excessive noise, and potential injury from objects that may fly into the transport container.

A back-up plan should be in place in case a vehicle transporting bats breaks down, especially in remote areas, and personnel should carry cell phones for emergencies. In the event of transport delays such as a vehicle breakdown, backed-up traffic, or other unforeseen situations, food and water should be available in a cooler. Baby bats may require milk during transit, and water or juice may be needed for adults.

INTERSTATE TRANSPORT

All Australian bats are protected native fauna and permits are required to transport them across state borders (see Appendix 1-1). It may be necessary to transport bats across borders for rehabilitation, and all states stipulate that these animals must be returned for release to the point of rescue. However, in the case of highly mobile species like Australian flying foxes this rule is often relaxed, particularly for hand-reared orphans. Some Australian bats, like the little red flying fox (*Pteropus scapulatus*), move seasonally across large expanses of the continent, and rehabilitated individuals must be released when conspecifics are nearby.

Applicants seeking permits for transport must explain why it is necessary to import or export the bat or bats in question. Australian authorities are particularly mindful of the illegal wildlife trade and expect applicants to fully justify the proposed transport, and to have permits to keep a bat or bats in captivity in the state of origin. Permits for interstate transport are required to export from the state of origin as well as to import into the state of destination. Permits may sometimes be applied for online from the respective state agencies (see Appendix 1-1) for a minimal fee.

HEALTH PRECAUTIONS

As noted earlier, some Australian species carry Australian Bat Lyssavirus, and good workplace health and safety practices require that it be assumed that all Australian bats may have the disease. ABL is similar in many respects to rabies, but the incidence of transmission to humans is extremely low. All bats must be secured in appropriate containers so handlers can not be scratched or bitten. This is especially important for travel that may involve persons not vaccinated against ABL, such as airline personnel. It is recommended that all bat workers be immunized against ABL. This practice serves to protect handlers from the risk of disease and the need for vaccination after the fact, while protecting bats against euthanasia for testing, should a person be bitten.

PRODUCTS MENTIONED IN TEXT

Exo Terra® Faunarium: Rolf C. Hagen Inc., Montreal, Quebéc H9X 0A2 Canada; http://exo-terra.com/en/products/faunarium.php)

Exo Terra® Flexarium: Rolf C. Hagen Inc., Montreal, Quebéc H9X 0A2 Canada; http://www.exo-terra.com/en/products/flexarium.php.

Shade Cloth (approx. 70%): Australian Shade Wholesalers Pty Ltd., Unit 5/9 Combarton Street, P.O. Box 5221, Brendale, Queensland 4500 Australia; http://www.shadewholesalers.com.au/shadecloth.html.

ACKNOWLEDGMENTS

I thank Dietrich Schaaf for his helpful edits on the manuscript.

Considerations for Transporting and Shipping *Glossophaga soricina* and *Carollia perspicillata*

John J. Rasweiler, IV and Nilima K. Badwaik

Pallas's long-tongued bat (*Glossophaga soricina*) and Seba's short-tailed bat (*Carollia perspicillata*) adapt so readily to captivity that they have been held for as little as 3-4 days on Trinidad before being shipped successfully to North America. In the experience of the authors, with careful handling and feeding, few or no mortalities are now recorded over many months from the animals' original time of capture. The occurrence of more than a few mortalities during this period would suggest that some aspect of the animals' capture, husbandry or shipping is deficient. Mortalities due to disease must also be considered as a possibility; however, the authors have never had to contend with this problem, despite introducing many batches of both *Carollia* and *Glossophaga* bats into captivity.

In order to prepare the bats for transport, they have always been provided with food and water on the night immediately prior to shipping. They were then recaptured and packed as late as possible, generally during the early morning hours on the day of shipping. This is important because 1) the bats seem to have no interest in consuming food placed in their shipping containers during short periods of transit (<14 hours); 2) disturbances and confinement in unfamiliar containers during shipping may disrupt their normal physiological response to food deprivation the previous night (i.e., entry into diurnal torpor), and 3) they are going to be released into completely new surroundings upon arrival at their destination. Although the authors have never found it necessary, for long transit periods, canned fruit nectar could probably be offered to the bats in their shipping containers by means of plastic water dispensers used for small caged birds. For international shipments, packing fresh fruit in shipping containers with the animals also raises the possibility of problems with agriculture inspectors when national borders are crossed.

The authors ship bats in reusable, cylindrical, quart-sized food containers made of water-resistant cardboard (Fig. 2-10). Holes are punched in the top, side, and bottom with a paper punch and cork borer for ventilation and drainage. Plastic aquaculture netting (e.g., 1/8-in. [0.3-cm] Naltex®) is then stapled and glued to the inside of each container to provide a surface on which the bats can hang. Up to 4 bats have been shipped successfully per container. Fourteen of these containers are then placed in sturdy cages, constructed of water-resistant cardboard and metal house screening that were originally designed for shipping mice. The shipping method also includes a strong, but light-weight, outer container constructed of 15/16-in. (2-cm) square aluminum tubing (as frames), ¼-in. (0.6-cm) plywood top and bottom, and sides of ¼-in. (0.6-cm) galvanized mesh (Fig. 2-11). This is designed to eliminate any concern on the part of airline or health regulatory authorities that damage to the inner shipping containers might allow the bats to escape. Much higher quality, light-weight, all-plastic shipping containers for rodents are now commercially available at reasonable cost from some animal suppliers (see *Products Mentioned in Text*) or could probably be obtained for reuse from many institutional animal facilities. If desired, additional light-weight plywood or masonite siding with ventilation holes could be readily affixed to the aluminum tubing of the protective outer container with self-tapping screws. The aluminum frames can either be custom made by a cage manufacturer (see *Products Mentioned in Text*) or assembled with simple hand tools (Rasweiler, 1992).

This shipping system has several other important advantages. The bats are provided with adequate ventilation during transit, but are still buffered from exposure to extreme temperatures or disturbance by humans. Furthermore, their division into small groups in the food containers prevents the bats from attempting to fly when startled, or congregating into large, tight clusters. *C. perspicillata* gather in this fashion when stressed, and this can lead to mortalities if the animals are also exposed to moderate heat.

It is important to note that if the bats are to be shipped by air, their containers must meet Container Requirement 77 (see Fig. 2-12) of the International Air Transport Association (see Schwellenbach *et al.*, this Chapter). In essence this states that the containers

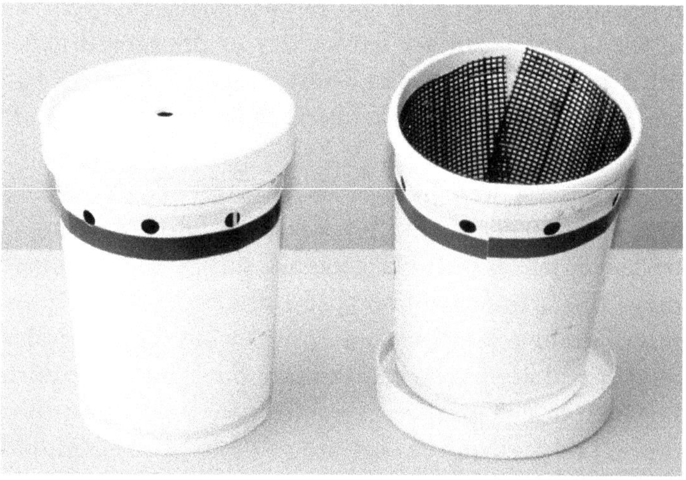

Fig. 2-10. Cardboard food containers lined with aquaculture netting used for transporting small nectarivorous and frugivorous bats. Photo by J. J. Rasweiler, IV.

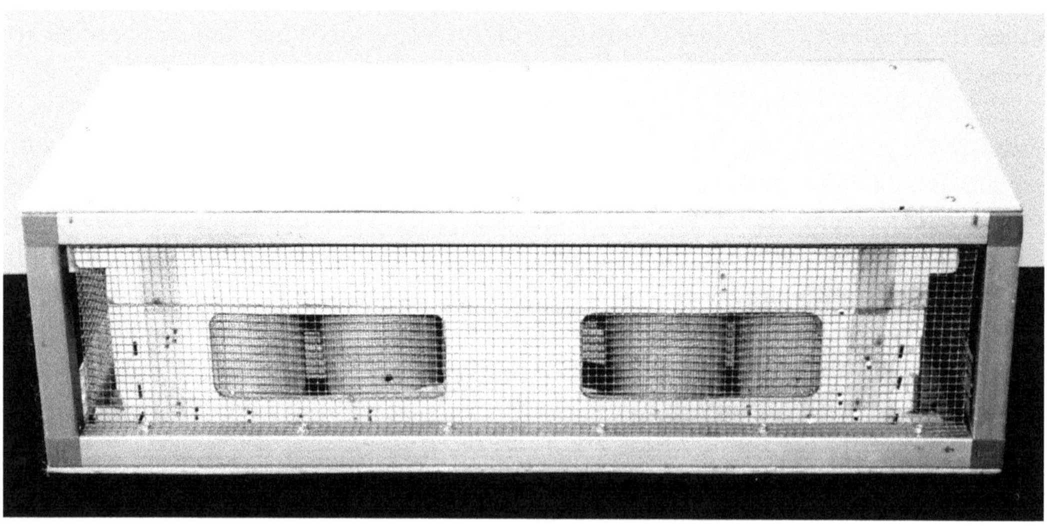

Fig. 2-11. System used for shipping small nectarivorous and frugivorous bats. Food containers holding the bats are loaded inside a cardboard mouse shipping box. The latter is then placed inside of a strong, but lightweight, outer container constructed of aluminum tubing, wire mesh, and plywood. Photo by J. J. Rasweiler, IV.

must be designed to withstand the likely rigors of transportation, prevent exposing workers to any animal contact, prevent escape of the animals during feeding and/or examination, and provide for the safety and well-being of the animals. Although one possible approach is illustrated in the regulations, these clearly recognize that other approaches may be acceptable and indeed advisable, depending upon the species.

PRODUCTS MENTIONED IN TEXT

Plastic Aquaculture Netting (Naltex® extruded square mesh): DelStar Technologies, Inc., 220 East St. Elmo Road, Austin, TX 78745 USA; http://www.delstarinc.com/naltex.html.

Plastic Rodent Shipping Containers or Transit Cages: JAX Mice and Services, 610 Main Street, Bar Harbor, ME 04609 USA; http://jaxmice.jax.org, or Taconic Farms, Inc., One Hudson City Centre, Hudson, NY 12534 USA; http://www.taconic.com.

Protective Aluminum Shipping Frames: Corners Limited, 841 Gibson, Kalamazoo, MI 49001 USA; http://www.cornerslimited.com.

REFERENCES

Rasweiler, J.J., IV. 1992. Practical approaches to housing bats for captive studies. *Bat Research News*, 33:8-9.

International Air Transport of Bats

Kathy Schwellenbach, Gregg Pittelkow, and Stephanie R. Ostrowski

INTRODUCTION

Arranging for international air transport of any animal species can be a time-consuming and detailed task. With respect to bats, it can also be complicated by the lack of knowledge that the general public, including cargo handlers, have about these animals. For anyone attempting to understand the regulatory and operational framework of how animals move by air internationally, it is helpful to identify resource manuals and points of contact. Shipment by boat or auto is much less common, but many of the same regulations apply with regard to permits, inspection, caging, and humane transport.

INTERNATIONAL TREATIES

International shipment of all species of wild animals is regulated by the Convention on International Trade in Endangered Species of Wild Fauna and Flora, or CITES. The International Air Transport Association (IATA) publishes Live Animal Regulations (LAR), which are revised and updated annually, and should be considered the "gold standard" for shipping instructions. IATA regulations are developed by the IATA Live Animals and Perishables Board, which consists of various airline representatives in consultation with a veterinary advisor and numerous animal conservation organizations around the world. Their mission is to provide guidelines and shipping container requirements that allow these special and fragile cargoes to be shipped safely, efficiently and humanely, while minimizing opportunities for injuries to cargo handlers and their exposure to infectious materials. Member states of the European Union and the Council of Europe use these regulations as a basis for their own minimum standards. The Office International des Epizooties (OIE) and CITES also accepts these guidelines for transportation of animals by air.

PORT OF ENTRY INSPECTION

Whether arriving by land, sea, or air, all live animal cargoes are subject to port of entry inspection by customs and/or public health, wildlife, or agriculture agents in the country of destination. Visual inspection of animals in containers may be required to verify species, numbers, and condition, and appropriate permit documents must be presented at that time. Lack of appropriate permits and other documentation, or containers that make it impossible to inspect the contents, may lead to significant delays and missed connecting flights. Such deficiencies may also put animals at risk of environmental stress, escape or seizure, and may result in fines and other criminal penalties, including charges of wildlife smuggling. Calling inspection officials at the intended port of entry in advance to provide courtesy notification of intent to clear customs can significantly expedite this process and allow for a final opportunity to remedy any deficiencies before travel.

CITES PERMITS

When transporting bats internationally, a CITES permit should be obtained from the wildlife conservation authority in the country of destination. In the United States, this is the U.S. Fish and Wildlife Service. If the species to be transported has been classified as endangered, it also requires a CITES export permit from the country of origin. Permit applications may take months to process, and shippers are advised to make permits an integral part of their advance planning.

OTHER PERMITS

Depending on the country, permits from the national agricultural, wildlife, and/or public health authorities may be required for export and subsequent entry of bats into the country of destination. Depending on the species and the perceived potential threat to public health or agriculture, some species may not be permitted entry, or may have strict quarantine or possession restrictions. Some countries may also impose significant restrictions, or may require licenses or permits for animals in transit to their final destination. It is the responsibility of the shipper to ensure that the shipment complies with regulations in each country along the shipping route. Shippers are advised to check thoroughly all such requirements when planning a shipment, well in advance of trapping or shipping.

IATA LIVE ANIMAL REGULATIONS

The international air transport of live bats is governed in large part by regulations developed by IATA. The regulations are updated as necessary, and published yearly in the LAR manual. IATA regulations have been incorporated into national legislation in a number of countries, including all member states of the European Union. In addition, the U.S. Fish and Wildlife Service has adopted LAR container requirements for transporting wild mammals and birds to the United States. Shippers who transport live animals with airline companies that are members of IATA, which includes most major air carriers, must comply with the LAR in their entirety, as well as any additional government regulations or air carrier restrictions that may apply. While IATA regulations apply to all international live animal shipments, many air carriers have elected to use the LAR as their standard reference for all live animal shipments, both domestic and international.

LAR regulations are in a standard format, which is consistent with airline regulations for all types of animals accepted for shipping. However, airlines vary in which animals they will accept. Thus, it is necessary to have the rules presented consistently for all animals, both wild and domestic, so that airline personnel can refer to them, and apply them easily and uniformly.

UPDATES

Compared to previous regulations, the new IATA LAR guidelines provide increased flexibility for individual species' needs. Based on feedback from a variety of bat experts (i.e., zoological, conservation, veterinary, research, and education), the *principles* of the revised regulations are written to allow shippers to design the interior of transport containers appropriately for each species, the number of animals being shipped, and the behavior of the species in question.

The current protocol was accepted and approved by the International Air Transport Association Live Animals & Perishables Board (IATA LAPB) in October 1998, following a presentation by a representative of the American Zoo and Aquarium Bat Taxon Advisory Group (Schwellenbach) with technical/regulatory advice from Northwest Airlines Cargo (Pittelkow). The conceptual recommendations were voted on and passed, with the precise language worked out subsequently in collaboration with the IATA LAR secretary.

The goals of the 1998 protocol were two-fold:

1. To respect the design function of the outer container, which is intended to provide sturdy handling and safe transport. Also, absolutely no contact between airline personnel and animals is permissible at any time during shipping. Finally, health experts in various ports need to be able to examine animals visually without risk of escape. Health ministries worldwide dictate these expectations, and implementation is determined by the operational needs of the airlines.
2. To make the inner container more "bat friendly" by allowing bat experts to consider the following: size and number of bats, social behavior, food and water requirements, temperature limitations, and any other information pertinent to the bat species being shipped (see also Appendices 2-1 through 2-3).

Shippers have been given more say in what is best for the bats being transported because they are the experts on the bats' needs, and airlines personnel are not. However, bat shippers need to understand and follow airline operational needs, which are dictated by Health Ministries and similar government organizations for all countries worldwide to assure successful operations of airline transport.

The following regulations for Container Requirement 77 were published in the October 1999 issue of the LAR (see also United States Government [USG] Exceptions in the current IATA LAR, Chapter 2). Bat shippers should be very familiar with these regulations and should check for any changes since the publication of the 1999 manual. In addition to specific container requirements, LAR Chapter 6 lists CITES categories for most species, and Appendix B provides an extensive pre-shipping question and answer checklist (see also basic transport considerations listed in Appendix 2-4).

CONTAINER CONSTRUCTION

(see also Exception QF-01 in IATA LAR, Chapter 3)

MATERIALS

Wood, plywood, wire mesh, non-toxic plastic mesh and screening materials, non-toxic plastic, aluminum, weather-resistant thermoplastic (e.g., Plexiglas®).

PRINCIPLES OF DESIGN

The following principles of design (Fig. 2-12) must be met, in addition to the General Container Requirements outlined at the beginning of Chapter 8 of the current IATA Live Animals Regulations.

Bats require very specialized handling, and the guidelines for Container Requirements are only general guidelines that can be applied to their shipment. Transportation of bats must only be undertaken by those shippers and carriers who can fulfill all the specialized needs of the species concerned.

The outer container must be designed to withstand the rigors of transportation, to prevent exposing workers to any animal contact, and to prevent escape of the animal(s) during feeding and/or examination. The inner container should be designed for the safety and well-being of the animal(s), while providing protection from outside elements.

DIMENSION

The container must be large enough for the bat(s) to move around freely and to hang normally from roosting material. Dimensions may vary with species, but they must

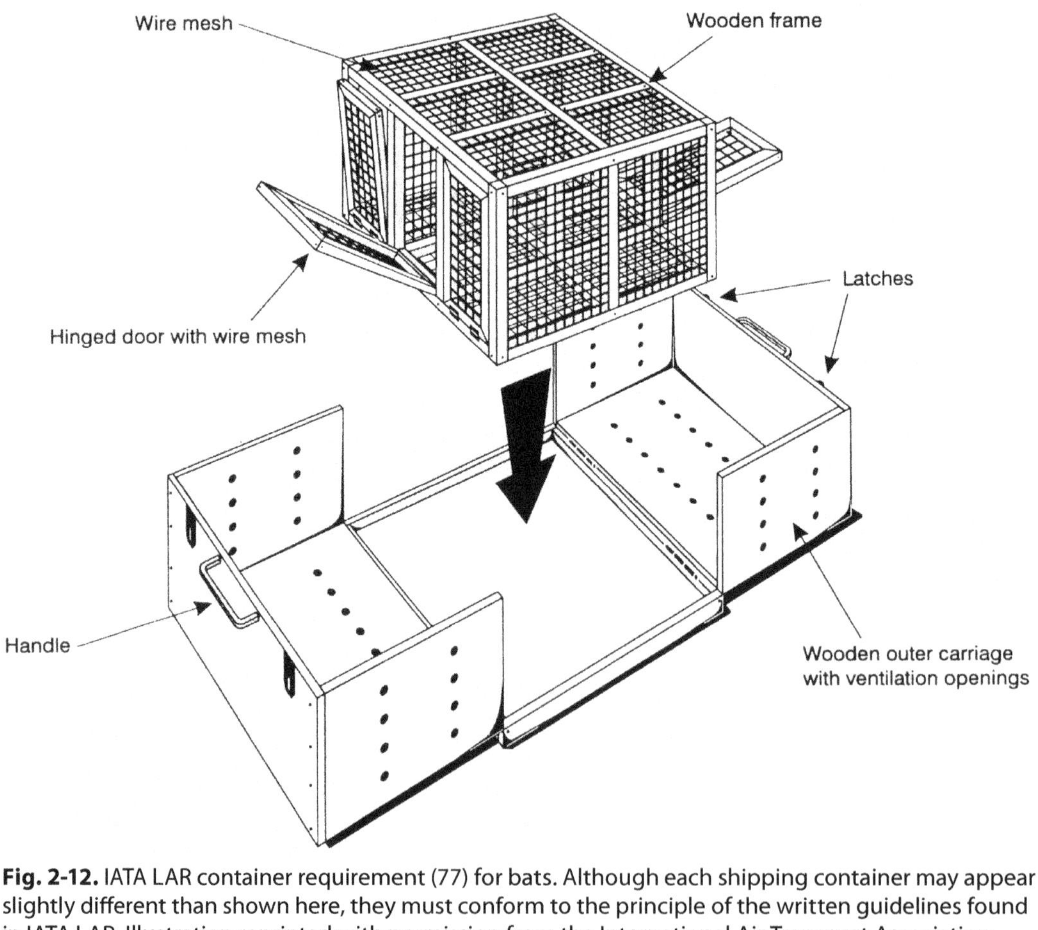

Fig. 2-12. IATA LAR container requirement (77) for bats. Although each shipping container may appear slightly different than shown here, they must conform to the principle of the written guidelines found in IATA LAR. Illustration reprinted with permission from the International Air Transport Association.

allow the animal(s) to hang with their heads extended, to flap their wings sufficiently to dissipate heat, and to position themselves comfortably in groups or separately.

OUTER CONTAINER

FRAME

The frame must be sufficiently strong to withstand the rigors of transportation. It should be impact-resistant, non-crushable, and weatherproof. Examples of appropriate outer containers include rigid plastic pet containers, hard plastic coolers, and solid, wood-frame containers securely fastened with screws.

SIDES

The sides must be made of wood, plywood, non-toxic plastic, aluminum, or weather-resistant thermoplastic (e.g., Plexiglas®).

FLOOR

The floor must be solid and waterproof.

ROOF

The roof must be made of water-resistant wood, plywood, non-toxic plastic, aluminum, or weather-resistant thermoplastic.

DOORS

A hinged door must be provided at one side, or on top of the outer container.

VENTILATION

Wire mesh ventilation is required on at least three sides, but preferably on all four sides. Required mesh-covered holes should be a minimum of 1/2 in. (1.3 cm) in diameter over a minimum of $1/10^{th}$ of the total surface area of the container sides.

SPACER BARS/HANDLES

Spacer bars must be made to a depth of 1 in. (2.5 cm), and must be present on three sides of the container as shown in Figure 2-12.

INNER CONTAINER

The inner container must be designed to prevent it from moving about within the outer container during handling.

FRAME

The frame must be made of wood, plywood, wire, or rigid non-toxic plastic mesh or other rigid screening material, non-toxic plastic, aluminum, or weather-resistant thermoplastic. The interior may be subdivided for individual animals or groups of animals, provided that feeding and/or examination for all animals in the container can be done safely.

Sides and Roof

The sides and roof must be made of wood, plywood, or screening material such as cloth and wire or non-toxic plastic mesh. Screening material should allow the animal(s) comfortable hanging and roosting positions, either together or individually. For species that normally roost in tight crevices, cloth or other safe, non-toxic material may be placed on the inside of the container, providing a padded roosting area for the bats. If necessary, insulation such as cloth or other safe, non-toxic material may be used between the interior and exterior containers, providing that it does not block any ventilation openings.

Floor

The floor may be solid, or wire mesh. Absorbent, non-toxic bedding (e.g., cloth, absorbent paper, newspaper or similar material) must be provided.

Door

A securely fastened hinged door must be provided.

VENTILATION

Unless constructed primarily of wire mesh, ventilation openings must be covered with wire or plastic mesh suitably sized to prevent any part of the animal from protruding out of the opening. All sharp edges must be covered with smooth material.

FOOD AND WATER CONTAINERS

Shippers must make provisions and supply equipment for providing food and water when necessary. Water should be given only in the event of a delay, and only in accordance with the instructions accompanying the shipment. Instructions for supplying food and water to the animal(s) (when necessary) must be provided by the shipper and attached to the outside of the container.

FEEDING AND WATERING GUIDE

(for emergency use only)

The need to provide food and/or water should not arise under normal conditions. Water delivery systems provided in containers must only be used in the event of a delay and only according to instructions provided by the shipper. Care must be taken to avoid spilling when using the water delivery system, as hypothermia and death can result if bats become wet. *At no time should the outer or inner container be opened to provide food and/or water.*

Feeding requirements vary by species. Many bats refuse to eat dead food items (e.g., dead insects, lizards, or fish) that they have not caught themselves. Fruit-eating bats may also be very fastidious about the quality of the fruit they eat. In addition, materials such as blood, fruit, live or dead insects, and other animals may not be permitted by agriculture or health ministry regulations in countries where a shipment may be in transit.

Many bat species become torpid during transport and require little or no food or water, as long as they are protected from environmental extremes of heat, cold, and humidity. Thus, rapid transit (< 24 hrs. duration) of bats from shipper to consignee is of the utmost importance.

- A plentiful food supply (subject to import restrictions at destination) may be placed in the container at the start of the journey for fruit-eating bats. If provided, these materials need to be accessible for removal and destruction by inspectors at the port of entry.
- Insectivorous bats require a live insect diet. Due to restrictions on importation of exotic insects, such food generally can not be supplied.
- Vampire bats require fresh mammalian or avian blood. These materials require special permits or other documents for import and must be appropriately contained. Therefore, they can not be handled as animal food items.

GENERAL CARE AND LOADING

(see also IATA LAR Chapters 5 and 10)

Most species of bats huddle together to maintain body heat and for security. If too warm, they space themselves and may fan themselves if they have adequate room in the container. Because of limitations for supplying food and water to bats, these animals should not be accepted for transportation if delays extending the itinerary more than 24 hours are anticipated. Extremes in temperature can be harmful to bats. Refer to shipper's instructions (IATA LAR Appendix C) for acceptable minimum and maximum temperatures, which vary by species. For all species of bats, ambient temperatures should not exceed 85 °F (29.4 °C).

The importer is responsible for supplying experienced personnel at the port of entry to open containers for agriculture and wildlife inspection, and for removing any material that is deemed inadmissible (fruit, blood, fecal material). For unaccompanied shipments, the importer should have expert personnel available to deal with animal emergencies at any intermediate transit points. Cargo personnel must not attempt to make contact with the bats for any reason.

***Warning**: Bats can inflict severe bites, and may transmit rabies and other diseases. Therefore, bat containers must never be opened except by experienced handlers. Public health authorities recommend that persons handling bats be vaccinated against rabies.*

PRODUCTS MENTIONED IN TEXT

Plexiglas®: http://www.eplastics.com/ or any local construction supply store.

REFERENCES

IATA Live Animals Regulations (LAR). 1999. International Air Transport Association (IATA), Montreal, Canada and Geneva, Switzerland. 368 pp.

Shipping Bat Specimens

Susan M. Barnard

INTRODUCTION

Bat specimens include a variety of materials such as whole carcasses or parts thereof, or any material collected directly from an animal like samples of blood, tissue, or urine. These materials may be transported as museum specimens, or as specimens or tissue samples for research purposes, or as samples for medical diagnoses that may lead to appropriate disease treatment.

Specimens may be transported by the postal service, commercial carriers, trains, or airlines. Most specimens are shipped by air to minimize the length of time that they are exposed to the hazards of transport. International shipments must comply with both the International Civil Aviation Organization (ICAO) and national regulations, which may differ from country to country. Shipments may also be required to meet International Air Transport Association (IATA) Dangerous Goods Regulations, along with any regulations that may be required by individual countries. In the United States, this includes requirements issued by the U.S. Department of Transportation (DOT).

Training is required to pack and ship infectious substances, dangerous goods, or hazardous materials according to DOT and IATA regulations. However, many diagnostic specimens do not contain infectious agents. Furthermore, dangerous goods or hazardous materials may be sent in such small quantities that the shipment is exempt from regulations. For information on dangerous goods training, visit http://www.iata.org/training/courses/Pages/tcgp03.aspx, and for shipping support information, visit http://www.unh.edu/ehs/shipping.htm.

It is beyond the scope of this paper to offer information on shipping regulations for every country. The information presented here covers United States international and domestic requirements for shipping bat specimens. Many of the U.S. regulations may also be applicable in other nations (Appendix 2-5). However, regulations are often changed, so the latest information should always be checked prior to making a shipment.

PACKAGING BAT SPECIMENS

The type of container used for shipping bat specimens depends on the type of specimen to be shipped (e.g., blood, culture swabs, or preserved or fresh carcasses). Important considerations include: 1) prevention of cross-contamination between specimens; 2) prevention of specimen decomposition; 3) prevention of fluid leakage, and 4) preservation of specimen information. The following packaging items and procedures are required for shipping specimens.

PRIMARY CONTAINER (A.K.A. INNER RECEPTACLE)

Labeled specimens must be packed in a securely sealed, crush-proof and leak-proof primary container. In most cases, the primary container will be specimen tubes (Fig. 2-13) or jars or bottles with screw-top caps (Fig. 2-14). As an added precaution against leaking, jar and bottle tops can be secured with Parafilm® (Fig. 2-15). Primary containers may also include plastic bags (Fig. 2-16) or metal cans.

SECONDARY CONTAINER

The purpose of a secondary container is to preserve and protect the specimen(s) if the primary container is damaged. The secondary container must contain enough absorbent material to absorb the entire liquid contents of the primary container if it should break in transit. In the case of glass or plastic tubes, the best secondary container is another plastic container with a screw-top lid (Fig. 2-17). All shipping containers must be packaged to fully contain leakage in accordance with applicable federal laws.

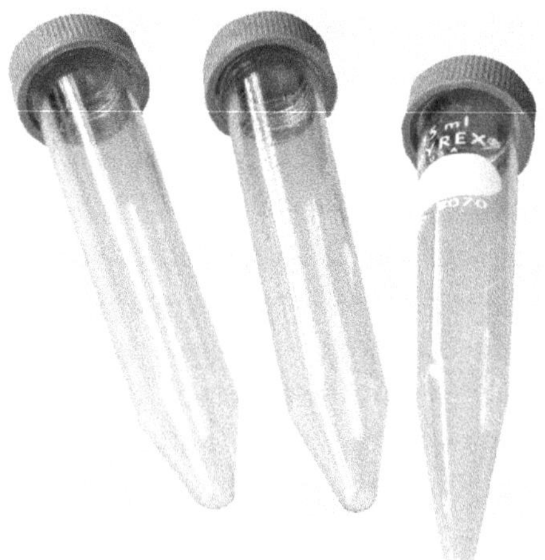

Fig. 2-13. Test tubes with screw-tops used as primary containers for transporting bat specimens. Photo by the author.

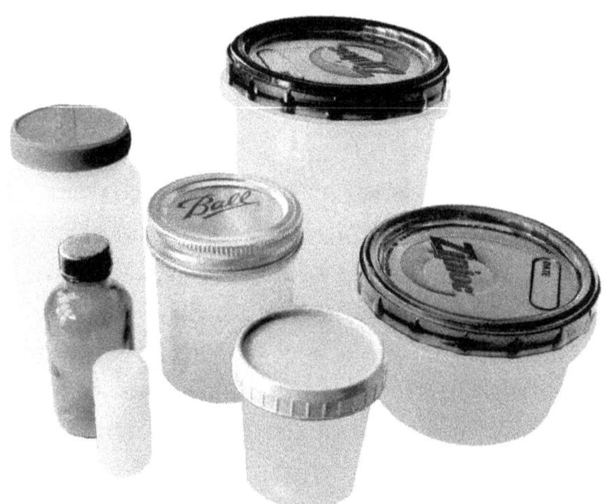

Fig. 2-14. Jars and bottles may be used as primary or secondary containers for transporting bat specimens. Photo by the author.

Fig. 2-15. Laboratory film (Parafilm®) stretched around jar and bottle tops helps to seal in liquids that might otherwise leak out of the vessel. Photo by the author.

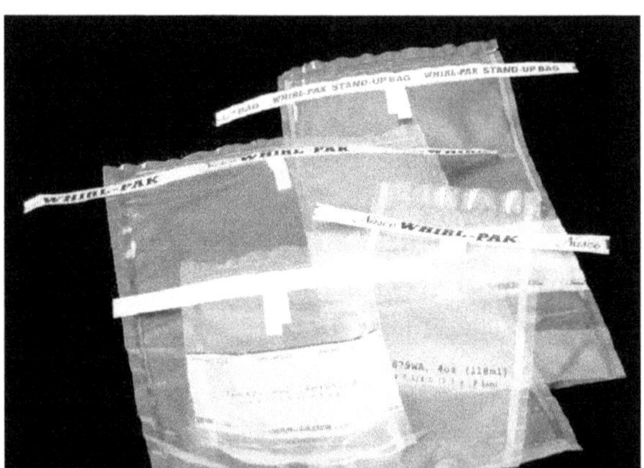

Fig. 2-16. Plastic bags may be used as primary containers for shipping bat specimens. Photo by the author.

OUTER PACKAGING

Each set of primary and secondary containers must be placed in a crush-proof shipping container made of fiberboard. The secondary container may be cushioned against damage with Bubble Wrap® or other packing materials (Fig. 2-18).

Fig. 2-17. Plastic jar with screw-top being used as a secondary container. Note the absorbent packing around the two primary containers. Photo by the author.

Fig. 2-18. Example of triple packaging and cushioning of secondary container inside a rigid outer fiberboard container. Photo by the author.

SHIPPING PRESERVED SPECIMENS

Diagnostic laboratories differ in their procedures for receiving specimens and should be contacted for shipping instructions. Specimens that are prepared with the appropriate preservative for transport will not contaminate, decompose, or spoil. Necropsy samples are typically transported in 10% formalin, whereas fresh carcasses are usually transported in coolants.

The most common substances used for preserving specimens are ethyl alcohol (ethanol), isopropyl alcohol (isopropanol), formaldehyde, and glycerin (glycerol). With the exception of glycerin, these substances are regulated as dangerous goods or hazardous materials. However, as mentioned above, exceptions are made for dangerous goods that are shipped in small quantities (see 49 CFR 173.4; http://setonresourcecenter.com/49CFR/Docs/wcd00009/wcd00925.asp).

Ten-percent formalin is the preservative of choice for necropsy samples. In some situations, whole carcasses may be transported in 10% formalin. In the United States, formalin in a concentration of 10% or less is exempt from regulation, and the DOT Pipeline and Hazardous Materials Safety Administration (PHMSA) no longer requires special labeling or packaging for specimens shipped in this preservative (see http://www.phmsa.dot.gov/staticfiles/PHMSA/DownloadableFiles/Files/Interpretation%20Files/2006/060267.pdf).

Fresh materials are often preserved temporarily with dry ice, wet ice, frozen-water ice packs, or gel packs. Dry ice (solid carbon dioxide) is considered a hazardous material. However, a dry ice label (Fig. 2-19) and a Shipper's Declaration for Dangerous Goods (Fig. 2-20) are not required when dry ice is used to cool non-dangerous goods that are transported domestically via the post office or commercial carrier, provided that all applicable requirements are met (see http://pe.usps.com/text/pub52/pub52c3_024.htm). The exception is that a dry ice label is required for shipments to and from Alaska, Hawaii, Puerto Rico, and Catalina Island because specimens must be sent as air shipments, even though the dry ice is used to cool non-dangerous materials (see http://www.ups.com/content/us/en/resources/ship/materials/coolants.html and http://safety.dri.edu/Forms/shippingnonhaz.pdf).

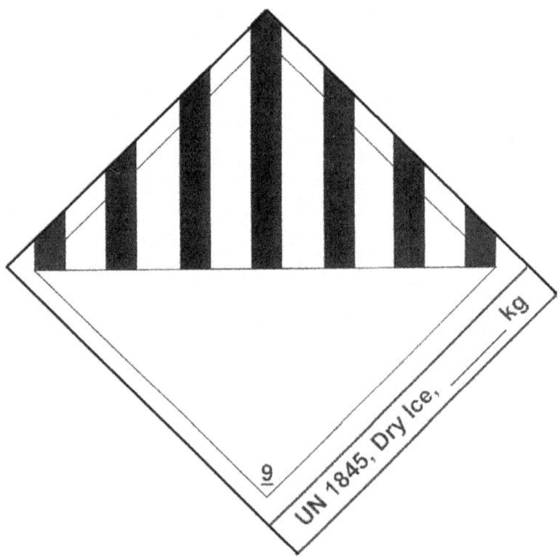

Fig. 2-19. Dry ice label. When shipping dry ice with dangerous goods, as the sole material, or with non-dangerous goods by air, a dry ice label must be on the package. In addition, the UN 1845 number and the weight of the dry ice in Kg must be included. Illustration by the author.

Wet ice (frozen water) has the disadvantages of adding weight to a shipment, and it requires leak-proof packaging. Corrugated and chipboard packaging can become weakened and may fall apart when wet. If water is found escaping a container, the shipment could be denied transport by a carrier. When using wet ice as a coolant, it is

SHIPPER'S DECLARATION FOR DANGEROUS GOODS

Shipper	Air Waybill No. Page 1 of 1 Pages Shipper's Reference Number (optional)
Consignee	
Two completed and signed copies of this declaration must be handed to the operator.	

TRANSPORT DETAILS

This shipment is within the limitations prescribed for: *(Delete non-applicable)*	Airport of Departure:
PASSENGER AND CARGO AIRCRAFT	CARGO AIRCRAFT ONLY
Airport of Destination	

WARNING

Failure to comply in all aspects with the applicable Dangerous Goods Regulations may be in breach of the applicable law, subject to legal penalties.

Shipment Type *(Delete non-applicable)*

NON-RADIOACTIVE	RADIOACTIVE

NATURE AND QUANTITY OF DANGEROUS GOODS

Dangerous Goods Identification						
UN or ID No.	Proper Shipping Name	Class or Division (Subsidiary Risk)	Pack-ing Group	Quantity and Type of Packing	Pack Inst.	Author-ization

Additional Handling Information

I hereby declare that the contents of this consignment are fully and accurately described above by the proper shipping name, and are classified, packaged, marked and labelled/placarded, and are in all respects in proper condition for transport according to applicable international and national governmental regulations. I declare that all of the applicable air transport requirements have been met.	Name/Title of Signatory Place and Date: Signature *(see warning above)*

Fig. 2-20. Shipper's Declaration for Dangerous Goods. This is a universal form used by all carriers worldwide. Image duplication by the author.

best to fill plastic jars with water and freeze them. Jar lids should be taped shut to prevent them from opening accidentally during transit.

The majority of gel-pack manufacturers use materials that are non-toxic and safe to handle. Frozen refrigerant packs may develop condensation around them during transit. Absorbent material wrapped around gel packs can prevent this moisture from damaging the packing material or leaking out of the shipping container. Condensation may also form on jars of frozen water, which should be similarly wrapped.

REGULATIONS FOR SHIPPING BIOLOGICALS

UNREGULATED BIOLOGICAL SPECIMENS

Bat shipments that do not contain infectious substances, or substances that are unlikely to cause disease in humans or animals, are exempt from regulation (see 49 CFR 173.134 Class 6, Division 6.2[b]) and packages may be sent with an "Exempt Animal Specimen" label (Fig. 2-21). A clear handwritten label is acceptable. No shipping paperwork is required, but samples must be packaged in appropriate leak-proof and cushioned secondary containers as shown in Figure 2-18. Containers with liquids that are being transported by air must be rated to an internal pressure of 95 kPa.

Fig. 2-21. Exempt Animal Specimen label. Bat shipments that do not contain infectious substances, or substances that are unlikely to cause disease in humans or animals, are exempt from regulation. Such shipments may be sent with the label shown here. Illustration by the author.

REGULATED BIOLOGICAL SPECIMENS

The classification of materials being shipped determines the packaging and training requirements for shippers. Materials known to contain infectious substances must be categorized and labeled appropriately.

CATEGORY A

The Code of Federal Regulations, 49 CFR 173.134 Class 6, Division 6.2(a)(1)(i), places in Category A any infectious substance that is in a form capable of causing permanent disability or life-threatening or fatal disease in otherwise healthy humans or animals upon exposure to it. Examples include confirmed infectious substances such as Hendra virus, Lyssavirus, and rabies virus (cultures only).

General packaging for Category A requires a watertight primary container for each specimen surrounded by absorbent material, a watertight secondary container with a list of contents attached to the outside, and a United Nations (UN) rated rigid outer container with proper UN labeling (e.g., UN 2814 for infectious substances affecting humans, and UN 2900 for infectious substances affecting animals).

Formal training and certification are required by law for anyone packaging and transporting Category A substances.

CATEGORY B

The Code of Federal Regulations, 49 CFR 173.134 Class 6, Division 6.2(a)(1)(ii), places in Category B any infectious substance that is not in a form capable of causing permanent disability or life-threatening or fatal disease in otherwise healthy humans or animals upon exposure to it. Examples include most blood, tissue, and urine samples that contain common microorganisms, and that are considered diagnostic specimens.

General packaging for Category B requires a leak-proof primary container for each specimen surrounded by absorbent material, a leak-proof secondary container with appropriate labeling on the outside listing the contents in the container, as well as appropriate hazard warnings (e.g., biohazard), and rigid outer packaging labeled with a UN 3373 label (see Fig. 2-26).

SHIPPING EXAMPLE

A bat suspected of dying of rabies must be sent to a public health laboratory for testing. Unlike many other mammals that may be tested for rabies, a bat carcass frozen prior to shipping will not delay testing because it is small and should defrost by the time it reaches the laboratory. However, many health laboratories do not recommend freezing a carcass. To ensure that the brains of such specimens are not damaged, check the local laboratory for shipping instructions. The following example for shipping a suspected rabid bat carcass meets DOT regulations.

PREPARATION

- Wrap the whole bat carcass in 2 plastic bags, one inside the other (Fig. 2-22).
- Place the double-bagged bat in a rigid, leak-proof container (secondary container), which may or may not be provided by the public health laboratory. Any appropriate rigid container may be used (see Fig. 2-23).
- Apply a biohazard label (Fig. 2-23) to the secondary container (not on the plastic bag holding the bat).

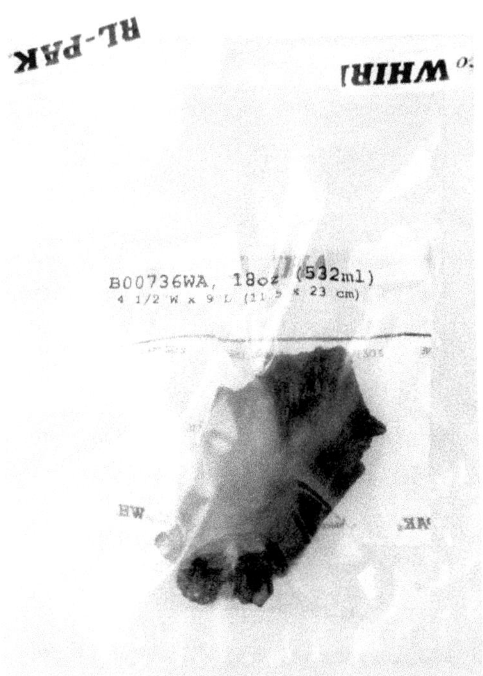

Fig. 2-22. Bat carcass in a plastic bag that has been placed inside another plastic bag. These bags serve as the primary container for shipping. Photo by the author.

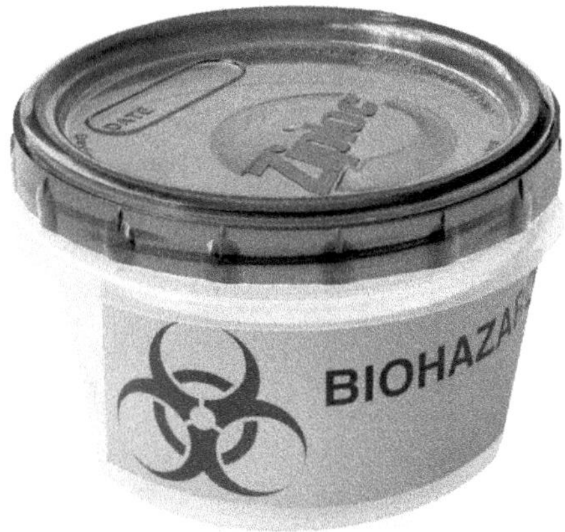

Fig. 2-23. Rigid, plastic secondary container with biohazard label. This container is suitable for transporting the double-bagged bat carcass shown in Fig. 2-22. Photo by the author.

FORMS

- Prior to preparing the bat carcass for shipping, obtain a submission form from the public health laboratory of destination. If submitting more than one bat, a form for each animal may be required.
- Place the completed form in a separate, sealed plastic bag (Fig. 2-24).
- Place the form inside the *outer* container (see Fig. 2-25), not inside the secondary container.

Fig. 2-24. Paperwork inside sealed plastic bag. The sealed bag protects written information against loss in the event a container should leak. Photo by the author.

PACKAGING

- Place the properly labeled secondary container into a polystyrene container (see Fig. 2-25).
- Place a frozen gel pack next to the secondary container. When a polystyrene container is used, it needs to be protected against damage by placing it in a rigid outer fiberboard container (Fig. 2-25).

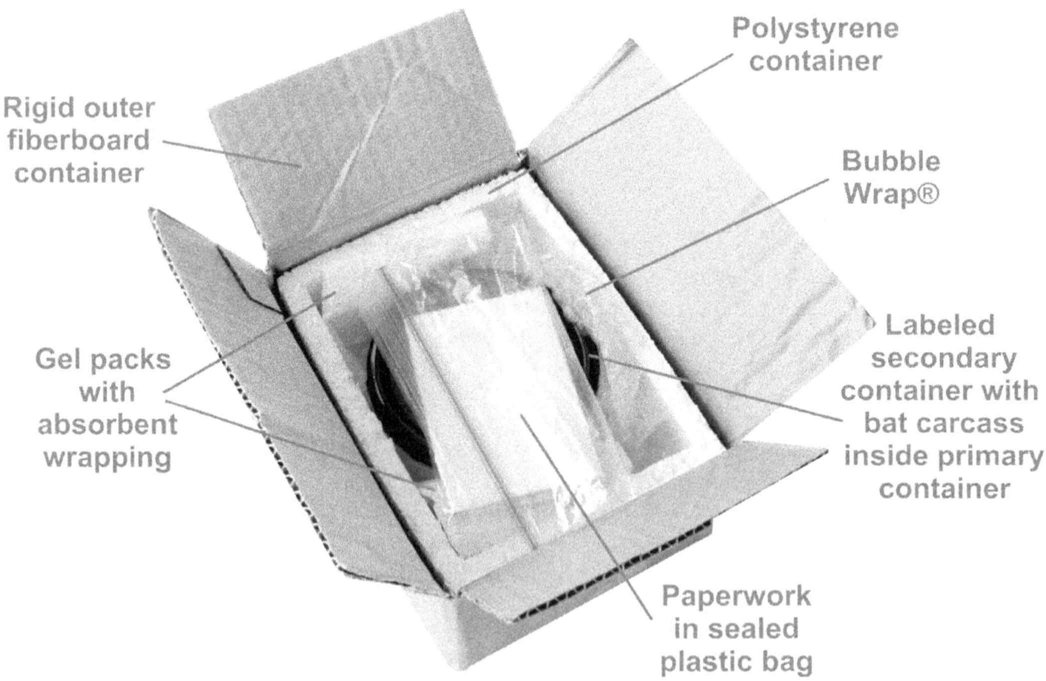

Fig. 2-25. Container contents that meet the Category B Code of Federal Regulations. Photo by the author.

LABELING

Fig. 2-26. Category B packaging requires a UN 3373 label as shown here. Photo by the author.

- Seal the outer packaging for shipping (see Fig. 2-26).
- Apply a UN 3373 label (Fig. 2-26) to the *side* of the shipping box.
- Write the shipper's name, address, and phone number on the outside of the box.
- Write the public health laboratory's address on the box.
- Be sure both addresses are correct.

SHIPPING

- Most laboratories prefer that shipments be sent via commercial carrier (e.g., FedEx, DHL or UPS), and not through the regular mail, because in many instances mail service centers delay specimens from arriving in a timely manner.
- However, when using the USPS, the package may be sent via first class mail, priority mail, or express mail.
- Be sure to obtain a tracking number.

PRODUCTS MENTIONED IN TEXT

Bubble Wrap®:Amazon.com; http://www.amazon.com/220ft-Rolls-110ft-Small-Bubble/dp/B000S5M9VK/ref=sr_1_3?s=office-products&ie=UTF8&qid=1293486479&sr=1-3, or most office supply stores.

Parafilm® M: *SPI Supplies Division, Structure Probe, Inc., P.O. Box 656, West Chester, PA 19381 USA;* http://www.2spi.com/catalog/supp/parafilm.php.

ACKNOWLEDGMENTS

I thank Cathleen A. Hanlon and Dietrich Schaaf for their helpful comments on the manuscript.

Appendix 2-1. Shipping container constructed for crevice-dwelling bats

(*Susan M. Barnard*)

Crevice-dwelling bats prefer tight places in which to roost, and they rarely roost in the open. By putting each bat into a bag before placing it in a shipping container, requirements for seclusion can be met, and the possibility of escape reduced. Food and water are not required for crevice-dwelling species while in transit. The shipping container depicted in Figure 2-27 was used to air ship insectivorous bats throughout the United States for nearly 20 years. During that time, no bats were lost.

Insectivorous bats are very small, and therefore shipping containers constructed for them may also be small, but sturdy. Using the IATA LAR container construction concept, a Keter® plastic toolbox (Fig. 2-27a), measuring 19.3 in. (49 cm) W x 10.2 in. (26 cm) D x 9.7 in. (25 cm) H, with an appropriate number of holes drilled into it for ventilation, makes an ideal outer container. It is able to meet the rigors of transportation, being impact resistant, non-crushable, weatherproof, and can be properly secured. These

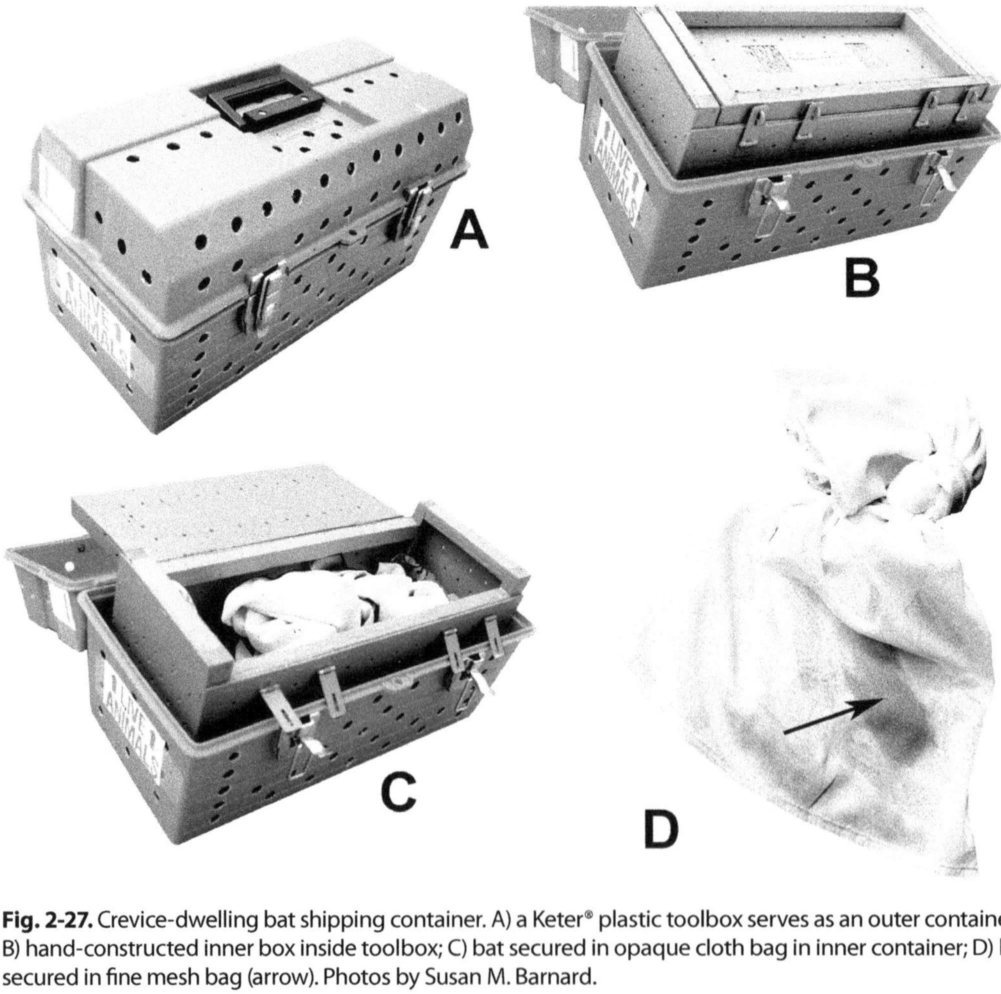

Fig. 2-27. Crevice-dwelling bat shipping container. A) a Keter® plastic toolbox serves as an outer container; B) hand-constructed inner box inside toolbox; C) bat secured in opaque cloth bag in inner container; D) bat secured in fine mesh bag (arrow). Photos by Susan M. Barnard.

Appendix 2-1 continued

toolboxes also have molded protrusions inside that serve as spacers that allow air circulation between the outer and inner containers.

To reduce weight, the inner container (Fig. 2-27b) may be constructed with a 1-in. x 1-in. (2.54-cm x 2.54-cm) wood frame and wood veneer (e.g., door skin) sides. Although insectivorous bats are highly carbon-dioxide tolerant (see Chapter 14, Volume 1), the inside container must also have an appropriate number of ventilation holes. For the safety of the occupants, it must fit tightly inside the toolbox. As safety is a priority for both the animals and airline cargo workers, clasps that can be opened should be used to allow inspectors to access the animals. In addition, bags containing the bats (Fig. 2-27c,d) offer the best seclusion if they are opaque (Fig. 2-27c). However, opaque material does not allow the bats to be inspected and counted. Therefore, to meet the IATA regulations fully, bags should be made of a material that allows health experts in various ports to examine the animal(s) (Fig. 2-27d, arrow) without risk of escape. Very fine cotton mesh material works well.

PRODUCTS MENTIONED IN APPENDIX

Keter® Plastic Toolbox: Available at Home Depot; http://www.homedepot.com/buy/storage-organization/garage-storage/keter-tool-storage/19-in-tool-box-99946.html.

Appendix 2-2. Shipping containers appropriate for fruit bats

(*Susan M. Barnard*)

The shipping container shown in Figure 2-28a-c was constructed for 4 giant Indian fruit bats (*Pteropus giganteus*) that were transported from Atlanta, Georgia to Albuquerque, New Mexico. With the exception of crevice-dwelling species, this design is also suitable for shipping all other bat species (e.g., phyllostomids and pteropodids).

The outer container (Fig. 2-28a,b) is made of plywood, and the 4-compartment inner container (Fig. 2-28c) has polyethylene mesh sides. The frames for both the outside and inner containers are made with 1-in. by 1-in. (2.54-cm by 2.54-cm) stock to reduce weight. The outer and inner containers are top loading, and two sides of the outer container open so bats can be inspected. The inner container is one unit, with each compartment opening separately to reduce the possibility of escape. The floor of each compartment can be lined with paper towels and there are 4 in. (10.2 cm) of clearance between the floor and the heads of large fruit bats. The compartments were sized to minimize the bats' movements, thus preventing injuries during transit. The bats were able to turn around, and also able to clutch the tops of the compartments with their thumbs, but they could not spread their wings fully. Therefore, containers of this design should not contain large bats for more than 8 hours.

The airline kennel (e.g., Petmate Pet Porter® [Fig. 2-28d,e]) is an excellent alternative to the container shown in Figure 2-28a-c, and it is familiar to all cargo handlers. Airline kennels are available in a variety of sizes, allowing for the transport of both small and large bats. Once a kennel of the right size is purchased, an inner container is then built to fit inside it. Shippers must be sure that all IATA live animals regulations have been met.

PRODUCTS MENTIONED IN APPENDIX

Pet Porter®: Amazon.com; http://www.amazon.com/Petmate-Porter-X-Large-Light-Gray/sim/B0009H56DA/2/ref=dp_cp_ob_k_sexpl or any pet supply store. ***Warning:*** *Product may not be accepted by all airlines. Check with the airline in question prior to purchase.*

Appendix 2-2 continued

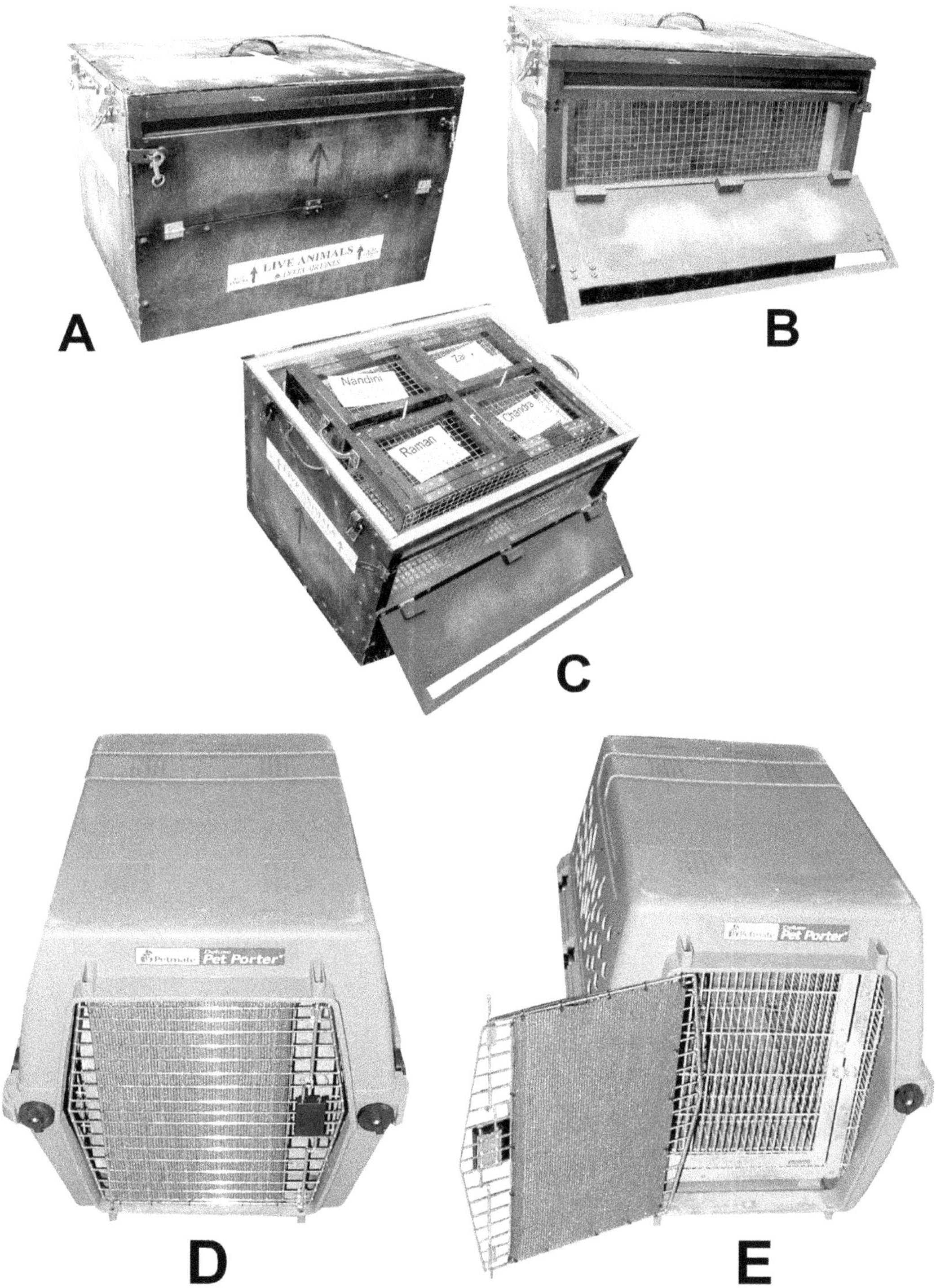

Fig. 2-28. Fruit bat shipping container. A) hand-constructed outer container; B) outer container with side panel open for inspecting bats; C) outer container with top open showing four-compartment inner container; D) Petmate Pet Porter®; E) pet porter with door open showing inner container. Photos A-C by Susan M. Barnard; D and E © Dana LeBlanc, Lubee Bat Conservancy.

Appendix 2-3. Shipping container constructed for vampire bats

(updated and reprinted with permission from Efird and Barnard, 1991)

Thirty vampire bats (*Desmodus rotundus*) were shipped from Atlanta, Georgia to Tokyo, Japan in the shipping container shown in Figure 2-29. Two outside containers (i.e., carrying cases) were used, with no more than 8 bats in each inside container (Fig. 2-29b-d); all bats survived the long trip. The most important design feature is the ice cube trays (Fig. 2-29d) built into each of the inside containers. Before transit, paper towels drenched in water were placed in the trays, so that water would not spill into the containers during flight, but if the bats were thirsty, they could obtain moisture. More important, blood could be placed in the trays without the bats escaping, and upon arrival at Narita International Airport, the trays were immediately filled with blood. Because vampire bats can not go without food for much more than 48 hours (Wilkinson, 1984), and these bats still had to travel more than 3 additional hours to Yokohama City University, providing a blood meal as quickly as possible was imperative.

The shipping container is composed of an outer container (Fig. 2-29a) and two inner mesh containers (Fig. 2-29b). General construction consists of a frame made of 1-in. (2.5-cm) stock, sheathed with ⅛-in. (0.32-cm) plywood or plastic laminate. The inside

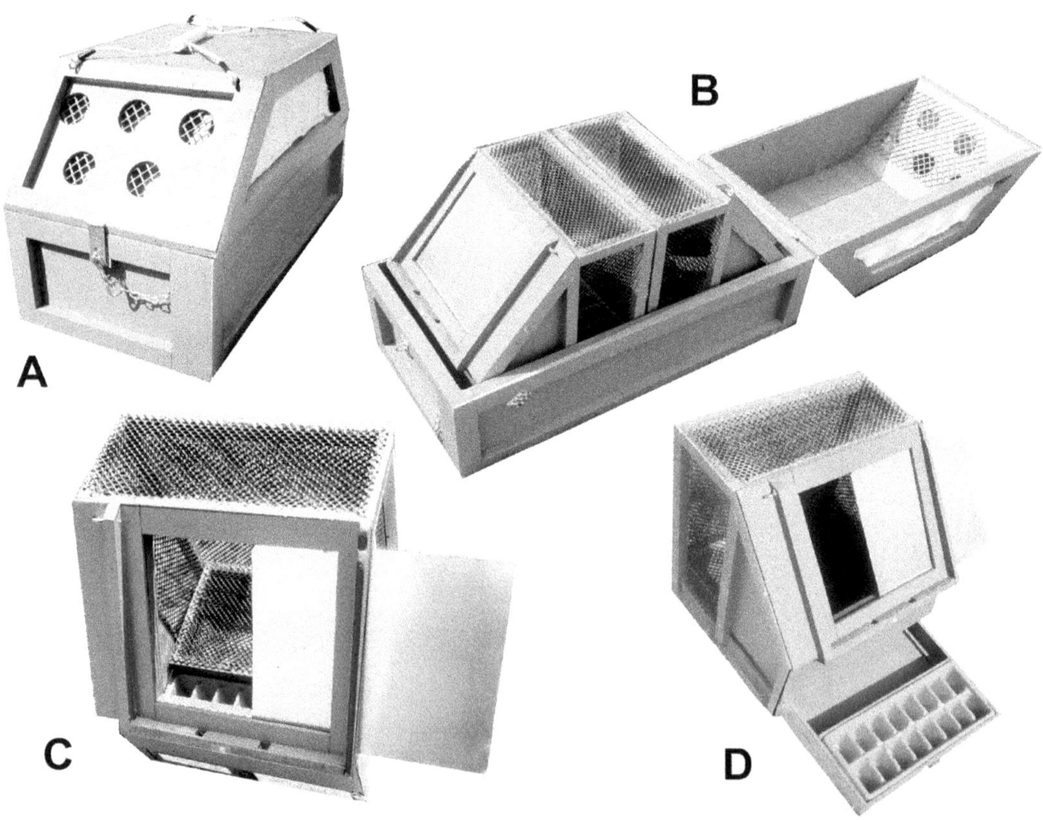

Fig. 2-29. Hand-constructed vampire bat shipping container. A) outer container; B) outer container open showing inner containers; C) inner container showing smooth sliding door; D) inner container showing food tray. Photos from Barnard (1995).

Appendix 2-3 continued

containers should be constructed first to ensure that they fit properly into the outer container or carrying case.

Ventilation holes are 2 in. (5.1 cm) in diameter and covered with ¼-in. (0.64-cm) polyethylene mesh, placed inside the outer container to protect the mesh from damage during transit. Holes should be placed only on the sides and sloped top part of the outer container, and not on the upper-most top. To comply with transport regulations, the bottom of the outer container must have a solid floor and sides to prevent fluids from leaking outside the container. In addition to affixing ice cube trays in the inner containers, three other important design features include 1) a sloping side over the ice cube trays (Fig. 2-29c,d)) and 2) a sliding door with a smooth surface (Fig. 2-29c,d), both of which prevent the bats from hanging over and contaminating their food, and 3) a raised mesh floor that protects the bats from direct contact with feces and urine.

One major improvement could be made to this design. A solid top could be placed over the mesh tops of the inner containers, so that when the outer container top is opened, the bats have protection from light and disturbance. This problem was solved temporarily for the shipment from Atlanta to Tokyo by placing a black cloth over the tops of the inner containers.

MATERIALS AND CONSTRUCTION

The shipping container shown in Figures 2-30 through 2-36 may be assembled by anyone possessing a moderate amount of woodworking skill and a general collection of household tools. Although it is not required, a table or radial-arm saw would greatly facilitate accurate construction. The general construction consists of a frame made of ¾-in. (1.9-cm) stock, sheathed with ⅛-in. (0.32-cm) plywood (e.g., Luan door skin) or plastic laminate (e.g., Formica®). Pine is a good wood for the framing because it is lightweight and strong. Luan door skin is three-ply, giving adequate strength when glued to the supporting frame. The inside containers should be built first to ensure that the outer carrier fits properly. This is especially important if dimensions are changed from those listed here.

MATERIALS

Below is the *List of Materials* needed for constructing the vampire bat shipping container. The list provides slightly more materials than is needed for the construction of one carrying case (see Fig. 2-29a) and two inside containers (see Fig. 2-29b) because the wood working skill of each individual varies; in addition, the kerf (saw cutting width) of the individual's saw varies, and the actual dimensions used in constructing the carrier differ from stock sizes. For example, 1-in. (2.5-cm) by 4-in. (10.2-cm) stock has actual dimensions of ¾ in. (1.9 cm) by 3¾ in. (9.5 cm). If stock is used that is wider than the suggested 4-in. (10.2 cm) boards, adjust the lengths to compensate. Also remember to allow for the thickness of your saw blade. A kerf of ⅛ in. (0.32 cm) allows for the cutting of three 1-in. (2.5-cm) frame members with ⅜ in. (0.95 cm) of waste.

Appendix 2-3 continued

List of materials needed to construct a shipping container for vampire bats*

ITEM	QUANTITY
Luan door skin or plastic laminate (e.g., Formica®), 1/8-in. x 36-in. x 80-in.	2 sheets
Pine stock, 1-in. x 4-in.	16 lin. ft.
¼-in. polyethylene mesh	850 sq. in.
Brass piano hinge, 1¼-in. x 12-in.	1 strip
Galvanized screw eyes, ¾-in.	4
Galvanized safety hasp, 2½-in.	1
Round eye snap hook, small	1
Chain, small	8 in.
Polypropylene rope, ¼-in.	5 ft.
PVC pipe, ½-in.	6 in.
Aluminum angle, 1/8-in. x ¾-in. x ¾-in.	6 ft.
Aluminum sheet, 8 in. x 9¾-in. x 20-22 ga.	2
Water-based, exterior paint (e.g., Kelly-Moore™)	1 qt.
Galvanized screws (2½-in.), brads, staples, wood glue, hot melt glue sticks, paint brushes, sandpaper, and other necessary items.	As needed

* See also *Products Mentioned in Appendix.*

CONSTRUCTION

Although the subject for another publication, the shipping container discussed in this paper contains modifications (p and q in Fig. 2-33) for receiving a purpose-built shift/transfer box (see Fig. 9-33, Volume 3). The box was designed to transfer vampire bats quickly and with little or no handling from their enclosure to the shipping container. This modification can be eliminated, or other modifications can be included to meet a shipper's particular needs.

Figures 2-32 through 2-36 detail the construction of one inside container.

The basic construction employs the use of ½-in. (1.3-cm) by ¾-in. (1.9-cm) frame members joined with simple butt joints. Solid portions may be either plastic ⅛-in. (0.32-cm) laminate or ⅛-in. (0.32-cm) Luan door skin. The three-ply door skin is lighter and easier to work than plastic laminate, but it requires two to three coats of paint to achieve a surface that cleans easily. On the other hand, the plastic laminate is easy to clean but difficult to cut and attach. When using plastic laminate attach it with water-based contact cement to avoid the toxic fumes from petroleum-based products.

Cut the six uprights (b, d, and f in Fig. 2-32, left side) to the specified length. Notch the bottom of the uprights to receive the solid bottom side panels (a in Fig. 2-32 left side, and m in Fig. 2-33 back). Use ½-in. (1.3-cm) brads and water-resistant glue to attach the panels. Fasten the horizontal frame members with glue and four-penny finish nails. Pre-drill the nail holes with a 1/16-in. (0.16-cm) bit to prevent the wood from splitting. Before attaching the front and front-side panels, attach the polyethylene mesh using staples and a hot-melt glue gun. Try to fill the mesh with glue where it lies against the wood. This will facilitate the cleaning of the shipping container. Install the raised floor by stringing the inside mesh support (s in Fig. 2-35) across the inside so that the back edge of the ice cube tray is located behind it. Attach the floor frame with brads or glue. Fasten the mesh to the floor frame, bottom side, and back panels with hot-melt glue. Next, attach the front and front-side panels.

Appendix 2-3 continued

Part 0, shown in Figures 2-33 and 2-36, is a 10-in. (25.4-cm) by 12¼-in. (31.1-cm) rectangle of ⅛-in. (0.32-cm) Luan door skin, with an 8-in. (20.3-cm) square hole cut ¾ in. (1.9 cm) in from the side. To provide space for the sliding door, cut strips of ⅛-in. (0.32-cm) Luan door skin ⅜ in. (0.95 cm) wide (part u in Fig. 2-36 side view); glue these to the base. This leaves a ⅜-in. (0.95-cm) space around the hole in which the door can rest. The aluminum angle can be cut accurately with a carbide-toothed blade on a table or radial saw. Miter the joints, then drill and countersink two 3/16-in. (0.48-cm) screw holes in each rail. Position the holes so the screw will go through the spacer underneath and not impede the movement of the sliding door. Fasten the sloping front to the sloping frame members with brads and glue. The shutter door should be made from light-gauge aluminum (e.g., 20-22 ga.). The door should be cut 8½ in. (21.6 cm) wide and 9¾ in. (24.8 cm) long. Trim ¾ in. (1.9 cm) off the long side to make a handle, and bend it 90 deg. to prevent it from interfering with the closing of the carrying case.

The drawer (Fig. 2-34) should be constructed with butt or miter joints. This component catches most of the fecal material, urine, and spilled blood. It must be as tight and moisture-proof as possible. Use ¼-in. (0.64-cm) square fillets to reinforce the joints. Be sure they do not interfere with the ice cube tray. Although most ice cube trays are approximately the same size overall, the shape of the bottom and reinforcing webs vary greatly. Caulk along the joints and seams before painting the finished drawer. For best results, use hot-melt glue for this purpose. Avoid using other caulking compounds because they may contain arsenical compounds meant to inhibit the growth of mildew. Use two or three coats of a high-quality, water-based paint, sanding between coats.

Refer to Figure 2-30 for dimensions of the various carrying case (outer container) components. Frame members a, f, i, k, and 1 provide the vertical support. The horizontal frame members rest on top of the vertical members so that the weight-load is borne by the end grain. All members are ¾-in. (1.9-cm) by l-in. (2.5-cm) stock, except one, which is ¾-in. (1.9-cm) square. The panels, five in all, use simple butt joints that should be fastened with galvanized decking screws. All joints and skins should be glued with water-resistant glue, as well as being fastened. For example, ¾-in. (1.9-cm) brads can hold the skins in place while the glue sets. Place the skins on the inside of the frames to provide a smooth interior surface. Such a surface allows the carrying case to be cleaned properly. Extend the skins to the outer edges of the frames (as shown in the corner details in Fig. 2-31) to provide greater stability.

Install the piano hinge at the narrow end of the carrying case's top and bottom components. At the opposite end, use a safety hasp with a round eye snap hook. To prevent misplacing the snap hook, connect it to the box with a short piece of small chain. Put a screw eye into each corner as shown in Figures 2-30 and 2-31. Tie or splice a ¼-in. (0.64-cm) diameter by 28-in. (71.1-cm) long section of polypropene rope to the eyes on each side of the carrying case top. Pass the two sections of rope through a piece of ½-in. (1.3-cm) by 6-in. (15.2-cm) PVC pipe for use as a handle. Make sure the rope's knots or splices are **secure** to prevent the carrier containing the bats from dropping accidentally to the ground during handling.

Ventilation holes, shown in Figure 2-30, can be cut into the carrying case with a fly cutter or hole saw. The holes should be approximately 2 in. (5.1 cm) in diameter. They

Appendix 2-3 continued

should be cut ONLY on the sides and sloped ends of the carrying case top. To prevent placing bats in close proximity with cargo handlers, or exposing them to drafts and unfavorable weather conditions, do not cut holes in the upper-most part of the carrying case.

Holes should not be cut into the bottom of the carrying case. The solid sides of the carrier bottom prevent fluids from spilling to the outside, thereby adhering to federal regulations governing animal transport.

By attaching Luan door skin to the inside of the carrying case frame, as described above, ventilation holes can not be blocked by other cargo during transport. In addition, by attaching the mesh to the inside of the carrier, damage to the mesh can be avoided. The mesh covering ventilation holes should be attached with ⅜-in. (0.95-cm) to ½-in. (1.3-cm) staples, with the legs clinched.

After construction of the carrier is complete, remove the hardware and sand it inside and out. When the surfaces are smooth, caulk around all inside joints, using hot-melt glue. Apply two to three coats of a water-based, exterior paint (e.g., Kelly-Moore™) on all surfaces, sanding lightly between coats. After the last coat of paint dries, re-install the hardware.

Appendix 2-3 continued

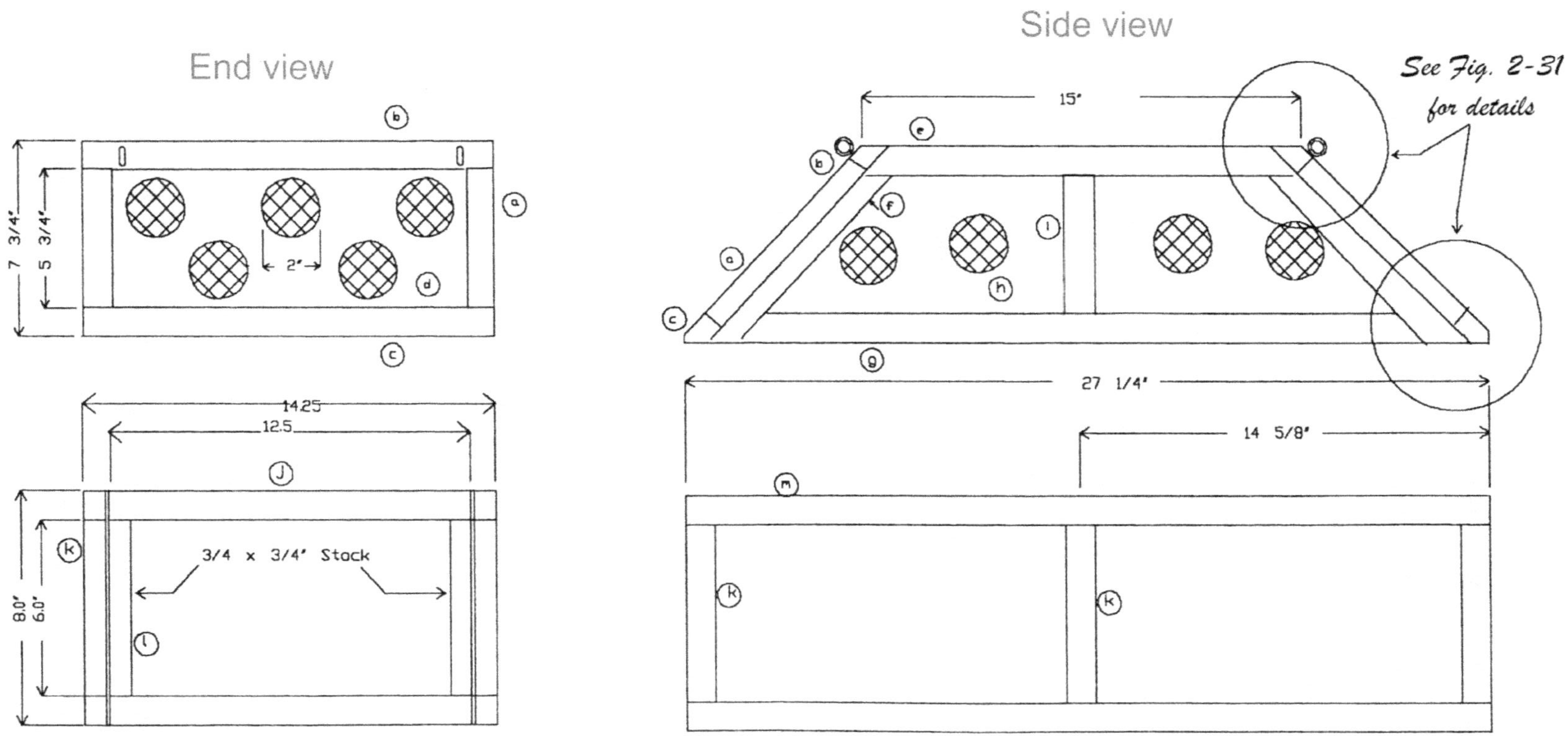

Fig. 2-30. Carrying case (end and side views). Ventilation holes should be located on sides and ends only; none on flat top. Parts, materials and quantity: a – ¾ in. x 1 in. x 7½ in. (4); b - ¾ in. x 1½ in. x 14¼ in. (60 deg. bevel on top) (2); c - ¾ in. x 1½ in. (double bevel on bottom) (2); d - 1/8 in. x 10 in. x 14¼ in. (2); e - ¾ in. x 1 in. x 16 in. (60 deg. bevel on ends) (2); f - ¾ in. x 1 in. x 10½ in. (60 deg. bevel on ends) (4); g - ¾ in. x 1 in. x 23¾ in. (60 deg. bevel on each end) (2); h - 1/8 in. x 25¾ in. x 7½ in. (2); i - ¾ in. x 1 in. x 5 5/8 in. (2); j - ¾ in. x 1 in. 12½ in. (4); k - ¾ in. x 1 in. x 6 in. (6); l - ¾ in. x ¾ in. x 6 in. (4); m - ¾ in. x 1 in. 27¼ in. (4). Illustrations reprinted with permission from Efird and Barnard (1991).

Appendix 2-3 continued

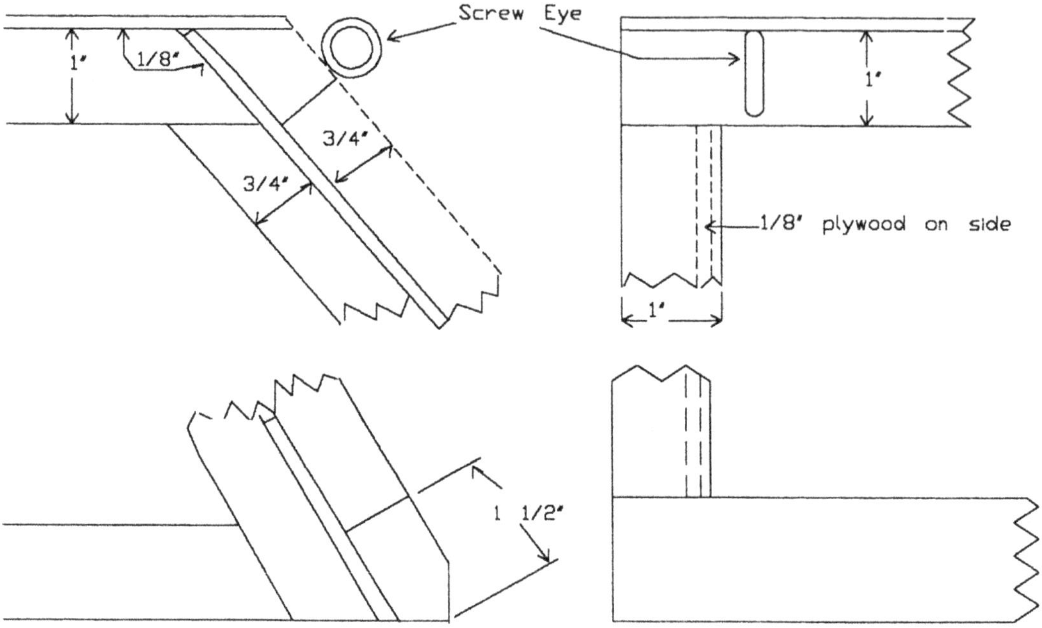

Fig. 2-31. Corner detail (see also Fig. 2-30). Illustrations reprinted with permission from Efird and Barnard (1991).

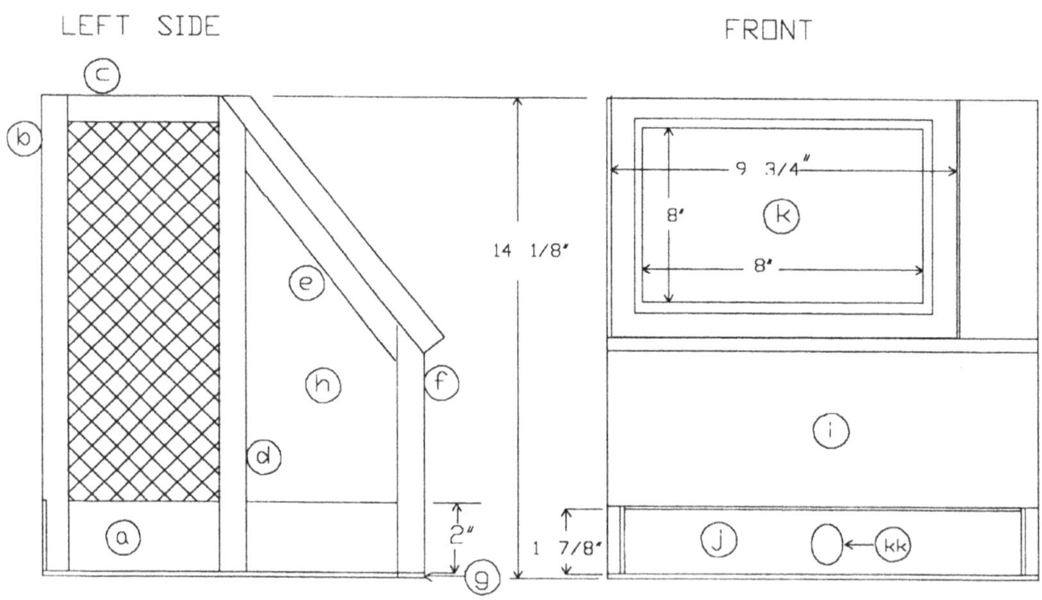

Fig. 2-32. Inside container (left side and front views). Parts, materials and quantity: a – bottom side panel, 1/8 in. x 2 in. x 11¼ in. (2); b – back upright, ½ in. x ¾ in. x 13 in. (2); c – top bar, ½ in. x ¾ in. x 4 3/8 in. (2); d – center bar, ½ in. x ¾ in. x 13 in. (38 deg. bevel) (2); e – front top, ½ in. x ¾ in. x 8 5/8 in. (38 deg. bevel) (2); f - front support, ½ in. x ¾ in. x 7 1/8 in. (38 deg. bevel) (2); g – bottom sheet, 1/8 in. x 11 3/8 in. x 12¼ in. (1); h- front side panel, 5 5/8 in. x 10¾ in. laminate (2); i – front panel, 12¼ in. x 4¼ in. laminate (1); j – drawer, see Fig. 2-34; k – door shutter (20-22 ga. aluminum), 8½ in. x 9¾ in.; kk – leather pull tab for drawer. Illustrations reprinted with permission from Efird and Barnard (1991).

Appendix 2-3 continued

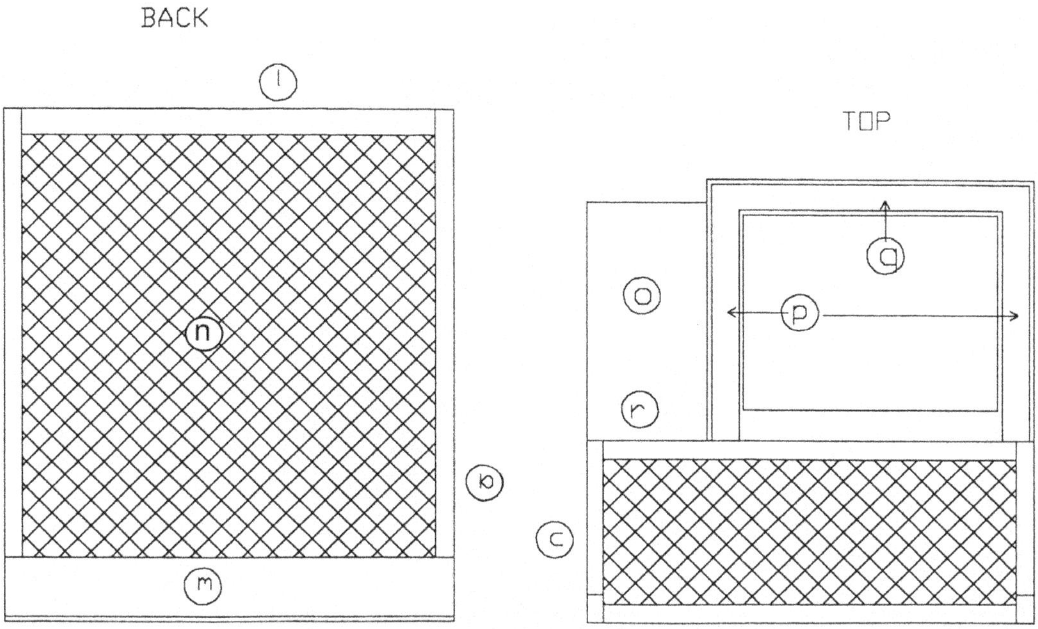

Fig. 2-33. Inside container (back and top views). Parts, materials and quantity: l – top back bar, ½ in. x ¾ in. x 11¼ in. (1); m – back bottom panel, 1/8 in. x 2 in. x 12¼ in. (1); n – back mesh (¼-in. polyethylene), 12 in. x 12 1/5 in.; o – slope panel, 1/8 in. x 12¼ in. x 9 7/8 in. (see also Fig. 2-36, detail a); p – transfer box side rails, ¾ in. x ¾ in. x 9 7/8 in. (2); q – transfer box top rail, ¾ in. x ¾ in. x 9 7/8 in. (1); r – top mid bar, ½ in. x ¾ in. x 11¼ in. (1). Illustrations reprinted with permission from Efird and Barnard (1991).

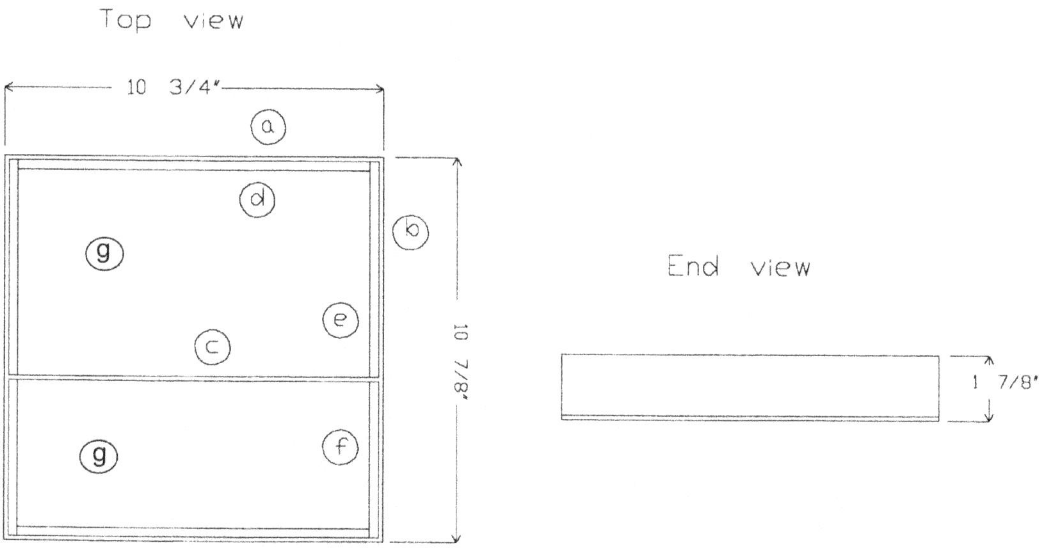

Fig. 2-34. Drawer (top and end views). Parts, materials and quantity: a – end panel, 1/8 in. x 1¾ in. x 10¾ in. (2); b – side panel, 1/8 in. x 1¾ in. x 10 7/8 in. (2); c – divider, 1/8 in. x 1¾ in. x 10½ in. (1); d – end fillet, ¼ in. x ¼ in. x 10 in. (3); e – long side fillet, ¼ in. x ¼ in. x 5 7/8 in. (2); f – short side fillet, ¼ in. x ¼ in. x 4½ in. (2); g – bottom, 1/8 in. x 10¾ in. x 10 7/8 in. (1). Illustrations reprinted with permission from Efird and Barnard (1991).

Appendix 2-3 continued

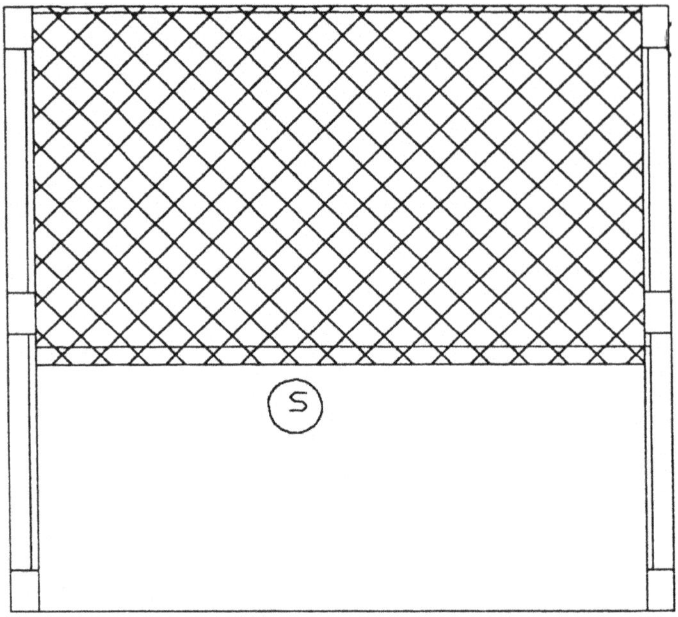

Fig. 2-35. Cross section view above drawer (elevated floor). Part s – inside mesh support, ¼ in. x ¼ in. x 11 3/8 in. (1). Illustration reprinted with permission from Efird and Barnard (1991).

Appendix 2-3 continued

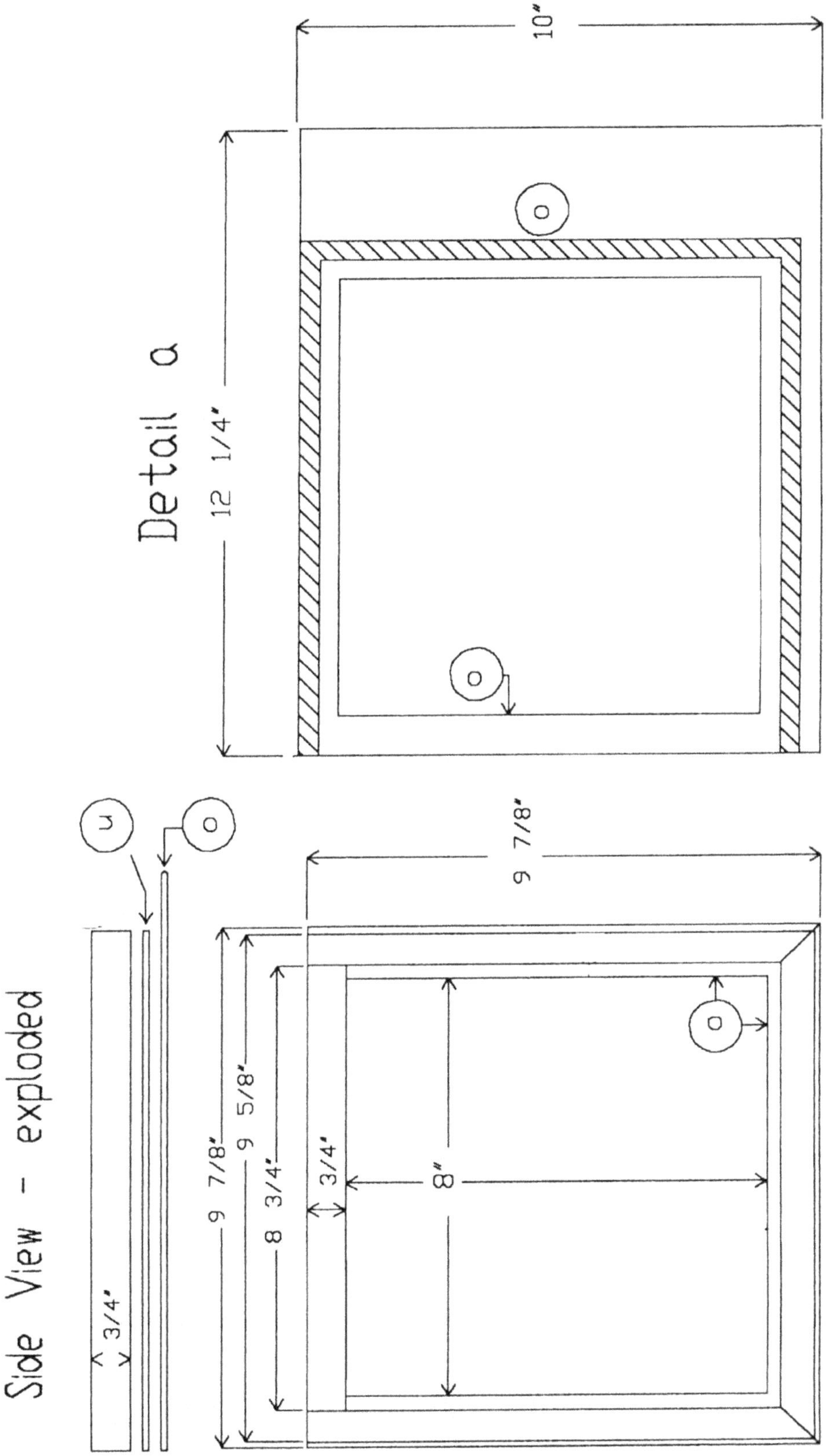

Fig. 2-36. Detail of sliding door assembly. Part u represents the 1/8-in. spacer strip. Detail a shows the front of the slope panel, o, with 1/8-in. x 3/8-in. spacer strips (shown hatched). The aluminum angles are not shown. The sliding door enters from the left of the illustration and is held down by the aluminum angles. Illustrations reprinted with permission from Efird and Barnard (1991).

Appendix 2-3 continued

SELECTED PRODUCTS MENTIONED IN APPENDIX

Aluminum Angle (¾-in. x ¾-in. x ⅛-in.): http://www.metalsdepot.com/products/alum2.phtml?page=sangle&LimAcc=$LimAcc.
Aluminum Sheet (20-22-ga.): http://www.metalsdepot.com/products/hrsteel2.phtml?page=sheet&LimAcc=$LimAcc.
Hot-Melt Glue Sticks: http://www.acehardwareoutlet.com/ProductList.aspx?ProductGroupID=2&MerchandiseClassId=212%20&CommodityGroupId=20.
Luan Door Skin (⅛-in. plywood): http://www.brittlaminate.com/products.htm#lp.
Paint (e.g., Kelly-Moore™): http://www.kellymoore.com/products/peah_1250_acry_shield.
Piano Hinge (1¼-in.; brass): http://www.baltimorehardware.com/index.cfm/product/374/continuous--piano-hinge-brass-plated-1-14w-x-48l-x-040t.
Plastic Laminate (e.g., Formica®): http://www.formica.com/en/home/TradeLanding.aspx.
Polyethylene Mesh (¼-in.; product XV 1170): http://www.industrialnetting.com/home.html.
Round Eye Snap Hooks: http://www.cdwplus.com/singleRDsnap.html.

REFERENCES

Barnard, S.M. 1995. *Bats in Captivity.* Wild Ones Animal Books, Springville, CA. 194 pp.
Efird, M. and Barnard, S.M. 1991. A proven shipping crate design for transporting select bat species. *Handbook of Live Animal Transport*, Supplement No. 30. Silesia Companies, Fort Washington, MD.
Wilkinson, G.S. 1984. Reciprocal food sharing in the vampire bat. *Nature*, 308:181-184.

Appendix 2-4. Basic transport considerations *(from Barnard, 2009)*

1. Ship bats at each species' thermal preferences. Allowable shipping temperatures range from a low of 45 °F (7.2 °C) to a high of 85 °F (29.4 °C). When temperatures are close to allowable limits, it may be necessary to include heating or cooling packs (e.g., Gempler's® warm packs or Uline® cold packs) in a transport container. If transporting bats by automobile during hot weather, always leave the air conditioner running when leaving the vehicle. During winter, leave the heater running for tropical species.
2. Avoid unnecessary transport delays by checking long-range weather forecasts for all stops along the shipping route.
3. Transport only during regular working hours. Government officials worldwide do not work routinely on weekends or holidays. Check the holiday calendar for all countries on the shipping route, and be aware of time differences.
4. Notify colleagues along the shipping route of flight plans and the species of bats being transported, and request their assistance if needed. Check periodically with all air carriers for delays or possible re-routings.
5. Feed bats prior to shipping, and ensure that a food source is available at their destination. When allowable, and depending on species and method and duration of transport, provide food and/or water in transit.
6. Construct the appropriate inner container for the safety and comfort of the species being transported.
7. Never ship mixed species in the same container. Separate the sexes during transport. In some cases, juveniles should be separated from adults, and males separated from each other.
8. When transporting bats directly from the wild, ship without delay. If a delay is unavoidable, hold the animals long enough to recover from the stress of capture.
9. To avoid respiratory problems, never use aromatic substrates in shipping containers (e.g., pine bark and cedar chips).
10. Whenever possible, transport bats as carry-on luggage.

PRODUCTS MENTIONED IN APPENDIX

Gempler's® Instant Warm Pack: Gempler's®, P.O. Box 44993, Madison, WI 53744 USA; http://www.gemplers.com/product/144707/6x10-Instant-Warm-Pack.

Uline® Cold Packs: Uline, 2105 S. Lakeside Drive, Waukegan, IL 60085 USA; http://www.uline.com/BL_2158/Cold-Packs.

REFERENCES

Barnard, S.M. 2009. Maintaining bats for captive studies. In: *Ecological and Behavioral Methods for the Study of Bats*, 2nd ed. (Kunz, T.H. and Parsons, S., eds.). The Johns Hopkins University Press, Baltimore, MD. Pp. 329-372.

Appendix 2-5. Helpful websites for shipping bat specimens

Classification, Packaging, and Labeling of Dangerous Substances		
United States	Code of Federal Regulations (CFR)	http://setonresourcecenter.com/49CFR/Docs/wcd00002/wcd00276.asp
	United States Postal Service	http://pe.usps.com/text/pub52/pub52c3_021.htm
	United States Geological Survey National Wildlife Health Center	www.nwhc.usgs.gov/mortality_events/shipping_instructions.doc
Canada	Transport Canada (TC)	http://www.tc.gc.ca/eng/tdg/clear-tofc-211.htm
Europe	Europa	http://europa.eu/legislation_summaries/consumers/product_labelling _and_packaging/l21276_en.htm
Australia	National Transport Commission (NTC)	http://www.ntc.gov.au/viewpage.aspx?AreaId=35&DocumentId=1147
Transporting Infectious Substances		
All countries	International Air Transport Association (IATA)	http://www.emro.who.int/stb/pdf/SAMPLEISSG7THED.pdf
United States	US Department of Transportation Pipeline and Hazardous Materials Safety Administration (DOT/PHMSA)	http://www.phmsa.dot.gov/staticfiles/PHMSA/DownloadableFiles/Files/ http://www.phmsa.dot.gov/staticfiles/PHMSA/DownloadableFiles/Files/ Transporting_Infectious_Substances_brochure.pdf
Canada	Public Health Agency of Canada (PHAC)	http://www.phac-aspc.gc.ca/publicat/lbg-ldmbl-04/ch10-eng.php
	Transport Canada	http://www.tc.gc.ca/eng/tdg/moc-infectious-menu-484.html
Europe	European Biological Resource Centres Network (EBRCN)	http://www.eccosite.org/transport.html
Australia	NTC	http://www.ntc.gov.au/filemedia/Publications/ADG7_Information_Guide.pdf

Appendix 2-5 continued

Other Helpful Websites		
Air transport	IATA	http://www.iata.org/about/Pages/contactus.aspx
	International Civil Aviation Organization (ICAO)	http://www.thecompliancecenter.com/publications/icao.htm
	World Health Organization (WHO)	http://www.colorado.edu/ehs/training/WHO_guidance.pdf
	DHL Express	http://www.dhl-usa.com/resources/Excepted_Quantity_Restrictions.pdf
Road and rail transport	US Department of Transportation (HazMat Regulations)	http://hazmat.dot.gov
	United Nations Economic Commission for Europe (UNECE)	http://www.unece.org/trans/danger/publi/adr/adr_e.html
	International Carriage of Dangerous Goods by Road and Rail (ADR, RID)	http://www.roadtransport.com/RoadLegal/11948/adr-hazardous-goods.html http://www.traceca-ltss.eu/admin/editor/uploads/files/RID%202005%20English.pdf (This website is an example of RID regulations)
	Department of Infrastructure and Transport	http://www.infrastructure.gov.au/transport/australia/dangerous/index.aspx
	Hazardous materials shipments to Canada from the United States by road or rail.	http://www.tc.gc.ca/eng/tdg/training-us-276.htm
	WHO	http://www.colorado.edu/ehs/training/WHO_guidance.pdf
Postal transport	Universal Postal Union (UPU)	http://www.upu.int/en/the-upu/the-upu.html
	United States Postal Service Domestic and International Mail Manuals	http://pe.usps.com/text/dmm300/dmm300_landing.htm http://pe.usps.gov/text/imm/welcome.htm

Section II

Public Education

Public Education

Patricia A. Morton

Just a few decades ago there were very few animals that suffered more from bad press than bats. Something all bat biologists still agree on is that one of the greatest barriers to bat conservation is misunderstanding driven by ignorance, myth, and superstition. Factual information that focuses on how bats benefit the environment and the miniscule risk of ever being harmed by a bat is fundamental to conservation planning. Working to change public opinion about bats is something all 'bat people' do for two reasons: 1) everyone recognizes that through greater awareness and understanding, bats will be appreciated and ultimately conserved, and 2) informing people about bats is fun! The public both fears and is fascinated by bats. It is quite easy to educate and entertain people because bats are such fascinating animals. There is a real "oooh-aaaah" result that comes from a well delivered program as people are quickly transformed from being very hesitant about bats to becoming enthusiastic advocates for their conservation, and many can not wait to share new facts and perspectives with family and friends. An effective bat presentation can multiply positive attitude changes far beyond the original audience.

In the past 20 years there has been an impressive number of excellent education resources produced, for example, books for all ages, classroom curricula, games and toys, and extraordinary audiovisual programs documenting bat activities and life styles that few people would ever see in the wild. Using live bats in public talks is a powerful education tool, but also a "double-edged sword" in being able to balance the amazing "up close and personal" experience with a "don't try this at home" warning. Bat houses are now popular ways to attract bats to private property where owners can provide roosting space and use the bat house in the yard for educating neighbors and friends.

Due to the growing popularity of bats, several bat conservation organizations and initiatives now exist in North America, Europe and beyond, and public education programming is fundamental to their missions. Through these organizations and state and federal government agencies, the creation of accurate information and implementation of bat conservation actions are increasing. Thanks to greater awareness among conservation professionals, bats and their habitats are now being included as part of comprehensive ecosystem management plans. With new threats to bats, for example the devastating white-nose syndrome disease that was first documented in North America in 2006, never has there been a more important time to build awareness and advocacy for bats.

Despite the success that has been realized, there is still a long way to go. Many bat conservationists are now working to tailor helpful information to key audiences, including the news media, public health and animal control agencies, pest control companies, veterinarians, state and federal land managers, and private landowners who own significant bat roosts and habitats. Focused education materials and technical assistance directed here can pay off with big results. Natural resource and public health agencies often contribute to bat conservation by developing best management practices that balance the needs of bats and people.

The seven chapters included in this section cover valuable information on a wide variety of topics that will help readers develop effective and responsible educational activities. No bat conservation action is more important than educating the public, and there are few animals more intriguing than bats. Please join the growing number of educators who are helping to make the world a more accepting and friendlier place for bats, and HAVE FUN!

Chapter 3

Suggestions for Developing Traveling Trunks and Loan Boxes

Photo courtesy of Bat Conservation of Wisconsin

Suggestions for Developing Traveling Trunks and Loan Boxes

Norma Lewis

INTRODUCTION

Bats are popular subjects for demonstrations in schools and for presentations to public audiences. Although a bat in the hand is always a nice addition to a program, it is certainly not a necessary one. Hands-on items commonly used for demonstrations in traveling trunks and loan boxes can illustrate simple concepts, and they never need to be fed or cleaned; they are easy to acquire and can liven up a program, all without the use of living animals.

Traveling trunks fill two different needs. One type of trunk is typically used by someone already familiar with bats. Another type is a loan box designed for use by teachers or other educators who are less familiar with bats and desire educational tools about these animals. The items in the boxes may vary, depending on the needs of users and the funds available to create and stock them.

TEACHING AIDS FOR TRAVELING TRUNKS

When putting together a traveling trunk for someone already familiar with bats (Fig. 3-1), one should think about what to cover in the course of a talk, and what parts of the presentation might be illustrated with simple interactive tools (Appendix 3-1). Watch the audience. A concept or idea may seem clear to one person, but may not be to another. One should not assume that adults have more knowledge than children about bats, ecology, or basic biological concepts. In fact, it is often the other way around. For this reason, when a presenter sees confusion or lack of comprehension during a program, he or she should consider adding interactive or demonstration items to the trunk to help illustrate the concept in question. This is an opportunity to be creative, and adults appreciate the opportunity to have a hands-on learning experience as much as children do.

Fig. 3-1. A small- to medium-sized, wheeled suitcase makes an ideal traveling trunk, especially for personal use. This small suitcase contains nearly everything listed in Appendix 3-1, including a bat costume and a few preserved bats in a Riker mount. Craft and activity supplies may be included, or packed in a separate container, such as a sturdy plastic box, to prevent paint and glue spills on demonstration materials. Heavy-duty plastic boxes with latching lids incur less damage than loan boxes, and can be used in the field (e.g., when shipping to foreign countries). Photo courtesy of Vicky Beckham Smith.

Whenever possible, an audience (especially children) should be involved in the program, sometimes by doing something as simple as holding up an item for everyone to see. For example, when discussing the great differences in body sizes among bats, participants can hold up life-sized cut outs of several bat species (Fig. 3-2). It is helpful to include the smallest and largest bat species with a couple of sizes in between. Be sure to include at least one bat that represents a local species.

Bat skulls (Fig. 3-3) are items that audiences of all ages find fascinating. The author uses skulls purchased from Skulls Unlimited (http://www.skullsunlimited.com/) and resin casts from BoneClones® Osteological Reproductions (http://www.boneclones.com/). These products are not only excellent illustrations of body-size differences, they are also useful for demonstrating the differences in dentition in animals that eat different foods. Resin skulls are very durable. Even tiny skulls hold up well if placed in clear plastic boxes with cotton or foam to prevent breakage. With a magnifying glass, program participants are able to see and be amazed by the dentition of the tiny skulls of insectivorous bats, compared with the dentition of large fruit bats.

Biological concepts such as pollination might be difficult for some people to understand. Showing how pollen is transferred from one flower to another as a bat feeds on nectar can be illustrated easily and inexpensively with silk flowers, corn starch, and a stuffed bat.

Fig. 3-2. Life-sized cut-outs of several bat species offer program participants a realistic view of the body sizes of bats throughout the world. Cut-outs can be made from felt, fabric, vinyl, or whatever is convenient, durable, and foldable. Suggested body sizes for life-sized cut-outs include the bumblebee bat (*Craseonycteris thonglongyai*), large flying fox (*Pteropus vampyrus*), and a native insectivorous species. In addition, wing shapes can be demonstrated by using life-sized cut-outs of the Mexican free-tailed bat (*Tadarida brasiliensis*) and pallid bat (*Antrozous pallidus*). The body sizes of the latter two species fall in between the bumblebee bat and the large flying fox. Photo by the author.

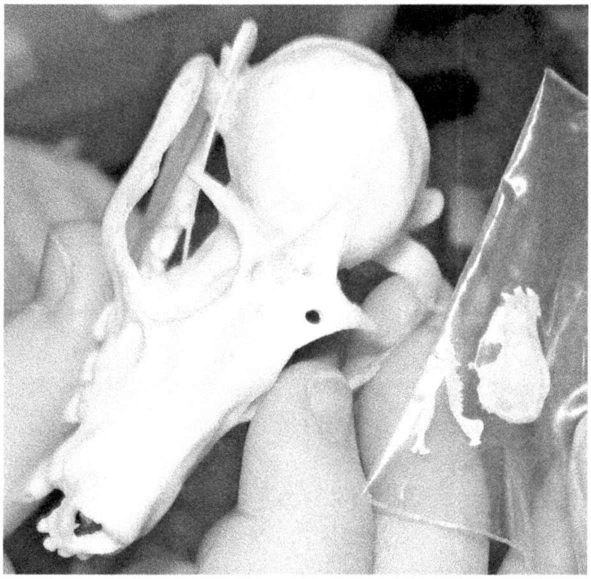

Fig. 3-3. Comparison of a *Pteropus vampyrus* skull (left) and an insectivorous bat skull (right). Having skulls available for program participants to examine stimulates discussions about animal sizes, food habits based on dentition, and comparisons of bone thickness. Bat skulls also help to demonstrate the differences among mammals (e.g., compare bat skulls with those of rodents, cats, or dogs). Photo by the author.

Anyone having difficulty with ways to illustrate concepts should seek help from a classroom teacher, or environmental educator at a zoo, museum, or nature center. Educators enjoy helping people develop new teaching strategies, and everyone can learn something new while thinking of engaging ways to teach about bats.

LOAN BOXES

Loan boxes or traveling trunks designed for use by someone not familiar with bats require more planning. Most loan boxes are designed for teachers or environmental educators and are often used in classroom settings. They can also be designed for informal use with family or adult groups. These boxes usually contain background material, lesson plans, books, games, activities, demonstration items, slides, audio and video tapes, craft ideas, and other items, such as bat detectors.

A binder with lesson plans and other information about bats should be included, along with information on how to use the items in the box. The contents of the binder should not exceed a dozen or so pages. The binder is also a good place for brochures, information from the organization developing the loan box or various related organizations, as well as bibliographies and a form for evaluating the quality and effectiveness of the loan box (Appendix 3-2).

Whoever creates a loan box about bats for teachers may wish to seek advice from a local environmental educator or a classroom teacher knowledgeable about bats. Unless a teacher already spends time on bats in the classroom, he or she will appreciate help on fitting bats into the curriculum, especially when tailoring lesson plans to meet a particular state's curriculum standards or national science standards.

Books are favorite items in loan boxes. The types of books included may range from bat curriculum books (e.g., AIMS' *Bats Incredible* and BCI's *Educators' Activity Book About Bats*) to general information books (e.g., M. Brock Fenton's *Bats*) to children's books (e.g., *Extremely Weird Bats, ZooBooks: Bats,* and Merlin D. Tuttle's *America's Neighborhood Bats*) and storybooks such as *Stellaluna* by Janell Cannon. For a list of books, see *Suggested Books for Loan Boxes* below.

Activities may vary too, depending on the audience or school grade level, but many activities can be used for a variety of groups and grade levels. Educational games can be made easily from simple items such as index cards and markers, or pictures cut from magazines and glued to index cards. These items are all that is needed to produce a memory game, played by laying out all the cards face down. Two cards are turned over in an attempt to make a match. Matched cards are removed from the game. Cards that are different are again placed face down. The object of the game is to remember what was seen in order to match cards. Because the cards are handled often, they last longer if they are laminated. For additional bat game ideas, see one of the activity books listed under *Suggested Books for Loan Boxes* below.

A loan box should include games and activities that require participants to stand up and move around. Project WILD and NatureScope books have excellent ideas for games and activities, as well as the recommended grade levels for such activities. For example, the 'Bat and Moth' game (Fig. 3-4; see http://lnr.cambridge.gov.uk/uploads/bat_and_moth_game.pdf) is a fast-moving activity designed to teach program participants about echolocation.

One should not be misled by the 'name' of a game in a book. Instead, focus on the concepts that a game illustrates. Many games can be adapted to suit a bat theme, and for different age groups. For example, the name "Quick Frozen Critters" (Project WILD) may not sound like a bat game, but by substituting the words 'bats and insects' for the words 'predator and prey', as cited in the directions, the game can teach about the predator-prey relationship between bats and insects.

Fig. 3-4. The 'Bat and Moth' game is excellent for teaching the concept of echolocation. Here, two people representing bats are blindfolded and are trying to locate two or more other people playing moths. Bats and moths are encircled by the rest of the program participants to keep them from straying. Each time the bat says "bat," the moth must reply "moth." The bat follows the sound of the "echo" from the moth and tries to catch it. If the bat touches the moth, then the moth is "eaten," and a new bat and moth are chosen. Photo by the author.

The task can be challenging when an organization attempts to develop its own game, or adapt a game to meet a specific need. It may be possible to ask a teacher or an environmental educator to field test a new or adapted activity with one of their groups. Such pretesting can help greatly in becoming comfortable with a new activity before including it in a loan box.

Whenever activities from books are used, or when such activities are adapted for a particular need, it is important to credit the source. Furthermore, published activities are often copyrighted. Therefore, it may be necessary to request permission from the publisher(s) before printing photocopies for a loan box binder.

Many organizations providing loan boxes charge a deposit that is refunded after the loan box has been returned and inspected to ensure that all items are in good condition. If they are not, the cost of replacements is taken from the deposit. This can be a problem if the cost to replace an item is more than the deposit. A borrower can be charged the additional amount, but if payment is not forthcoming, the cost may have to be absorbed.

Some organizations charge a small fee plus a deposit. The fee helps replace consumable and worn out items that need to be replaced.

Because loan boxes may contain expensive items such as books and equipment, information should be obtained about the borrower before a loan is made, in case a box has to be traced. This also makes the loan "formal," so that borrowers are less likely to take their responsibilities to care for loan box items lightly. A complete list of items contained in the box (Appendix 3-3) is necessary, and this checklist should be reviewed before the box is sent out. When the box is returned, it should be inspected carefully against the checklist. Any missing or damaged items should be noted so that the borrower can attempt to locate or replace them. However, some wear and tear is to be expected in the course of normal use, and borrowers should not be expected to pay for damage like dog-eared book pages or bent index cards.

An evaluation form (see Appendix 3-2) is very useful in maintaining the quality of a loan box and ensuring that materials included meet the needs of users. The form should ask pertinent questions and allow for a variety of answers from the different borrowers who will use the box. Some items will be appropriate for some users, but not for others, and one should try to include items that are appropriate for the particular audience or school grade level using it.

SUGGESTED BOOKS FOR LOAN BOXES

The following is a list of books that teachers have found useful and fun for bat programs, but there are many more excellent titles available, and new ones are continuously being published. Many of these books are available at the sources listed under *Where to buy "Bat Stuff"* below.

CURRICULUM AND ACTIVITY BOOKS

Bats: Complete Cross-Curricular Theme Unit That Teaches About This Amazing Mammal (Robin Bernard, Scholastic Professional Books)

Bats: A Complete Thematic Unit (Evan-Moor Educational Publishers)

Bats: A Creativity Book for Young Conservationists (Jane F. G. Jennings, Bat Conservation International)
Bats: A Science Discovery Book (Annalisa McMorrow, Monday Morning Books)
Bats Incredible (AIMS Instructional Media)
Bats, Butterflies and Bugs: A Book of Action Toys (S. Adams Sullivan, Little, Brown and Company)
Educators' Activity Book (Bat Conservation International)
Discover Bats! (Bat Conservation International, book and video tape set)
Let's Explore Caves and Caverns: A Young Explorers Activity Book (Connie Toops, Explorer Press)
Swift Shadows in the Twilight (Ann C. Cooper, Denver Museum of Natural History)

INFORMATION AND STORY BOOKS FOR CHILDREN

A New True Book: Bats (Susan Heinrichs Gray, Children's Press [CT])
Amazing Bats (Eyewitness Juniors-13, Frank Greenaway, Alfred A. Knopf, Inc.)
Bats: Creatures of the Night (Joyce Milton, All Aboard Reading, The Putnam & Grosset Group)
Bats: Mysterious Flyers of the Night (Dee Stuart, Carolrhoda Books, Inc.)
Bats for Kids (Kathryn T. Lundberg, Northword Press, Inc.)
Bats Undercover (Geosafari®, Educational Insights)
Cactus Cafe (Kathleen Weidner Zoehfeld, The Nature Conservancy)
Extremely Weird Bats (Sarah Lovett, John Muir Publications)
Eyes on Nature: Bats (Celia Bland, Kidsbooks, Incorporated)
Misunderstood Bats (Art L'Hommedieu, Child's Play International Ltd.)
One Small Square: Cave (Donald M. Silver, McGraw-Hill)
Shadows of Night: The Hidden World of the Little Brown Bat (Barbara Bash, Sierra Club Books for Children)
Stellaluna (Janell Cannon, Harcourt Brace and Company)
Taking Katy for a Nightride (Elin Menzies, Blue Cat Press)
The Magic School Bus: Going Batty (Joanna Cole, Scholastic, Inc.)
Zipping, Zapping, Zooming Bats (Ann Earle, Harper Collins Publishers)
ZooBooks: Bats (Wildlife Education, Ltd.)

GENERAL INFORMATION BOOKS

America's Neighborhood Bats (Merlin Tuttle, University of Texas Press)
Bats (M. Brock Fenton, Facts On File)
Bats in Question (Don E. Wilson, Smithsonian Institution Press)
Bats of the United States (Michael J. Harvey, J. Scott Altenbach, Troy Best, Arkansas Game and Fish Commission and the US Fish and Wildlife Service)
Bats of the World (Gary L. Graham, Golden Press)
The World of Bats: The Flying Goblins of the Night (Klaus Richarz and Alfred Limbrunner, T.F.H Publications, Inc.)
With Wings on their Fingers: An Intimate View of the Flying Fox (Pamela Conder, Angus & Robertson: An imprint of Harper Collins Publishers)

WHERE TO BUY "BAT STUFF"

In addition to the organizations listed below, remember to check out the local zoo, nature center, or natural history museum to find other 'cool' bat things.

Bat Conservation International (BCI)
http://www.batcon.org
P.O. Box 162603
Austin, TX 78716
Phone: 512-327-9721

BoneClones® Osteological Reproductions
http://www.boneclones.com
21416 Chase Street #1
Canoga Park, CA 91304
Phone: 818-709-7991

National Park Service
http://www.nps.gov/index.htm
1849 C Street NW
Washington, DC 20240
Phone: 202-208-3818

Organization for Bat Conservation
http://www.batconservation.org
BatZone at the Cranbrook Institute of Science
39221 Woodward Avenue
P.O. Box 801
Bloomfield Hills, MI 48303
Phone: 800-276-7074

Skulls Unlimited International
http://www.skullsunlimited.com
10313 South Sunnylane
Oklahoma City, OK 73160
Phone: 800-659-SKULL; if calling from outside the United States, dial 405-794-6985

Speleobooks
http://www.speleobooks.com
P.O. Box 10
Schoharie, NY 12157-0010
Phone: 518-295-7978

PRODUCTS MENTIONED IN TEXT

Books: see *Where to Buy "Bat Stuff"* above.
Games: see *Where to Buy "Bat Stuff"* above.
Heavy-duty Plastic Box with Latching Lid: Target Brands, Inc., 1000 Nicollet Mall, Minneapolis, MN 55403 USA; http://www.target.com/Sterilite-Gallon-Ultra-Tote-Blue/dp/B004ARMNV8/ref=br_1_16?ie=UTF8&id=Sterilite%20Gallon%20Ultra%20Tote%20Blue&node=13825601&searchSize=30&searchView=list&searchPage=2&sr=1-16&qid=1305636313&rh=&searchBinNameList=subjectbin%2Cprice%2Ctarget_com_primary_color-bin%2Ctarget_com_size-bin%2Ctarget_com_brand-bin&searchRank=salesrank&frombrowse=1.
Riker Mounts: Ward's Natural Science, 5100 West Henrietta Road, P.O. Box 92912, Rochester, NY 14692 USA; http://wardsci.com/product.asp?pn=IG0005349&cm_mmc=Mercent-_-Google-_-NULL-_-303310&mr:trackingCode=BD60975B-6581-DE11-8C0A-000423C27502&mr:referralID=NA.
Skulls: see *Where to Buy "Bat Stuff"* above.
Slinky: Available in toy stores; see also http://www.amazon.com/Poof-Slinky-SLINKY-Original/dp/B00000IZKX.
Wheeled Suitcase: Amazon.com; http://www.amazon.com/Samsonite-Aspire-Expandable-Carry-Upright/dp/B004P883DE/ref=pd_sbs_a_.

ACKNOWLEDGMENTS

I thank Dietrich Schaaf for his helpful comments on the manuscript.

Appendix 3-1. Suggestions for a Personal-use Traveling Trunk

(Norma Lewis and Vicky Beckham Smith)

The following is a list of useful items to use when presenting bat education programs. Once a program has been established, the only major trunk changes are craft and activity items, which depend on each group and the planned activities.

1) A box with examples of what bats eat (e.g., insect, scorpion/centipede, fish, frog, mouse or other small mammal, bird, flower [includes pollen and nectar], and cow blood).
2) Examples of what eats bats (e.g., spider, bat, owl/hawk, snake, cat, raccoon, bull-frog, human).
3) Life-sized silhouettes of bats (see Fig. 3-2). Use examples of the smallest and the largest bats, as well as sizes in between. Include one similar in size to a common local species.
4) Bat skulls (e.g., one each from a fruit- and an insect-eating bat) (see Fig. 3-3).
5) Inexpensive magnifying lens to view bat skulls.
6) Supplies needed for planned games and activities (e.g., bandanas or taped-over toy glasses for a person to wear while playing a bat during the "Bat and Moth" game) (see Fig. 3-4). Other activities can be found in resource manuals such as AIMS *Bats Incredible* (http://store.aimsedu.org/aims_store/books/bats-incredible-plus-dvd.html) and Bat Conservation International's *Educator's Activity Book About Bats* (http://www.utexas.edu/utpress/books/tute2p.html).
7) Bat masks (Fig. 3-5) and costumes representing different types of bats. Suggestion: Masks last longer if they are laminated.
8) Slinky to demonstrate echolocation (Fig. 3-6). Demonstrates how sound waves are emitted from a bat, then bounce back from objects (e.g., the slinky goes from the hand holding it, while it hits the other hand, and then bounces back to the original hand).

Fig. 3-5. Children wearing masks to represent different bat species. Before donning the masks, they learn about the variety of food items that different bats eat. Most children want to be a vampire bat or a fruit bat because those are the ones they hear the most about. These masks were produced on card stock, colored with pencils, and laminated for durability. Photo courtesy of Vicky Beckham Smith.

Appendix 3-1 continued

9) Bear claw hair clip or banana hair clip. These items help demonstrate how bat feet work.
10) Craft supplies needed for planned craft activities (e.g., paper, scissors, glue, crayons, bat stamps and pad [Fig. 3-7]), or anything that might be needed for children to make bat masks or other take-home crafts during a program (see the list of activity books in the text).
11) Inexpensive bat detector for evening "bat walks" (e.g., http://www.batsound.com/; http://www.econvergence.net/batdet.htm; http://www.batbox.com/products.html; http://www.ultrasoundadvice.co.uk/pages/Mini3home.html; http://www.semafor.co.uk/bat-detector.html#Nottop; http://www.scientificsonline.com/belfry-bat-detector.html?gclid=CMGa8rWdx6gCFYi8KgodEWI4sw)
12) Assortment of bat books for display (see *Suggested Books for Loan Boxes* above).
13) Bat motif table cloth on which to display items.

Fig. 3-6. A slinky helps to demonstrate how echolocation works. The slinky bounces from one hand to the other, the way a bat emits calls that go out and return to its ears after bouncing off an object. Photo courtesy of Vicky Beckham Smith.

Fig. 3-7. Arts and crafts can extend a bat lesson into other areas of a school curriculum. Children can create works of art or illustrate their writing journals with items as simple as rubber stamps and colorful stamp pads. Providing a few basic art supplies can inspire children to write and illustrate poetry, stories, and science journals. Photo by the author.

Appendix 3-2. Loan Box Evaluation

(courtesy of the Environmental Education Association of Alabama [EEAA])

We need your help to ensure that our programs remain of excellent quality and continue to meet your needs. Please take a moment to complete this evaluation form and leave it in the loan box when you return it, or mail it to EEAA at P.O. Box 2152, Huntsville, AL 35804.

Circle box on loan: Take a Hike Owl Pellets Bats Fossils
Water Quality Trees Butterflies

Age group or grade level this box was used for: ______________________________________
(e.g., families, first grade, middle school)

Number of participants: ____________

Box was used to supplement what type of program?

Condition of box when received: Excellent Good Fair Poor

I noticed that ____ was missing/broken but not noted on the check list or loan form. Please note any other problems with the loan box here:

What I liked best in this loan box was: ____

Why? ____

What the program participants liked best was: _____________________________________

Why? ____

What changes or additions would make this box more useful to you?

What other loan boxes would you like to see EEAA offer?

Appendix 3-3. Loan Box Checklist – BATS

(courtesy of the Environmental Education Association of Alabama)

Date loaned out: _________________ Date returned: ______________________

Check in	Check out	Item	Condition
_______	_______	Bat detector	__________
_______	_______	Bats, the True Story (video)	__________
_______	_______	Very Elementary Bats (video)	__________
_______	_______	Caves (video)	__________
_______	_______	Bat Chat (audio tape)	__________
_______	_______	Threatened and Endangered Bats (slides)	__________
_______	_______	Insectivorous bat skull	__________
_______	_______	Fruit bat skull (bone clone)	__________
_______	_______	Plush bat toy	__________
_______	_______	Stellaluna finger puppet	__________
_______	_______	3 flannel-backed bat silhouettes	__________
_______	_______	3 silk flowers	__________
_______	_______	Corn starch or baby powder	__________
_______	_______	6 bat stamps and one stamp pad	__________
_______	_______	Insectivorous bat game w/fuzzy die	__________
_______	_______	Bat twister game w/cards and spinner	__________
_______	_______	Matching game, 24 cards and envelope	__________
_______	_______	Bats in Question (book)	__________
_______	_______	Bats of the United States (book)	__________

Appendix 3-3 continued

Date loaned out: _________________ Date returned: _____________________

Check in	Check out	Item	Condition
_______	_______	ZooBook: Bats (book)	__________
_______	_______	Extremely Weird Bats (book)	__________
_______	_______	Bats of the World (book)	__________
_______	_______	Bats: A New True Book (book)	__________
_______	_______	Bats for Kids (book)	__________
_______	_______	Zipping, Zapping, Zooming Bats (book)	__________
_______	_______	Eyes on Nature: Bats (book)	__________
_______	_______	Cactus Cafe (book)	__________
_______	_______	Stellaluna (book)	__________
_______	_______	Bats, A Creativity Book (book)	__________
_______	_______	Bats, A Complete Thematic Unit (book)	__________
_______	_______	BCI Educators' Activity Book (book)	__________
_______	_______	Swift Shadows in the Night (book)	__________
_______	_______	Bat House Builders' Handbook (book)	__________
_______	_______	Guide to Bat House Designs (pamphlet)	__________
_______	_______	Bat facts (laminated sheet)	__________
_______	_______	Wooden bat house	__________
_______	_______	Binder with activities and information	__________

Chapter 4

The Use of Ultrasonic Mobility Devices in Education Programs

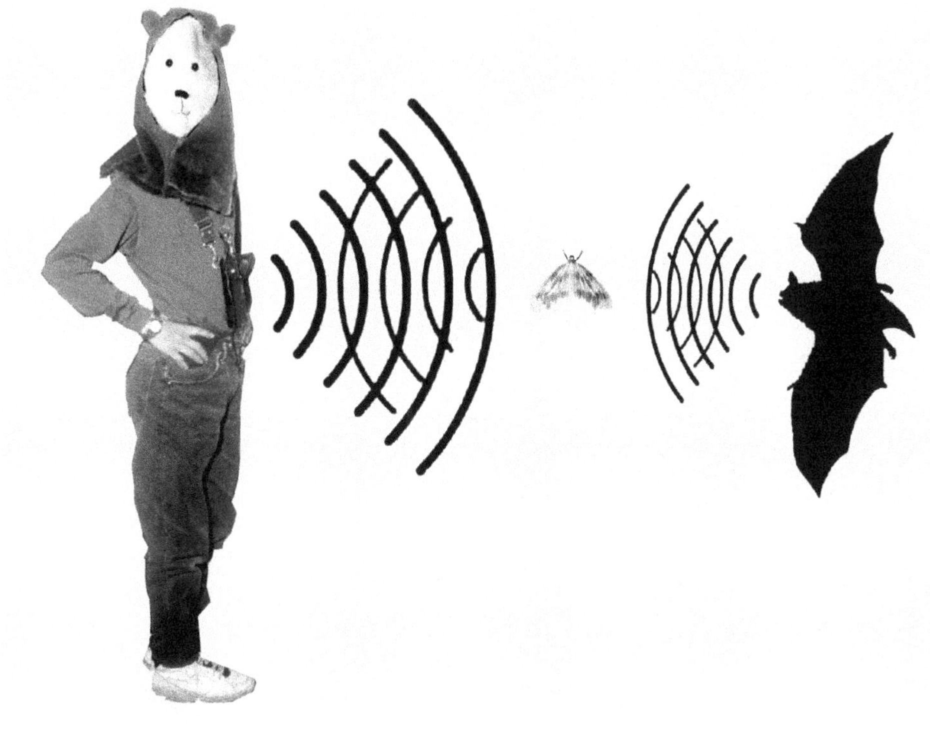

Image by Susan M. Barnard

The Use of Ultrasonic Mobility Devices in Education Programs

Suneth Attygalle and Susan M. Barnard

INTRODUCTION

Bat detectors (e.g., Magenta Bat4, Mini-3, and Pettersson D100) are extremely useful teaching aids to demonstrate the ultrasonic calls of bats. However, an education program can be enhanced by allowing members of an audience to experience the concepts of echolocation from the bat's point of view. In echolocation, bats send out high-pitched sounds that bounce off objects. By "calculating" the time it takes sounds to return (or the frequency of returned sounds), bats can determine how far objects are in space. With the use of ultrasonic devices, either designed to assist the blind with mobility, or merely for educational purposes, members of an audience can be given an opportunity to transmit ultrasonic waves, and to avoid objects by detecting the pattern of waves that return.

THE ULTRASONIC DEVICE IN BAT EDUCATION

The manufacturing of ultrasonic mobility devices for the blind has occurred over many years, including such devices as K-Sonar, Sensory 6, Miniguide, Sonic Glasses, Sonic Guide, and Sonic Pathfinder. Because they are expensive, many blind individuals could not afford them, forcing several manufacturers to discontinue making the devices. This has led to the need for bat educators to make their own. The homemade manufacturing of one such device is described below (see *Batgoggles*).

SENSORY 6

In approximately 1994, S. M. Barnard had the idea that using an ultrasonic device for the blind would add an extra dimension to her bat education programs. She visited the Center for the Visually Impaired in Atlanta, Georgia and purchased the Sensory 6. Although, like other ultrasonic devices, the Sensory 6 was expensive and is no longer manufactured, it is still being used in programs, allowing area school children the joy of experiencing echolocation.

The Sensory 6 was made with a stereophonic headset, and it had transmitters and receivers mounted on eyeglasses (Fig. 4-1). However, this instrument was expected to be shared among a large number of children. To eliminate the possibility of transmitting eyeglass-associated infections such as bacterial or viral conjunctivitis (i.e., pink eye), the transmitters and receivers were removed from the glasses and mounted onto the base unit (Fig. 4-2). To prevent children from seeing their surroundings with their eyes, a hood was placed over their heads (Fig. 4-3).

Before the ultrasonic device was placed on a child, the "instructor bat" demonstrated how to avoid obstacles by listening to the pitch of the signals returning to her. The equipment was then placed on the children. Blinded with the bat hood, each child was asked to walk up to a wall in the classroom, without walking into it. The closer the student was to the wall, the higher the sounds became. Most students caught on quickly and performed well.

The transmission frequency of the Sensory 6 was a steady 50 kHz. The class was not given this information. While one student was being a "bat," the others were asked to figure out the frequency at which the student-bat was echolocating with the use of a bat detector.

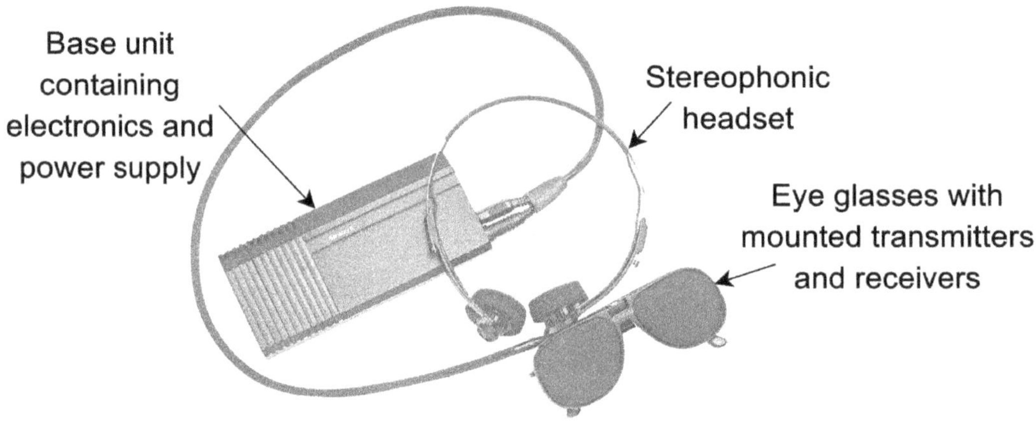

Fig. 4-1. Sensory 6 ultrasonic mobility device for the blind as manufactured by Brytech®. Photo courtesy of E.L. Bryenton & Associates, Inc., Ontario, Canada.

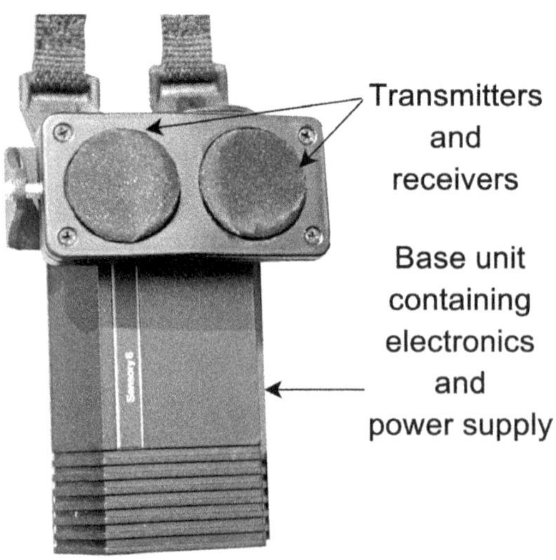

Fig. 4-2. Sensory 6 ultrasonic device modified for school children. Instead of having to wear the transmitters and receivers as eyeglasses, the children can wear the device more comfortably around their necks. Photo by Susan M. Barnard.

Fig. 4-3. Student with an embroidered hood over her head to block her vision. She is wearing the Sensory 6 around her neck. Photo by Susan M. Barnard.

BATGOGGLES

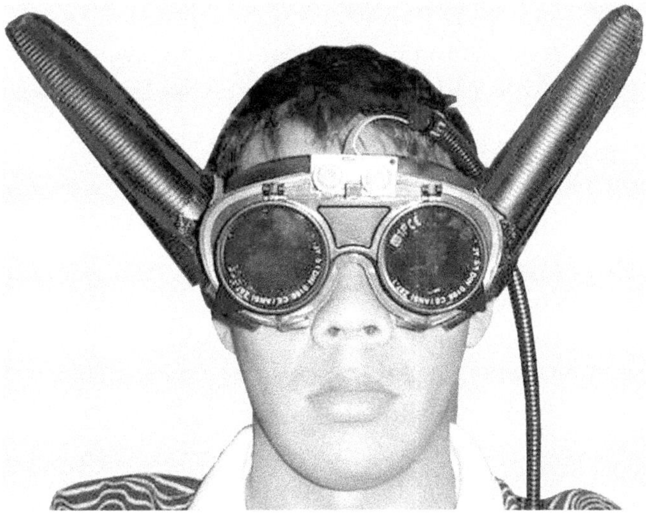

Fig. 4-4. Suneth Attygalle wearing Batgoggles. Photo is self-portrait.

The sight-blocking Batgoggles (Fig. 4-4) were constructed by S. Attygalle for the "Dynamic User-centered Research and Design" course in the Arts, Media and Engineering program at Arizona State University. They are inexpensive to make and incorporate an ultrasonic sensor that measures distances to objects in its path. Because ultrasonic sounds are inaudible to humans, the distance information is mapped to audible "pings" played by a buzzer in the goggles, allowing the wearer to hear what a bat would when an object is within a certain range. Thus by moving closer or further from objects in a classroom, a student can experience alternative senses for determining distances to objects. In addition, the goggles have "bat ears" to encourage children to imagine themselves as bats.

The following is a step by step method for constructing the Batgoggles. A *Glossary of Terms* is also provided below.

Materials

- Welding goggles with strap (e.g., Neiko brand)
- Devantech SRF05 ultrasonic sensor
- 9V battery connector
- Plastic stacking cones (for bat ears)
- 4-in. x 2.5-in. plastic electronics enclosure with 9V battery compartment
- 3/8-in. split seam, flexible, black convoluted tubing (to hide connecting wires)
- Piezo buzzer (any model that runs on 5v-9v)
- Assorted wires
- Plasti Dip spray (black)

Microcontroller Electronics (these components can be skipped if using a pre-built Arduino Nano controller)

- Arduino Uno
- PC board
- 7805 5v voltage regulator
- 16 MHz crystal
- [2] 22pF capacitors
- 10 microfarad (μF) electrolytic capacitor
- 1 microfarad (μF) electrolytic capacitor

- 28-pin DIP socket
- Female and male header connectors
  Note: (see Products Mentioned in Text below).

Tools and Supplies

- Soldering iron
- Hot glue gun
- Dremel rotary tool
- Newspapers
- Masking tape
- Sandpaper
- Wire strippers

Making the Bat Ears

There is no *one* way to start making the Batgoggles. In particular, the physical form of the goggles can be tailored to any specific application. In this example, Attygalle began with the bat ears. He used plastic stacking cones used in physical therapy to create the ears (Fig. 4-5). To begin constructing the bat ears, draw an oval with a sharpie felt pen and cut it out with a Dremel rotary tool. Using the Dremel rotary tool, shave the two cut-off pieces of the cones to fit inside the large pieces (Fig. 4-6). Then glue them in place with hot glue. If care is not exercised while using the hot glue gun, accidental melting of the plastic cones may occur.

The ears should be sanded and spray painted with at least three coats of black Plasti Dip. Allow 10-15 minutes between coats. It is important to protect the lungs by wearing a mask and safety glasses while sanding and spray painting with Plasti Dip. When wet, the paint appears glossy, but it dries to a matte finish.

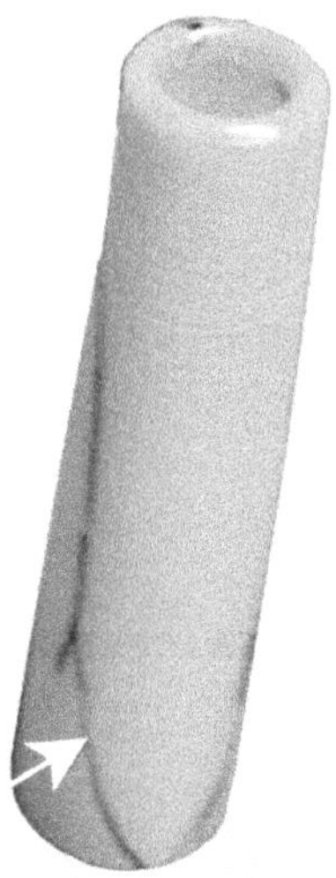

Fig. 4-5. The marked area (arrow) on this plastic stacking cone shows where the cone is to be cut with a Dremel rotary tool to fashion bat ears. Photo by Suneth Attygalle.

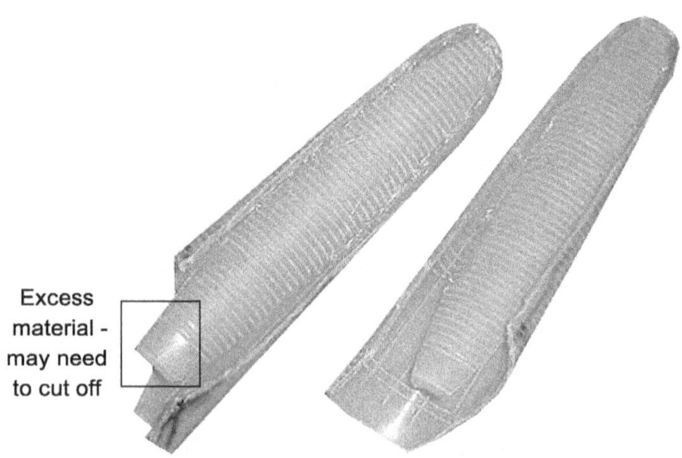

Fig. 4-6. Two plastic stacking cones cut up for bat ears. The two pieces are trimmed so they fit inside the other two pieces. Photo by Suneth Attygalle.

PREPARING THE GOGGLES

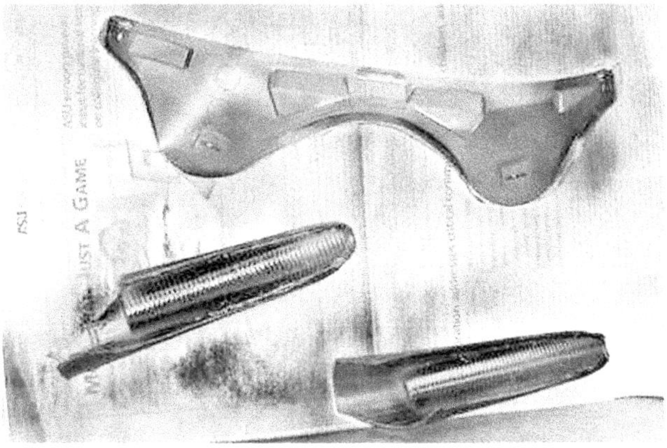

Fig. 4-7. Shown here are the welding goggles taken apart and laid flat for painting. The bat ears are also painted at the same time. Photo by Suneth Attygalle.

The Neiko welding goggles used for this project were a shiny aqua-blue color, bearing minimal similarity to a bat. This matter was remedied by coating the goggles with black Plasti Dip spray (Fig. 4-7). To proceed, first remove the nose-piece and lenses and lightly sand the remaining frame. Cover the interior of the goggles and all parts that touch the skin with masking tape. Following similar steps to the ears, paint the frame with Plasti Dip spray to give the frames a matte leathery texture (Fig. 4-7). The nose-piece is not painted because the paint reduces the flexibility of the goggle material: The nose piece is necessary to hold the goggles together. If desired, the lenses can be painted as well to block vision.

Reminder: When spraying with Plasti Dip, wear a mask and safety glasses.

UPLOAD CODE TO ARDUINO

Before assembling the electronics, the Arduino microcontroller must be programmed with the code that processes the ultrasonic sensor data and generates sound. Begin by downloading and installing the Arduino software and USB driver from http://www.arduino.cc/, following the instructions on the webpage. After it is installed, connect a USB cable from the computer to the Arduino. Next, download the file named "batgoggles.pde" code to your computer from http://suneth.com/downloads/batgogglescode.pde.zip. Unzip the downloaded file. Next, run the Arduino software on your computer and open the downloaded file from the "File" menu. Once the file is open, ensure that the correct Arduino board and serial port is selected from the "Tools" menu. Once those have been set, press the upload from the "File" menu. This will transmit the code from the computer to the microcontroller. Specific instructions for using and setting up the Arduino software for different platforms (OS X, Windows, Linux) are available at http://arduino.cc/en/Guide/HomePage.

The code (also shown in Appendix 4-1) does the following:

- triggers the ultrasonic sensor to detect distance,
- calculates the distance to an object in the sensor's path, and
- generates sounds.

Specifically, as objects are brought closer to the sensor, the frequency of beeps increases. The scaling in the code is set so that for every 1 inch, the frequency of the beeps changes by 100hz.

Assembling the Electronics

Basic electronic soldering skills and an ability to read a schematic are necessary for this project. Arrange the electronics components as shown in Figure 4-8, and solder all of them. The specific layout of components can vary, as long as schematic (Fig. 4-9) is followed. The Atmega168 chip is the one that is on the Arduino microcontroller. It needs to be removed carefully and placed on the PC board. A 28-pin DIP socket placed underneath the chip allows the Atmega168 to be removed and replaced, so that the code can be updated. To reduce the difficulty of this project, the Arduino Nano board can be purchased instead of the Arduino Uno. In this case, follow the simpler schematic shown in Figure 4-10.

Preparing the Buzzer and Ultrasonic Sensor Wires

Because the buzzer and ultrasonic sensor will be placed at a distance from the rest of the electronics, they need to be connected with long wires (Fig. 4-11). Specifically, the ultrasonic sensor requires 4 wires and the buzzer requires 2 wires. Ribbon cable that is trimmed to 4 and 2 wires is convenient for this purpose, but any type of flexible wire can be used. Solder the four-wire cable to the ultrasonic sensor as indicated on the schematic (see Fig. 4-9). Solder a row of female headers to the end of the two-wire cable. Attach the Piezo buzzer into these female headers. On the remaining ends of both wires, solder male header pins (Fig. 4-12).

Putting the Electronics into an Enclosure

The convoluted tubing should be cut so it is the right length from the goggles to one's hand or pocket. The wires connecting the ultrasonic sensor and Piezo buzzer should be placed inside the split-seam, convoluted tubing. An appropriate-sized hole is drilled in the electrics enclosure to fit the convoluted tubing. The wires are run through the hole,

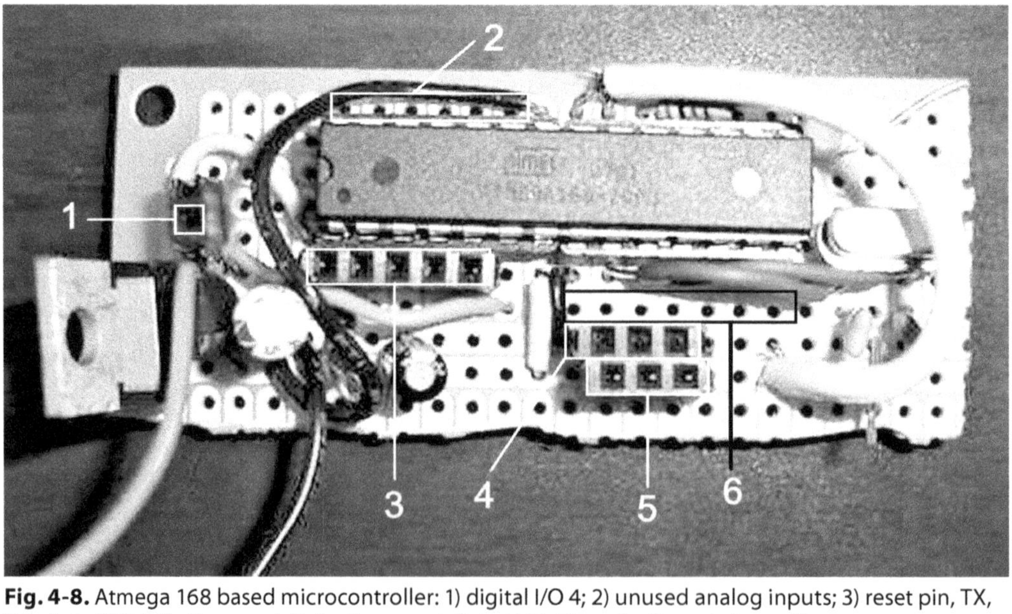

Fig. 4-8. Atmega 168 based microcontroller: 1) digital I/O 4; 2) unused analog inputs; 3) reset pin, TX, RX, digital I/O 2/3; 4) ground; 5) 5v; 6) unused digital I/O. Photo by Suneth Attygalle.

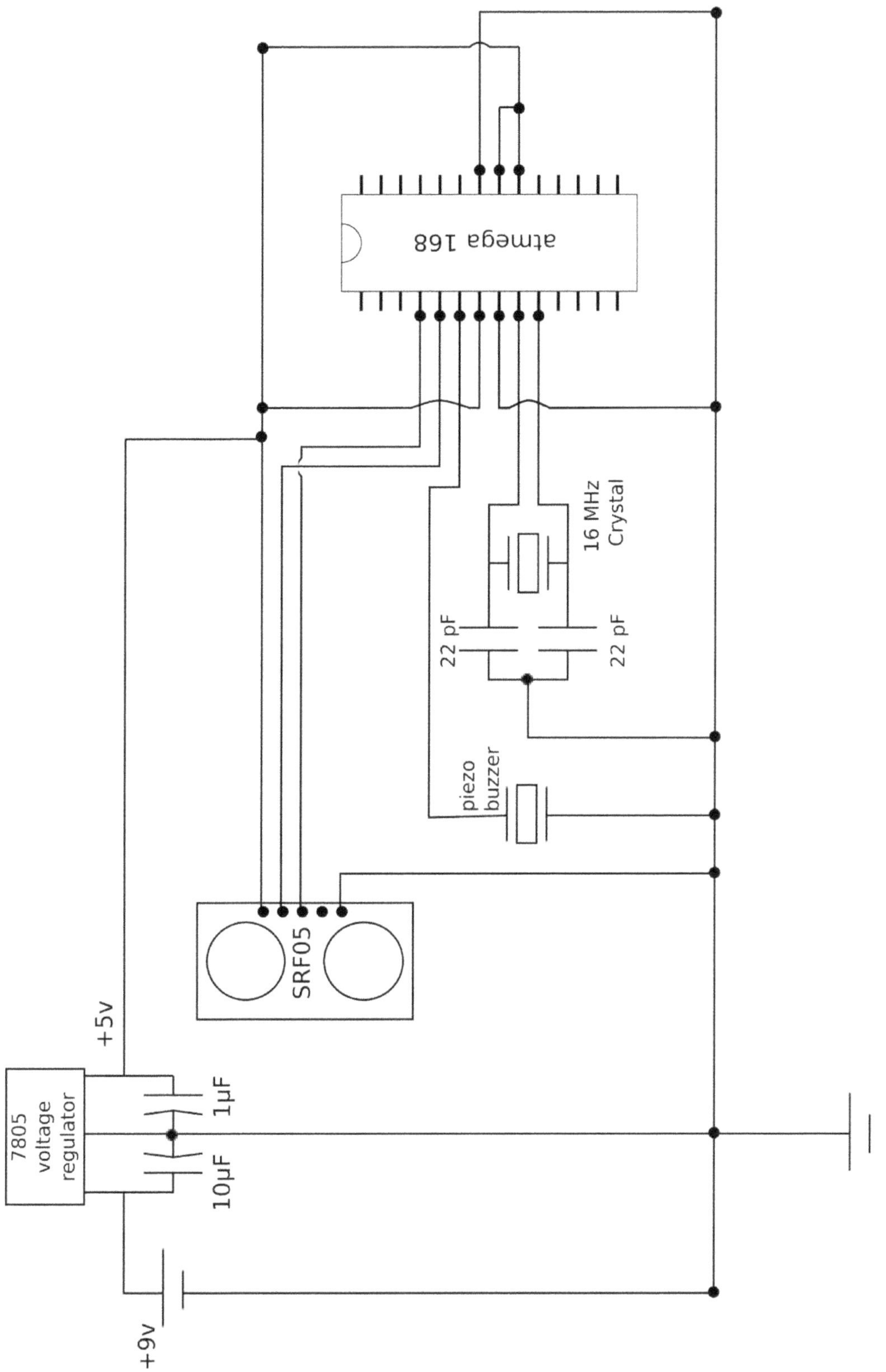

Fig. 4-9. Electronics schematic for the Arduino Uno board. Illustration by Suneth Attygalle.

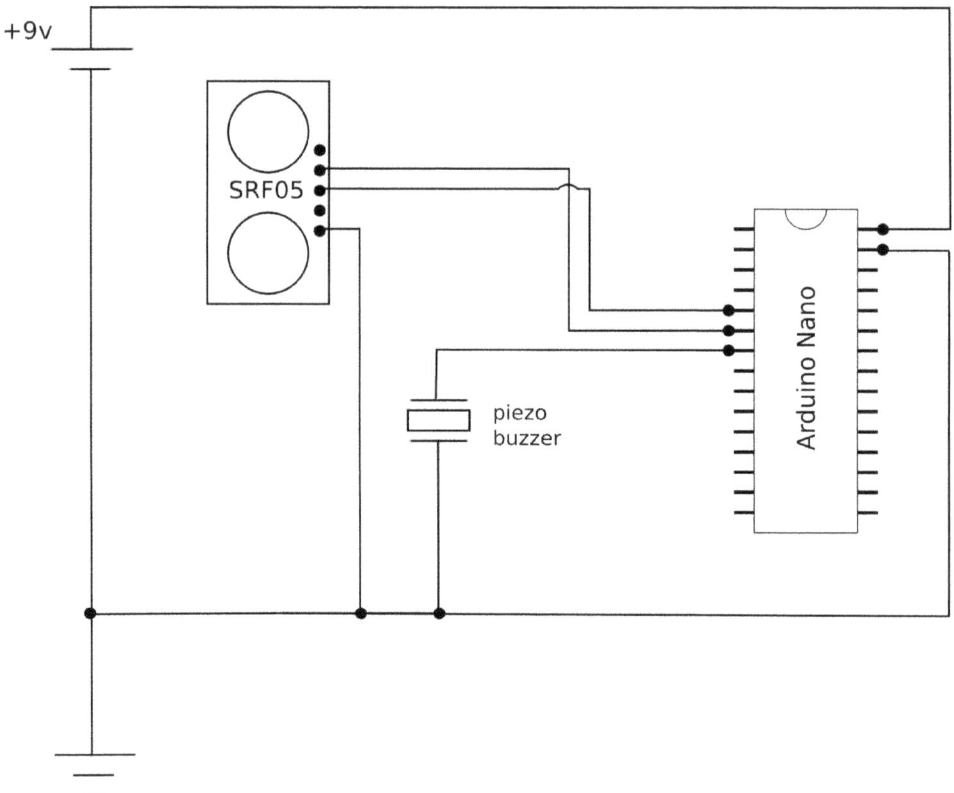

Fig. 4-10. Electronics schematic for the Arduino Nano board. Illustration by Suneth Attygalle.

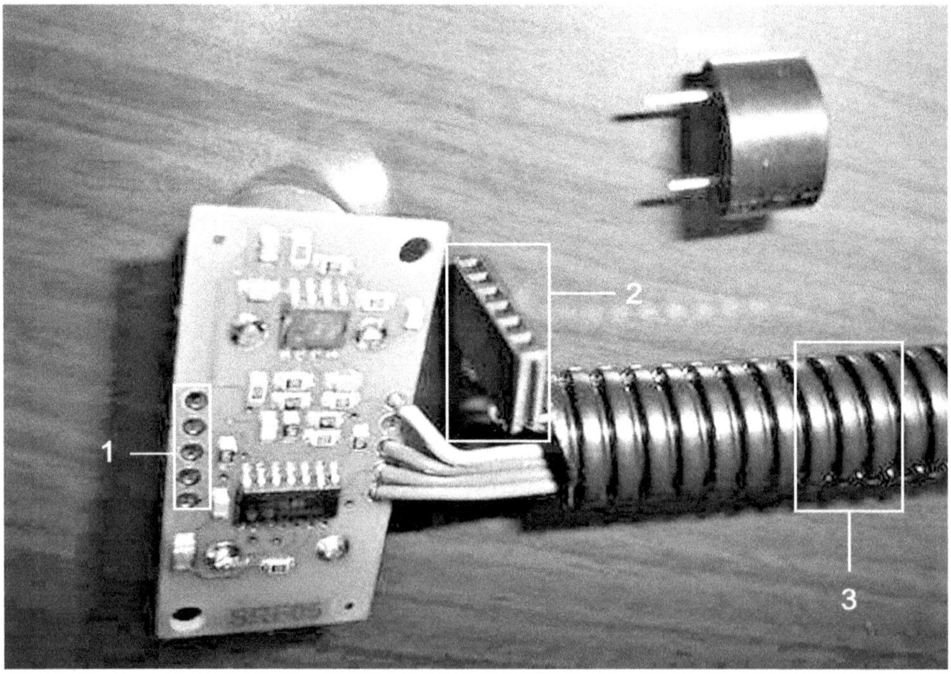

Fig. 4-11. Ultrasonic sensor and Piezo buzzer: 1) do not solder to this side; 2) buzzer fits into these headers; 3) this flexible, convoluted tubing can be added in this step, or as discussed in a later step. Photo by Suneth Attygalle.

followed by the convoluted tubing (Fig. 4-13). Velcro® can be used to hold the circuit board to the electronics enclosure.

Following the schematic, the male header pins at the ends of the wires can be connected to the appropriate pins on the PC board (Fig. 4-14). The electronics enclosure may now be closed (Fig. 4-15).

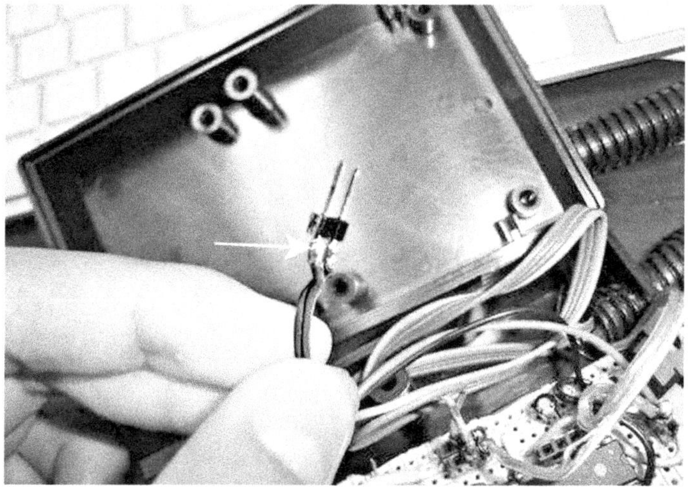

Fig. 4-12. Shown here is a male header pin soldered to wires (arrow) that connect to the microcontroller.

ATTACHING THE EARS TO THE GOGGLES

To attach the ears to the goggles, a vertical slot (Fig. 4-16) must be placed in each ear using the Dremel rotary tool. This allows a strap to pass through each ear. If the strap does not pass through easily, then it can be fed through the slot with the aid of a flat-headed screwdriver.

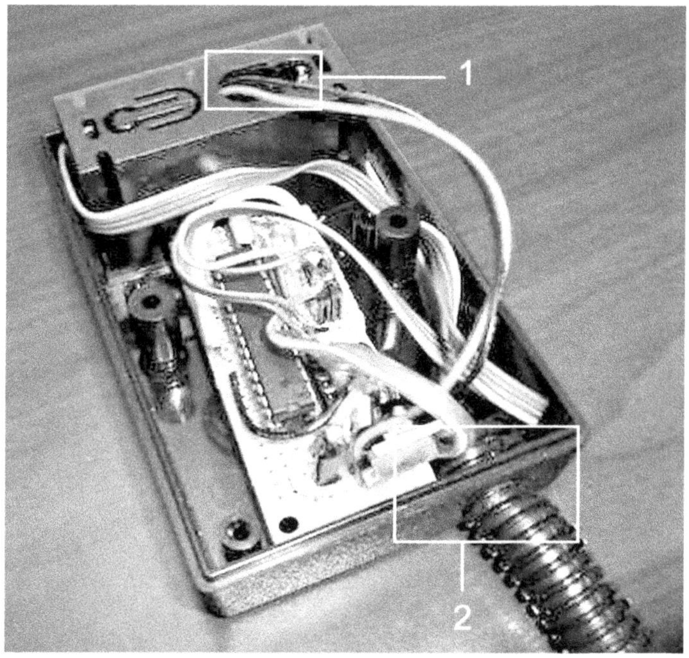

Fig. 4-13. Electronics in enclosure: 1) plastic piece may have to be snapped off to fit the 9v connector in the battery compartment; 2) drill hole to fit tubing. Photo by Suneth Attygalle.

After running the straps through the ears, Velcro® can be used to affix each ear to the goggles. Although slightly unstable, Velcro® is highly adjustable, making it easy to point the ears in the right direction. By using glue, the ear attachments can be made permanent.

The goggles have a flip-up component (see http://www.amazon.com/Neiko-Welding-Safety-Goggles/dp/B002GWE0VA), which is perfect for fitting in the ultrasonic sensor. However, the rubber goggles frame must be pulled out of the plastic lens piece slightly at the top of the goggles to make space available for the sensor. Furthermore, the sensor may pop out of the space if the pop-up feature is not glued to the goggles.

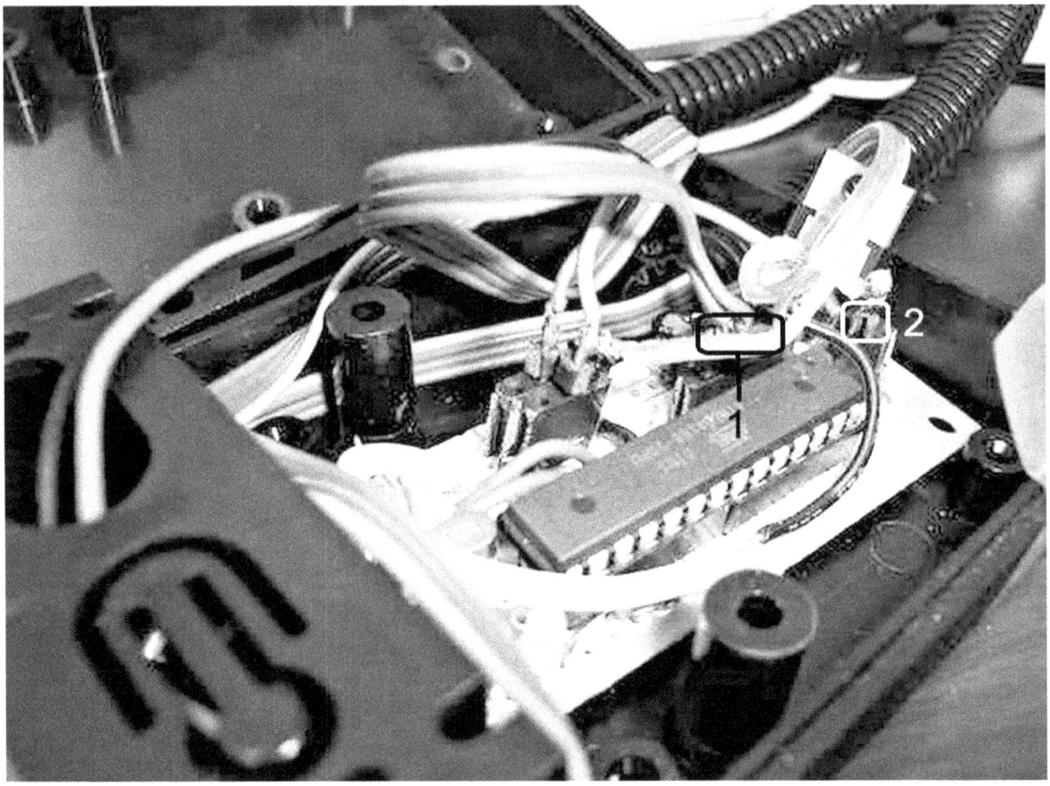

Fig. 4-14. Alternate view for electronics in enclosure: 1) I/O for sensor; 2) output for buzzer. Photo by Suneth Attygalle.

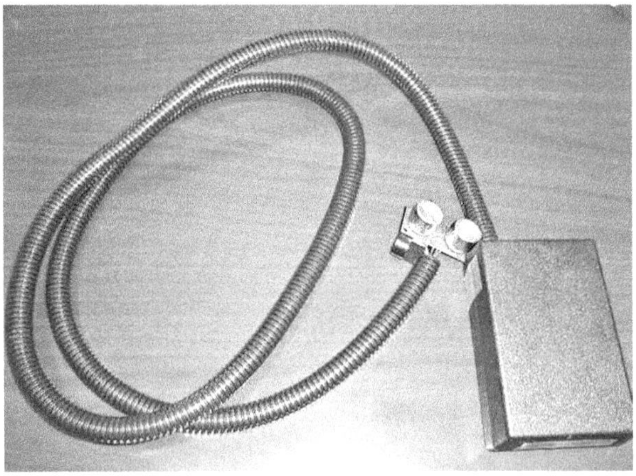

Fig. 4-15. Completed electronics for Batgoggles. Photo by Suneth Attygalle.

EXPERIENCING ECHOLOCATION

Connect a 9v battery; put the electronics enclosure in a pocket; and begin echolocating like a bat! See http://www.youtube.com/watch?v=GqOCJsqTkgY.

Authors' Note: For color photographs of the construction of the Batgoggles, visit http://www.instructables.com/id/Ultrasonic-Batgoggles/.

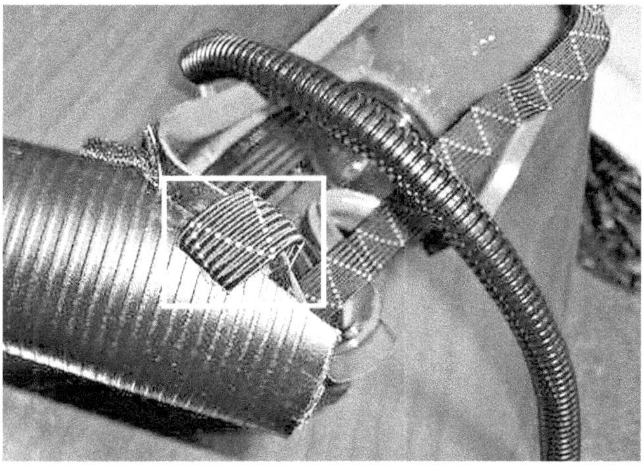

Fig. 4-16. Slot for strap made with a Dremel rotary tool. Photo by Suneth Attygalle.

TERMS USED IN TEXT

Analog Input: Refers to hardware interfaces that accept non-digital signals. For decades, all the plugs and sockets on traditional audio and video equipment connected analog lines.

Capacitor: A device used to store an electric charge, consisting of one or more pairs of conductors separated by an insulator.

Electrolyte: A liquid or gel that contains ions and can be decomposed by electrolysis (e.g., that present in a battery or capacitor).

Electrolytic Capacitor: An electrolytic capacitor is a type of capacitor that uses an electrolyte, an ionic conducting liquid, as one of its plates, to achieve a larger capacitance per unit volume than other types.

I/O: Short for *input/output*. The term I/O is used to describe any program, operation, or device that transfers data to or from a computer and to or from a peripheral device. Every transfer is an output from one device and an input into another.

PC Board: A board with a printed circuit.

Resistor: A device having resistance to the passage of an electric current.

Ultrasonic Sensor: An ultrasonic sensor (a.k.a. a transceiver when it both sends and receives) works on a principle similar to radar or sonar, which evaluates attributes of a target by interpreting the echoes from radio or sound waves respectively.

Voltage Regulator: A voltage regulator is an electrical device designed to automatically maintain a constant voltage level.

PRODUCTS MENTIONED IN TEXT

Arduino Nano: http://www.robotshop.com/arduino-nano-usb-microcontroller-v3-no-headers.html.

Arduino Uno: http://www.sparkfun.com/products/9950.

Assorted Wires: http://www.allelectronics.com/make-a-store/item/RCBL-9/9-CONDUCTOR-FLAT-RIBBON-CABLE/1.html.

Devantech SRF05 Ultrasonic Sensor: HVW Technologies, Solarbotics, Ltd., 3740D 11A Street NE, Suite #101, Calgary, AB T2E 6M6 Canada; http://www.hvwtech.com/products_view.asp?ProductID=943&utm_source=Google&utm_medium=Product+Search&utm_campaign=Product+Search+%28Apr11%29.

Dremel Rotary Tool: http://www.amazon.com/Dremel-300-1-24-Variable-Speed-Rotary/dp/B002BAHFBE/ref=sr_1_2?ie=UTF8&qid=1308131584&sr=8-2.
Enclosure for Electronics: http://nz.digikey.com/1/1/31594-9v-i-amd-box-3-25x4-38x1-5-231-i-almond.html.
Flexible, Convoluted Tubing (3/8-in./Black): Amazon.com; http://www.amazon.com/JT-Products-4305F-Flex-Guard-Convoluted/dp/accessories/B002UZW9H0.
Hot Glue Gun: http://www.amazon.com/Surebonder-Mini-Glue-Gun-Kit/dp/B001AQRLM8/ref=sr_1_8?ie=UTF8&qid=1308131200&sr=8-8.
Neiko Brand Welding Goggles: http://www.amazon.com/Neiko-Comfort-Fit-Welding-Safety-Goggles/dp/B000IBEBB0.
PC Board: http://www.radioshack.com/product/index.jsp?productId=2102845.
Plastic Stacking Cones: Sears; http://www.sears.com/shc/s/p_10153_12605_SPM1534297001P?sid=IDx20101019x00001a&ci_src=14110944&ci_sku=SPM1534297001.
Plasti Dip Spray: ACE Hardware; http://www.acehardwareoutlet.com/productdetails.aspx?sku=19507&source=GoogleBase.
Piezo Buzzer: http://www.radioshack.com/product/index.jsp?productId=2062403&CAWELAID=107596958.
Soldering Iron: http://www.amazon.com/Weller-WLC100-Soldering-Hobbyist--Yourselfer/dp/B000AS28UC/ref=sr_1_1?ie=UTF8&qid=1308131724&sr=8-1.
1 Microfarad (μF) Electrolytic Capacitor: Avrthing.com; http://avrthing.com/product.php?id_product=66.
9v Battery Connector: http://www.amazon.com/9V-Battery-Connector/dp/B001F39728.
10 Microfarad (μF) Electrolytic Capacitor: Amazon.com; http://www.parallax.com/Store/Components/Capacitors/tabid/150/ProductID/165/List/1/Default.aspx?SortField=ProductName,ProductName.
16 MHz Crystal: Sparkfun Electronics, 6175 Longbow Drive, Boulder, Colorado USA; http://www.sparkfun.com/products/536.
22pF Capacitor: Sparkfun Electronics, 6175 Longbow Drive, Boulder, Colorado USA; http://www.sparkfun.com/products/8571.
7805 5v Voltage Regulator: Amazon.com; http://www.amazon.com/7805-5V-Voltage-Regulator-TO-220/dp/B0002ZPXJ6.
28-Pin DIP Socket: http://www.solarbotics.com/products/dc-28_pin/.

ACKNOWLEDGMENTS

We thank Earl Bryenton of Brytech, Inc., Ontario, Canada and Alex Fink for their helpful comments on the manuscript.

Appendix 4-1. Batgoggles Code

This program gets the distance returned by the SRF05 ultrasonic sensor, converts it to inches, and creates a linear mapping of beeps related to the distance by changing the interbeep delay time. It accounts for the time taken to read the distance from the sensor.

```
#define echoPin 2 // pulse in
#define trigPin 3 // trigger
#define beepPin 4 // piezo
#define scalingfactor 10 // this sets the scaling from inches to milliseconds

int inches = 0; // variable for storing distance in inches.
unsigned long pulseTime = 0; // variable for storing pulse width from sensor
unsigned long maxtime = 1000; // variable for storing maximum delay in time
unsigned long i = 1; // counter
unsigned long time2 = 0; //for timer of readdistane()
unsigned long time1 =0; //for timer of readdistance()

void setup() { //initializes Arduino Pins
 pinMode(trigPin, OUTPUT); // set trigger pin as output
 pinMode(echoPin, INPUT); //set echo pin as input
 pinMode(beepPin, OUTPUT); //set beep pin as output
}

void readandmapdistance(){ // gets distance in inches and
 time1 = millis(); // time at start of getting distance
 digitalWrite(trigPin, HIGH); // sends a pulse to the sensor
 delayMicroseconds(10); // that is 10 microseconds long
 digitalWrite(trigPin, LOW); // end of pulse
 pulseTime = pulseIn(echoPin, HIGH); //reads resulting pulse width
 inches = pulseTime/148; // convert pulse width to inches
 time2 = millis(); //time at end of getting distance
 maxtime = scalingfactor*inches; // reset total time mapped distance in ms
}

void delayandcheckdistance(){ //this function delays the next beep
based on the distance, it checks distance every 60ms.
 i=1; // reset counter
 while(1){ // run this loop until counter exceeds maximum ime set in
         // readdistance()
 if (i % 60 == 0){ // check every 60 msecs (since sensor can only
         // be sampled every 50ms accurately
 readandmapdistance(); // get distance in inches
 }
 delay(1); // delay by delaytime ms
 ++i; // increase timer as long as maxtime not attained.
 if(i >= maxtime){ // if the counter gets to the max time
 break; //end while loop
 }
 }
}

void beep(){ //this functions makes the goggles beep. it also checks
         // the distance part way through the beep to prepare for the next
function
 digitalWrite(beepPin, HIGH); //start beep
 delay(90); //make beep 150ms long (the missing 50ms is added subsequently)
 readandmapdistance(); // check distance during the beep.
 if(time2-time1 < 30){ //this accounts for the time it took to check
         // the distance and makes sure the beep lasts 150ms (maximum delay is
         // 30ms after which it times out)
```

Appendix 4-1 continued

```
 delay(60-(time2-time1)); //this adds the amount of time it took to
         // sample the data.
 }
 digitalWrite(beepPin, LOW); //end beep
}

void loop() {
 beep(); // beep!
 delayandcheckdistance(); //delay by appropriate amount based on distance
}
```

Chapter 5

Conditioning and Training Bats for Public Demonstration

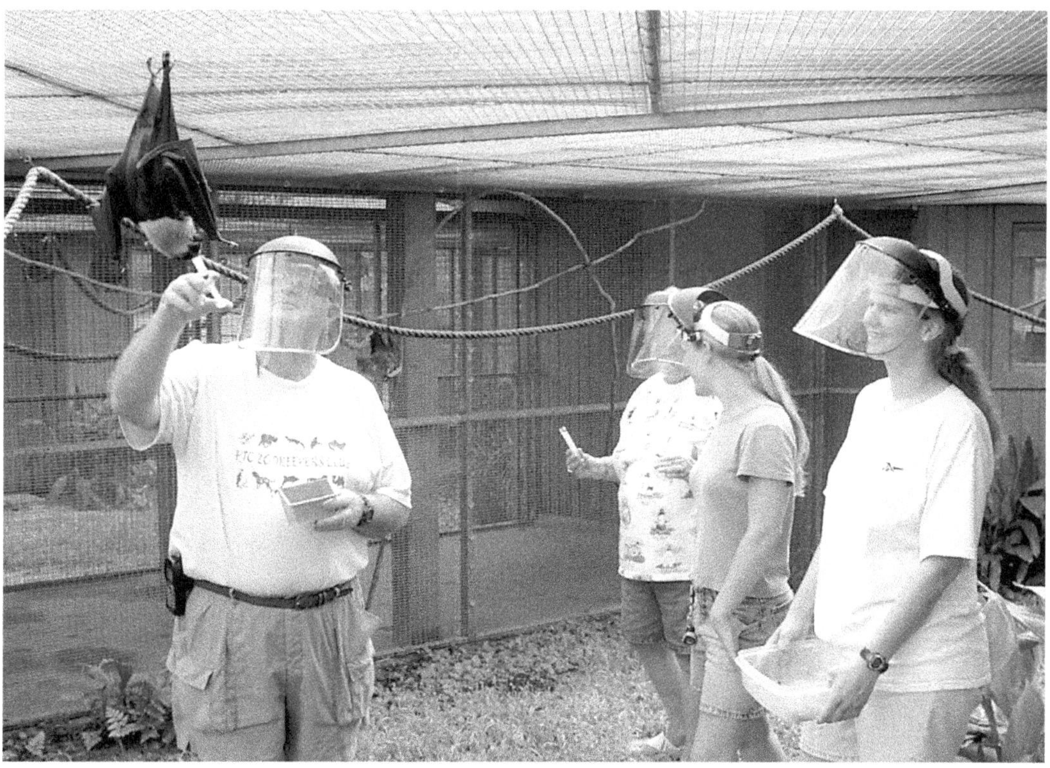

Photo © Dana LeBlanc, Lubee Bat Conservancy

Conditioning and Training Bats for Public Demonstration

John Seyjagat and Denise R. Tomlinson

INTRODUCTION

Using live bats in public education programs is one of the most effective ways to increase awareness of the benefits of these misunderstood creatures and to gain support for bat conservation. Some types of bats are more suitable than others for such programs, and all may require special training and conditioning.

Most bats that are used for demonstrations in Europe, Canada, and the United States are non-releasable, insectivorous, crevice-dwelling species that have been acquired from rehabilitation facilities. Crevice-dwelling species typically live in caves, mines, rock crevices, under bridges, behind loose bark, and in tree cavities. Many, however, prefer man-made structures, where they may be injured by people or their pets. These individuals are often presented to rehabilitation facilities, but even with the best care, their injuries may render them non-releasable. These bats can be ideal candidates for public demonstrations, particularly if they have lost the ability to fly. Healthy, captive-reared bats capable of flight are also displayed in outreach programs. However, because a bat's first defense against predators or stressful or uncomfortable situations is to fly away, these bats must be conditioned to being hand held. Conditioning helps them become comfortable with the handler and the surrounding environment. Their natural fears eventually subside, and the flight response is greatly decreased.

In the United States, Old World fruit bats are extremely popular when exhibited in zoos, and in some situations, a flying fox may go from a zoo exhibit to being featured in an outreach program. New World fruit bats and nectarivorous species are also very popular zoo-exhibit animals. Most of these species are too delicate to be displayed in outreach programs, although New World fruit bats, such as *Artibeus* spp, are occasionally displayed in portable enclosures.

In this chapter the authors discuss methods for conditioning and training bats for public demonstration, with particular focus on crevice-dwelling species and flying foxes.

CREVICE-DWELLING BATS

Conditioning and training crevice-dwelling bats for display in outreach programs is relatively easy, but it is important to keep in mind the animal's natural tendency to fly away or hide. The goal is to present these bats in a way that prevents them from flying off or hiding from the public, while making them feel secure.

Acclimating injured bats to being handled is best done after ensuring their wounds have healed fully. Throughout the course of rehabilitation, these animals have had to tolerate handling for medical treatment and general care, which greatly aids in the conditioning process. While restraining a crevice-dwelling bat, the handler becomes accustomed to how the bat prefers to rest in his or her hand. Some crevice dwellers are comfortable being restrained with a thumb over their rump or feet (Fig. 5-1a). Others prefer to feel the security of a hand or towel enclosing them with just their heads sticking out (Figs. 5-2 and 5-3). The preferences of each animal should be well known prior to presenting them to the public.

The best way to acclimate a bat to handling is to pick it up daily while gently stroking it. Eventually, the animal becomes accustomed to being held and should not display agitation. Once this is achieved, the length of time the bat is held can be increased. Start with a few minutes and progress to longer periods of time. Whenever the bat displays agitation, place it back in its cage and wait until the following day before trying again. A bat should never be restrained beyond the limits of its tolerance.

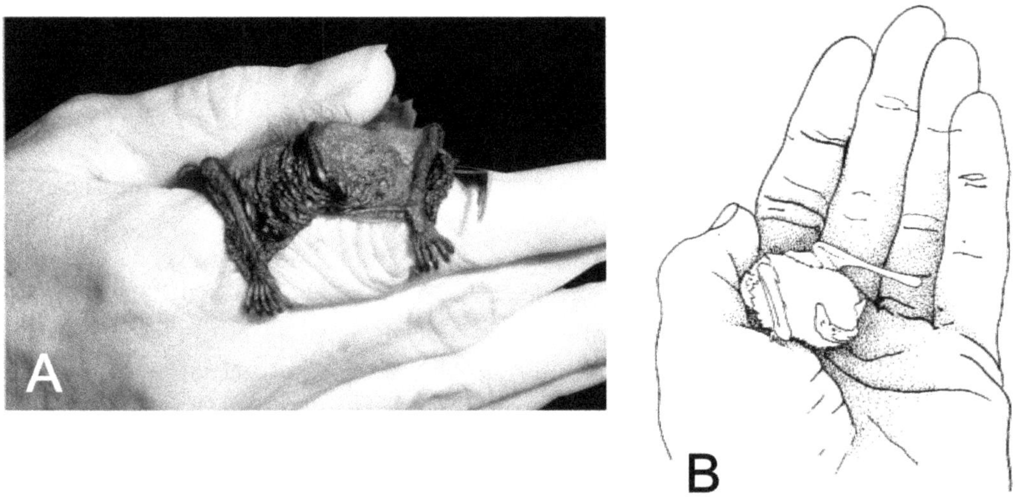

Fig. 5-1. Suggestion for holding a crevice-dwelling bat during a public demonstration. A) bat being "pinned" under the handler's thumb to prevent it from flying out of the hand; B) front view of a crevice-dwelling bat in hand. A) photo courtesy of Susan M. Barnard; B) illustration courtesy of T. P. McOwat.

Fig. 5-2. A crevice-dwelling bat may feel more secure when being held in the hand with only its head in view. Photo courtesy of Susan M. Barnard.

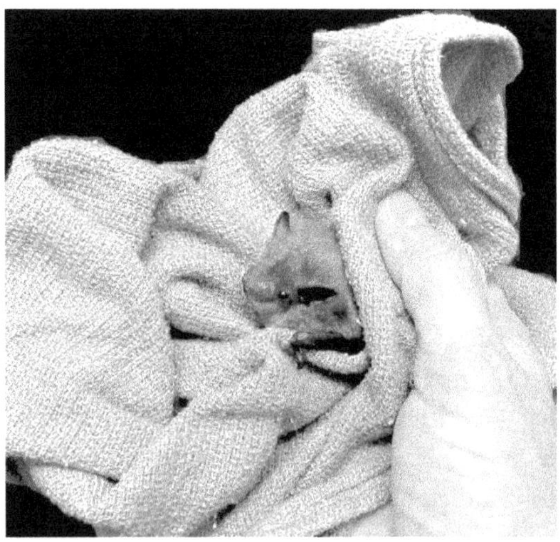

Fig. 5-3. Crevice dwelling bats can be controlled more easily when held in a cloth. The cloth offers more security to the bat than the method shown in Figure 5-1; however, the public can not see the animal's entire body. Photo courtesy of Susan M. Barnard.

There is no specific length of time required for a bat to acclimate to handling, and it should be understood that a few bats may never adjust to handling. When working with bats from the wild, handlers should protect themselves against bites by wearing gloves (see Fig. 5-2), and wild bats should never be displayed in programs until they have completed appropriate quarantine.

Hand-reared, long-term captives may never have been held in a gloved hand. In such cases, they must become accustomed to the feel of a glove before they are displayed in outreach programs.

Once a handler is comfortable holding a conditioned bat, the animal should be taken to public demonstrations, but not displayed. It should be introduced to new and distracting situations gradually, and should remain in its carrier while it becomes adjusted to the sounds of an audience. The carrier should be modified for the needs of each species (e.g., crevice dwelling or foliage roosting; see *Transporting and Temporary Holding*, Chapter 8, Volume 2). This helps to reduce any initial stress from the sounds of an audience. After being transported to educational programs a few times without being displayed, a bat is more likely to remain calm when it is handled during programs. If a bat exhibits signs of stress or agitation, it is best to place it back in its carrier. By conditioning more than one bat for public demonstration, the handler can immediately replace the agitated individual with another bat.

Not all states allow bats to be hand held in public. In such situations, an animal can be displayed in a viewing cage (see Barnard, Chapter 6). Unless a crevice-dwelling bat is very comfortable and acclimated to being viewed in a cage that exposes it on all sides, it is best to use cages that have front viewing only.

FLYING FOXES

Flying foxes, or Old World fruit bats, are extremely popular animals in education programs, and the larger the species, the more spectacular the demonstration. However, because of their large size, they can inflict painful scratches or escape easily when held. Only highly experienced handlers should display these animals. In the United States, flying foxes in the genus *Pteropus* are categorized as injurious wildlife under the Lacey Act (see *Legislation: United States*, Chapter 1). On rare occasions, a bat covered by the Lacey Act may be displayed in outreach programs by obtaining a special permit from the United States Fish & Wildlife Service, Office of Management Authority. The permit must show that the bat in question is permanently marked and identifiable, has been medically certified as being incapable of flight, and can be transported to and from programs in a double-enclosed, escape-proof carrier. For over a decade, S. M. Barnard demonstrated a well-conditioned giant Indian fruit bat (*Pteropus giganteus*) to the public (Fig. 5-4). In this case, handling began at an early age, when the bat was still a pup and wandering away from its mother on its own. When the pup became comfortable being hand held, it was introduced to the sounds of a television set. Thereafter, it was acclimated to a transport carrier.

There are several ways in which flying foxes may be selected for public demonstration. The animal in question may be a pup that was rejected by its mother and hand reared. In other situations, a severely harassed or injured juvenile or adult may require isolation from its group for its well-being. These animals can often be held in isolation in an enclosure next the main group to satisfy their social needs, while also being conditioned for public demonstration. An individual may be selected from a group based on its receptivity to handling, training, and proven ability to tolerate noisy humans. Three primary qualities to look for when selecting an individual for demonstration are:

Fig. 5-4. A giant Indian fruit bat (*Pteropus giganteus*) being displayed to an audience. Photo courtesy of Susan M. Barnard.

1. Approachability. If a flying fox appears calm when approached, it will most likely become comfortable with handling. If the bat backs away when approached, it is not likely to react well in front of a group of people.
2. Curiosity. If a flying fox appears interested in anyone entering its enclosure and exhibits a natural curiosity about the activities of people within its environment, it may do well in the presence of the public.
3. Gender. Although one sex is not necessarily preferred over another, there are extenuating circumstances to consider. Females in breeding situations should never be displayed because they may be pregnant. Males in breeding situations tend to be difficult to handle. Typically, the best choice is non-reproducing males that display the other criteria mentioned above.

Procedures for conditioning flying foxes for public demonstration differ, depending on the facility. Bats housed in institutions such as zoos may have to be acclimated to handlers other than their daily care givers. In some situations, the demonstrators are also care givers, although it is best for the bat if it knows more than one handler. This process should start in the bat's current environment. Flying foxes can be trained to approach a handler with fruit juice in a feeding syringe (see Chapter 5 divider), or with a small piece of their favorite fruit, such as papaya, mango, or melon. Such rewards help to make conditioning a positive experience, which hastens the process and helps to reduce stress on the bat.

Once an animal is accustomed to a person's presence and voice, it can be conditioned to being handled. An area should be selected to work with the animal, which is typically in its enclosure environment. When the same area is used each time, the bat learns what to expect while there, and will be less distracted by roost mates. To protect against bites and scratches, the handler should wear gloves and long-sleeved shirts during the conditioning process.

When first handling a flying fox, hold it for a few minutes, gradually increasing the duration as time passes. The goal is to condition the animal to be comfortable while being held for the time it takes to display it to an audience.

As with the crevice-dwelling bats discussed above, flying foxes should remain in a covered transport carrier and allowed to adjust to the sounds of an audience prior to being displayed. It is best to display a flying fox initially to a small, quiet group of people until it is ready for a regular audience. Each individual is different, and there is no set time for accomplishing a training program. It should be noted that a bat placed in front of the public before it is ready may react by biting the handler, thus sending a negative message to the audience.

In preparation for transporting a flying fox to outreach programs, the animal should be acclimated to an appropriate carrier (see *Transporting and Temporary Holding*, Chapter 8, Volume 2, and *Flying Foxes*, Chapter 6, this volume). To accomplish this task, gently place the bat in the carrier, and allow it to remain there for a brief time. This experience can be made more pleasant by offering the animal small pieces of fruit. Repeat this process daily, until the bat appears to be comfortable in the carrier. To reduce the chance of a flying fox attempting to fly out of a carrier or a demonstrator's hand during a public display, do not allow it to fly out of the carrier while releasing it in its regular enclosure after programs.

"FLIGHTY" BATS

Flighty bats belong to those species that typically dart skillfully through dense foliage. Unlike crevice dwellers, they take flight quickly when approached. Because some of these species are extremely delicate (e.g., nectarivorous species and many New World fruit bats), they may injure themselves in containers that are too small for continuous flight (Richard LaVal, pers. comm.). Furthermore, flighty bats should never be restrained in the hand for public demonstration. These animals are poor candidates for outreach programs.

CONCLUSION

If an educator's message is to be taken seriously and without distraction, total control must be maintained over the animal being displayed. This is achieved by making the animal as comfortable as possible during a demonstration, and by selecting bats that are receptive to handling, training, and ability to tolerate noise.

Because bats are vectors of rabies, all individuals from the wild must have completed appropriate quarantine, and presenters must always wear gloves to protect themselves against bites. By ensuring public safety and animal comfort, a program featuring live bats can be an unrivaled educational experience.

ACKNOWLEDGMENTS

We thank Valerie Lounsbury and Dietrich Schaaf for their helpful comments on the manuscript.

Chapter 6

Bats in Outreach Programs

Photo courtesy of Janna C. Kimel

Crevice-dwelling Bats

Susan M. Barnard

INTRODUCTION

Television is an excellent medium for wildlife educators, but it can not replace a personal encounter with living creatures. This is especially the case when the animals are secretive creatures-of-the-night, like bats. During their nightly forays, crevice-dwelling bats collectively feed on billions of insects, a behavior that is so important to the world's production of food crops. In addition to discussing the many benefits of bats, it is important that educators dispel the misunderstandings surrounding these animals.

Crevice-dwelling species are encountered worldwide, and throughout the world, misguided humans have long expressed fear of these beneficial animals. Audiences typically want to know if bats will get caught in their hair or infect them with rabies. Other important messages can be shared, such as what to do if a bat is found inside the home (see Finn and Finn, Chapter 8), or what to tell children when they see a bat on the ground.

Most educators start their programs by showing images of all types of bats while explaining their many benefits. An education program that includes a crevice-dwelling bat for a live demonstration does not mean that the lecturer is limited to discussing just these species. However, by displaying these animals during programs, people can see what bats they may encounter, at least for audiences in temperate regions of the world.

DISPLAYING CREVICE-DWELLING BATS

Only trained animal handlers should demonstrate live bats to the public. Bats displayed in outreach programs can be hand held or displayed in portable containers. With few exceptions, crevice-dwelling bats are hand held, perhaps because of their small size and secretive ways. This practice, however, has created some controversy about whether or not to wear gloves. Those who advocate hand protection agree that the 'flow' of a program can be interrupted if a demonstrator is bitten by a bat. Conversely, those who oppose wearing gloves may argue that it is safe to handle hand-reared bats because they are conditioned to being handled and tend not to bite, and there is little chance that they have been exposed to the rabies virus. Whether gloves are worn or not during public demonstrations, silent messages may be delivered. If gloves are worn, the message is that "bats are dangerous, and that is why the handler is taking care not to be bitten." If gloves are not worn, the message is that "bats are safe to handle," thus inviting a member of the audience to be bitten if picking up a wild bat sometime in the future. Therefore, the lecturer should always explain to the audience the reason for wearing gloves (or not). It may help to keep people from being bitten when attempting to handle found bats.

An alternative to handling crevice-dwelling bats in outreach programs is to place them in portable exhibits, such as the attic vent exhibit shown in Figure 6-1. Attic vents are familiar to most homeowners, and this exhibit allows a close-up view of a bat without handling it.

At Zoo Atlanta in Georgia, a mobile display (Fig. 6-2) was designed by the author to present both live and taxidermic crevice-dwelling bats to the public. Two bat boxes

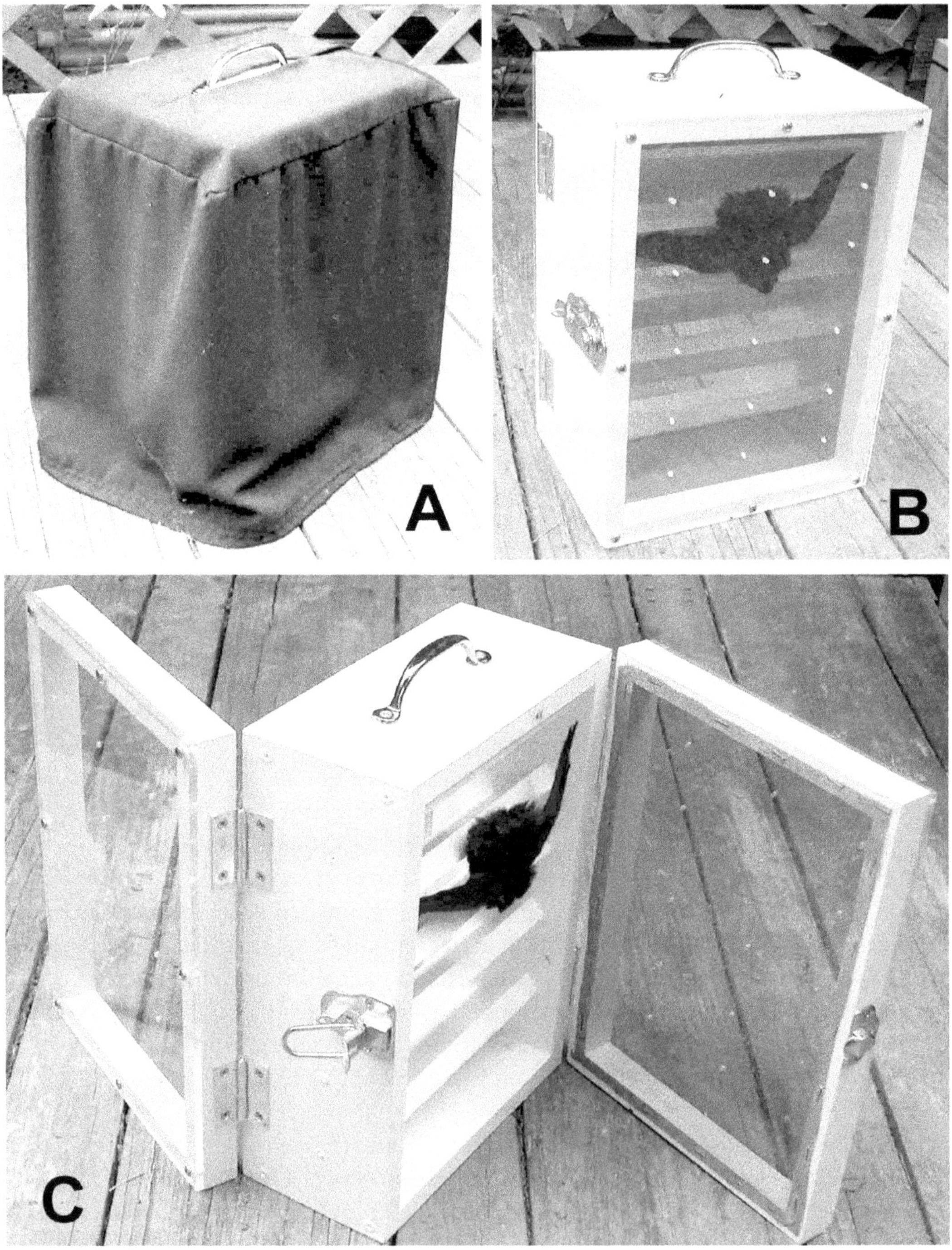

Fig. 6-1. Portable attic vent exhibit for crevice-dwelling bat species. Photos by the author.

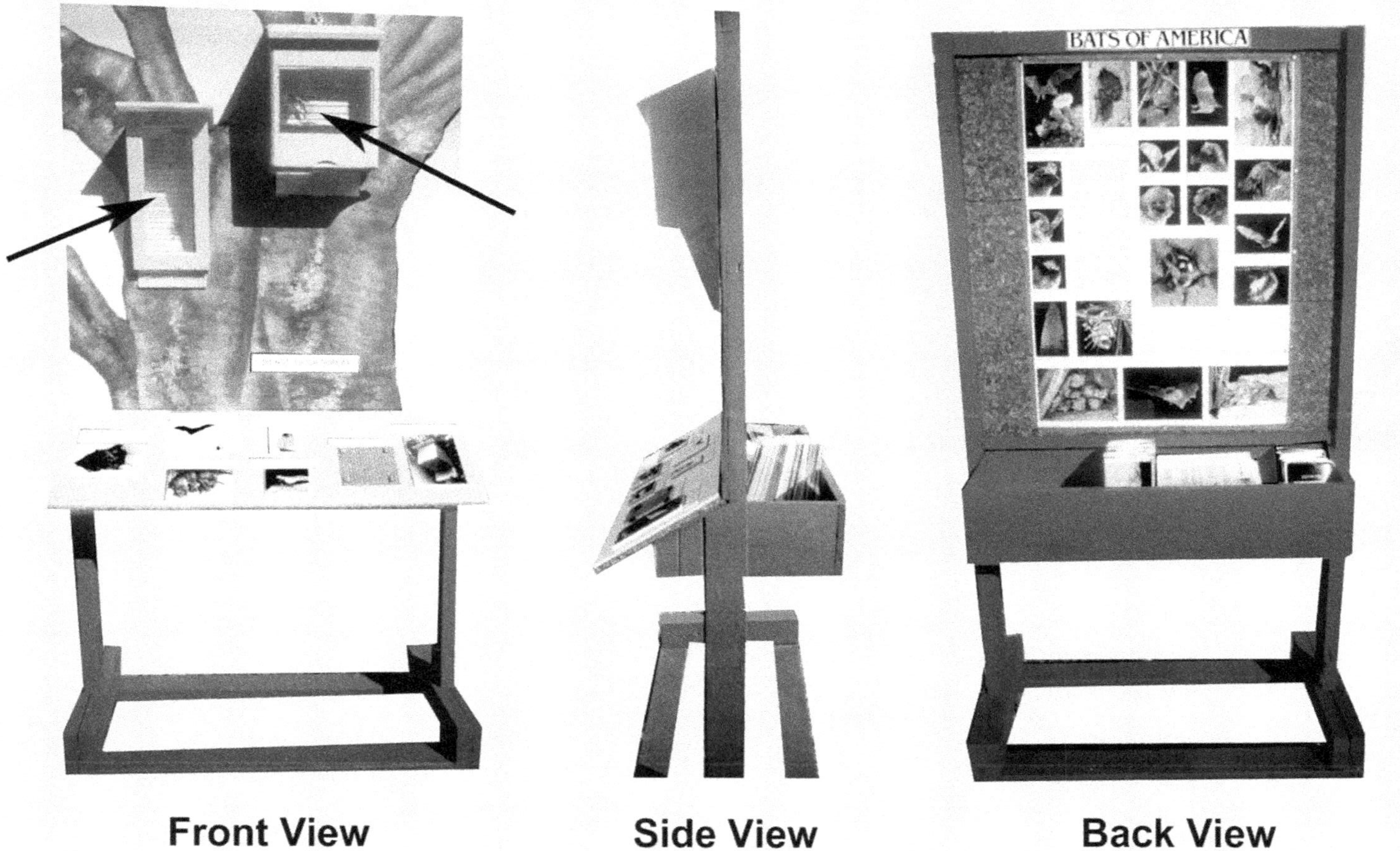

Fig. 6-2. Mobile bat exhibit used at Zoo Atlanta. The exhibit incorporates living and taxidermic bats, as well as educational materials. Front view: left arrow shows a bat house for displaying live bats; right arrow shows a bat house containing taxidermic specimens, including a female adult with pups. Below the bat houses is a graphics board. Back view: a Bat Conservation International poster of American bats is displayed. Below the poster are educational pamphlets for distribution to zoo visitors. Photos by the author.

were mounted onto a board on which a tree was painted. One box contained a live bat and the other, taxidermic specimens (Fig. 6-3). On the back side of the board a Bat Conservation International poster entitled *Bats of America* and various informative pamphlets were placed for public consumption.

The exhibits above were designed and used prior to the state of Georgia banning the display of rabies vectors such as bats, unless they are displayed in their permanent cages. Figure 6-4 shows how this regulation can be met, while at the same time displaying a live bat in outreach programs. This is achieved by making the permanent home of a display bat light weight for travel. It should also be designed for bat viewing without exposing the public to the possibility of disease, while also maintaining the bat's need for security.

Fig. 6-3. Close-up of bat box with taxidermic specimens shown in front view of Figure 6-2. Photo by the author.

Fig. 6-4. Lightweight housing for bats that can double as a portable exhibit (e.g., a simulated barn). A) removable cover (arrow) over "hay loft" bat roost. To assist a bat in climbing to the loft/roost, a mesh ladder is attached from the floor to the loft. B) cover over loft removed to allow audience to view the bat (arrow). Most bats climb to the highest point in their housing, making it reasonably predictable that they will be on display during programs. Photos courtesy of Janna C. Kimel.

People are very curious about the wings of bats. A demonstration is rarely completed without someone asking to see the animal's wings. Although spreading out a bat's wing does not injure the animal when done properly (Fig. 6-5), bats dislike it so intensely that if done often enough, a bat may begin biting whenever it is handled.

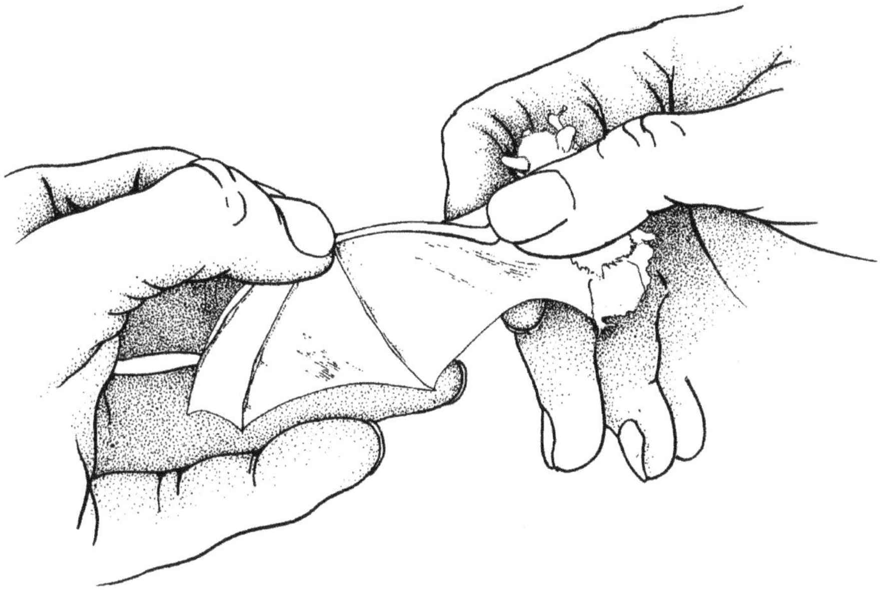

Fig. 6-5. Proper method for displaying a bat's wing. Illustration courtesy of T. P. McOwat.

THINGS TO AVOID WHEN FEATURING BATS IN OUTREACH PROGRAMS

- Avoid excessively long displays. As soon as a bat shows agitation or signs of stress, stop displaying it. If more than one bat is available, rotate them.
- Avoid holding a bat while lecturing. This can be distracting, resulting in the loss of control over a bat while displaying it. It may also distract the audience as people focus on the bat and not the message.
- Avoid escapes, but always be prepared to recapture an escaped bat using appropriate methods. This means keeping the audience calm and preventing injury to the animal.
- Avoid purposely flying a bat, unless it is in a secure flight cage.
- Avoid allowing members of the audience to pet or hold a bat.
- Avoid the use of aromatic products such as hand creams, perfumes, and shaving lotions. Bats may be allergic to them.

ACKNOWLEDGMENTS

I thank Dietrich Schaaf for his helpful comments on the manuscript.

Flying Foxes

John Seyjagat and Denise R. Tomlinson

INTRODUCTION

Flying foxes (Old World fruit bats) make exciting public demonstration animals. Their large eyes and dog-like faces appeal to humans and help to alleviate people's fear and negative opinions of bats. Flying foxes play a vital role in forest ecology as pollinators and seed dispersers. As biologically important species, some fruit bats have evolved a symbiosis with plants in their native habitats: The plants provide fruit for the bats, and the bats disperse their seeds. Recent studies have revealed that as pollinators, plant visiting bats are responsible, in part, for forest-based economies producing fibers, flavorings, food, and medicine.

Many lecturers open their programs by showing images of flying foxes, while at the same time explaining the variety of benefits derived from these animals. This is typically in the form of a slide show or power-point presentation. The presentation is usually followed with a live bat demonstration. Prior to closing a program, hands-on learning activities, such as drawing pictures or playing games (see also Lewis, Chapter 3), can be shared with the audience, especially when working with school children. By engaging the audience with a variety of learning activities, the program's messages are more likely to be remembered. However, before taking a live bat to an outreach program, some key animal behavior elements should be considered.

NOT ALL FLYING FOXES ARE TREATED EQUALLY

Flying foxes belong to the family Pteropodidae, which comprises 42 genera. Of these, bats of the genera *Cynopterus, Eidolon, Rousettus,* and *Pteropus* are the most commonly displayed to the public in zoos, and *Eidolon* and *Rousettus* are the genera typically displayed in outreach programs. Flying foxes of the genus *Pteropus* are considered injurious wildlife in the United States, and with rare exceptions, may not be featured in outreach programs (see Seyjagat, Chapter 1). Even though all flying foxes share similar physical features and feeding habits, only bats of the genus *Pteropus* are considered injurious under the Lacey Act.

FEATURING FLYING FOXES

There are many ways to present flying foxes in outreach programs. The object is to feature the animals in a manner that fosters acceptance and appreciation. In the opinion of the authors, a handheld flying fox has a great impact on its audience and almost immediately dispels any myth that bats are dangerous, scary creatures. Hand-held flying foxes may be featured in outreach programs (Fig. 6-6), or the animals may be displayed in portable enclosures (Figs. 6-7 and 6-8) that can be placed on a table. Some flying foxes are not comfortable in a totally transparent display carrier such as the one shown in Figure 6-7. In such cases, only one wall can be made transparent for viewing, especially when featuring a cave-dwelling species such as those in the genus *Rousettus.* Portable displays should always be designed to meet the individual needs of each species. Flying foxes, except *Rousettus*, can also be displayed in enclosures large enough to include both the presenter and bat (Fig. 6-9). Such displays work well at indoor or outdoor wildlife events. Because *Rousettus* bats live in caves, they must be kept away from direct sunlight.

Before a flying fox is displayed at public presentations, one must be certain that the animal is conditioned and ready for an appearance. Displaying unconditioned animals stresses and panics them, causing them to bite and scratch in a bid to escape. Such behavior may strengthen the perception of audiences that bats are dangerous animals. The object is to present bats in a manner that engages an audience and dispels the myths that give bats a bad name. For more information on conditioning and selecting flying foxes for public demonstrations, see Seyjagat and Tomlinson (Chapter 5).

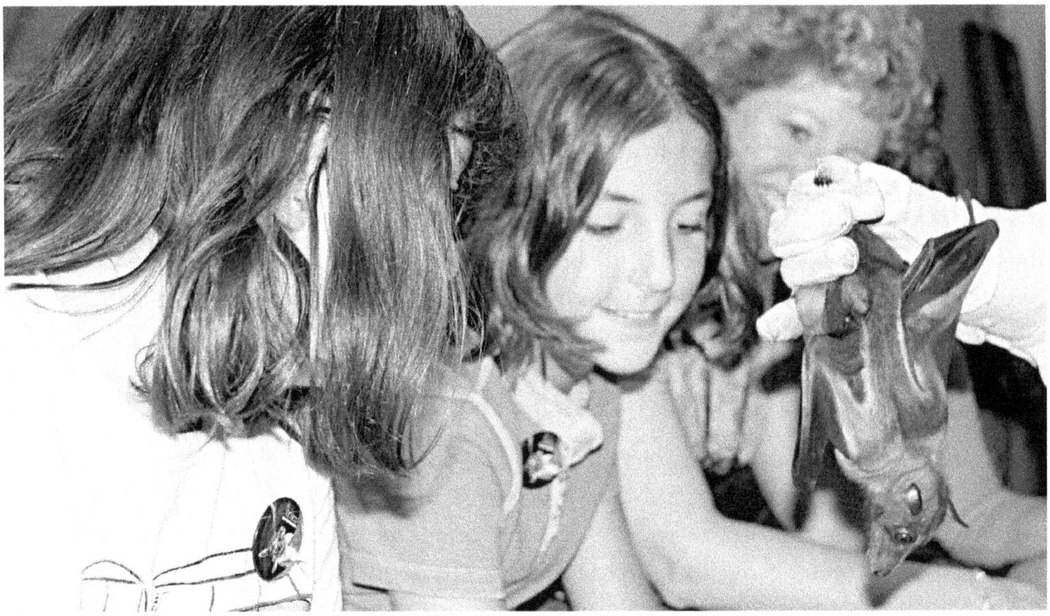

Fig. 6-6. A hand-held flying fox being shown to an audience. Photo courtesy of the Organization for Bat Conservation.

Fig. 6-7. Display carrier for flying foxes that are comfortable being viewed from all sides. Photos courtesy of Susan M. Barnard.

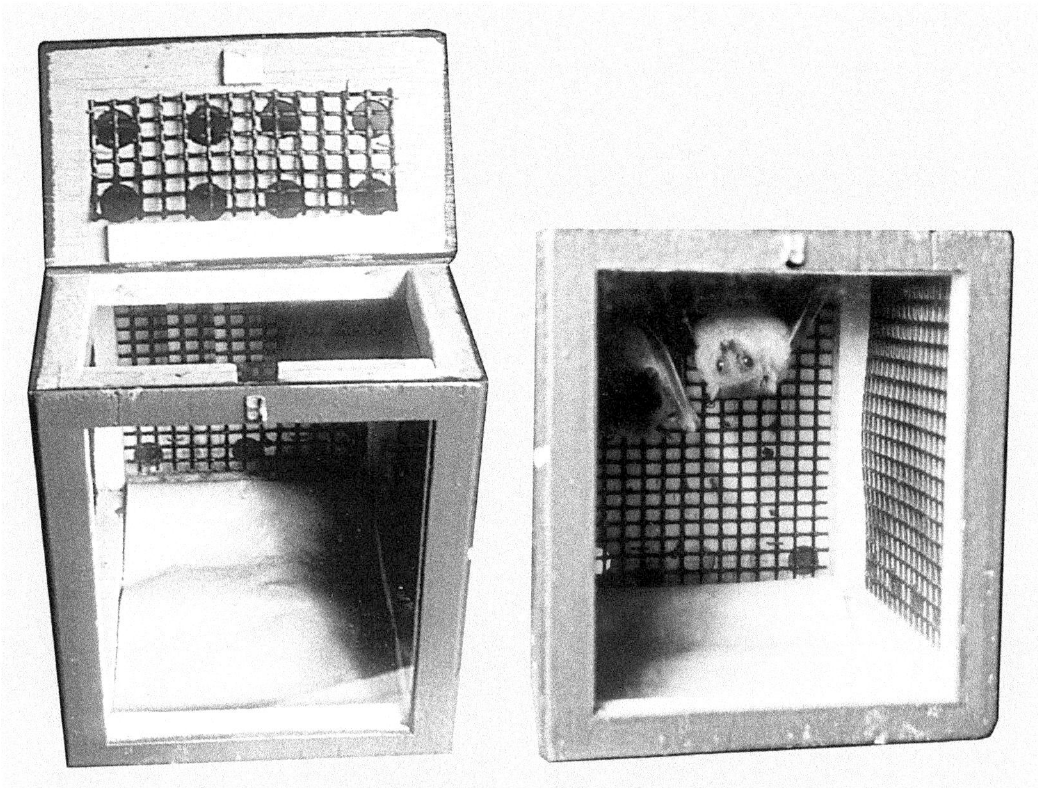

Fig. 6-8. Display carrier for flying foxes that are not comfortable being viewed from all sides (e.g., *Rousettus* species). Photos modified from Barnard (1995).

Demonstration time with flying foxes varies, depending on the individual animal. Once a bat becomes agitated, it should be returned to its carrier immediately, and not displayed again for the rest of the program. Not all bats react to an audience in the same way. Some are nervous and may urinate when removed from a transport carrier, and the audience should be warned that bats may void their bladders or bowels when handled. Flying foxes have a rapid gut transit time (in *Pteropus*, 12-34 min.; Tedman and Hall, 1985) and should not be fed for at least 2½ hours prior to presentation. Presenters should carry paper towels to clean up excreta, and the audience should be at least 6 feet from an animal and its handler. Audience members should never be allowed to touch a bat, or come into contact with its bodily fluids.

Audiences should always be cautioned that loud noises and sudden movements may frighten a flying fox, causing it to panic and to attempt to escape. In some cases, a large pteropodid may break away from the presenter. Audiences should be cautioned that this may occur, and that everyone should remain as still as possible to prevent injury to the bat or audience members.

Taking photographs of flying foxes during demonstrations has little effect on them, except when strobe flashes are used. Strobe flashes may frighten an animal, and therefore a specific time should be set aside for taking pictures. The presenter should take extra care while restraining a bat during this time, and keep photographers at a safe distance.

Fig. 6-9. Outdoor, walk-in portable exhibit appropriate for featuring flying foxes. Photo courtesy of Talking Talons Youth Leadership, Tijeras, New Mexico.

Unlike trained birds of prey that are taken to outreach programs in sports stadiums, flying foxes do not share the abilities of these animals, and should not be flown in open-air situations or over audiences. Moreover, bats should not be jessed because their stifle joints (knees) are weak. If a jessed bat attempts to fly, the stifle joints may be dislocated when the jess is pulled tight.

ACKNOWLEDGMENTS

We thank Valerie Lounsbury and Dietrich Schaaf for their helpful comments on the manuscript.

REFERENCES

Tedman, R.A. and Hall, L.S. 1985. The morphology of the gastrointestinal tract and food transit time in the fruit bats *Pteropus alecto* and *P. poliocephalus* (Megachiroptera). *Australian Journal of Zoology*, 33:625-640.

Common Vampire Bats (*Desmodus rotundus*)

Dianne Devison

INTRODUCTION

Unlike the attractive flying fox, common vampire bats (*Desmodus rotundus*) have few appealing attributes. They are parasitic on other mammals, lapping their blood and leaving bleeding divots on their hosts that become infested with screwworms. However, advertise that a live vampire bat will be on display, and there will be no trouble filling an auditorium because people enjoy the adrenalin rush of meeting a true "vampire."

Common vampire bats are long-lived, surviving in the wild for nearly 20 years. They live in structured societies, usually including an adult male and several females and their young. After feeding, the group returns to the roost where they meet face-to-face and groom one another. This contact aids a bat that has failed to feed to beg for blood from a roost mate (Fenton, 1992). Such reciprocal altruism may be an important factor contributing to the success of common vampire bats, but does this amazing behavior make them more appealing to people? Probably not, unless you are someone who has survived an ischemic stroke after being given vampire bat saliva, or vampire bat saliva-derived anti-clotting agent. Vampire bat saliva contains a potent anticoagulant that helps keep the blood of the host animal flowing while the bat feeds. In stroke patients, this anticoagulant dissolves blood clots without increasing the risk of internal bleeding. Unlike other "clot-busting" medicines that need to be administered within four hours of a stroke to be effective, the new drug called Desmoteplase can have a similar effect up to nine hours later (FoxNews.com, 2011).

PREPARING THE COMMON VAMPIRE BAT TO MEET THE PUBLIC

Hand-raised vampire bats make the best public demonstration animals. Hand-reared bats are typically handled frequently over the 9- to 10-month rearing and conditioning period. Because these bats are presented to an audience in the hand, baby vampire bats should be carried by care givers in the palm of the hand (Fig. 6-10) as much as possible. Common vampire bats tend to bond closely with their care givers, and it is best if several humans handle and feed an infant that is being prepared for outreach programs. However, vampire bats are wary, intelligent animals that will take advantage of a novice bat handler by biting and

Fig. 6-10. Juvenile vampire bat being held in the hand. Photo courtesy of Bob Johnson.

leaping from the hand. Unlike many bat species, common vampire bats can be clicker trained (the scientific term is operant conditioning), which means an escapee can be recovered if it is conditioned to receive a reward of a few drops of blood by returning to the handler.

Vampire bats are very sensitive to noises, odors, and sudden movements. Therefore, animals to be used in education programs should be conditioned, preferably by professional animal trainers, to as many people and locations as possible. S. M. Barnard conditioned a giant Indian flying fox (*Pteropus giganteus*) for public demonstration by exposing it to a variety of noisy television programs. The same can be done with vampire bats. For additional information on conditioning bats for public demonstration, see Seyjagat and Tomlinson (Chapter 5).

TRANSPORTING THE COMMON VAMPIRE BAT TO OUTREACH PROGRAMS

Keeping vampire bats at the right temperature is a critical concern. Vampire bats do not tolerate temperatures over 80-82 °F (26.7-27.8 °C), and a transport carrier should meet the temperature and roosting requirements of the bat or bats that it contains. The author prefers the Exo Terra® Faunarium, or a similar type of plastic container. The size of the container is determined by the length of time a bat is to be confined. Thus, the longer the confinement the larger the container should be. The snap-on lid of a Faunarium contains a small hinged access port that allows a presenter to remove and replace a bat easily and safely. Toweling or other cotton or fleece material can be placed inside the container. This unit is then placed inside a polystyrene cooler of the right size for safe transport.

Depending on the time of year, a cool pack or hot pack can be placed inside a cooler to maintain an appropriate temperature during transit. It is imperative that the temperature in the transport container be monitored closely. The ideal temperature in the box should be 75-78 °F (23.9-25.5 °C). Never place hot packs in a transport carrier with an adult vampire bat unless the ambient temperature is so low that it may be life-threatening to the animal. Common vampire bats are capable of generating a lot of heat in confined spaces, but they can not cool themselves.

DISPLAYING THE COMMON VAMPIRE BAT TO THE PUBLIC

Bats do not like to have their wings spread, but most audiences enjoy seeing the bat wing. Therefore, from a young age, the wings of bats that will be displayed to the public should be opened gently and held open for increasing periods of time. Two or three minutes is usually enough time for an audience to view a bat's wing.

A bat should never be held up to an audience by its wing tips, and no one from an audience should be allowed to touch a bat. If a bat becomes agitated while on display, return it to its transport carrier immediately, and do not display it again during the program. An agitated bat may bite the presenter, and because vampire bats have an anticoagulant in their saliva, if the presenter is not wearing gloves, a bite could become bloody and frightening to onlookers, thereby reinforcing an already negative opinion about bats.

Leather gloves should always be worn when handling bats in an outreach environment (Fig. 6-11), which means the bat requires conditioning to a gloved hand long before it is to be displayed. For a bat's comfort, it is best if a handler stays at least 18 in. (45.7 cm) away from any member of an audience. Some vampire bats become very nervous when presented too closely to unfamiliar people and can produce a blood-curdling scream when frightened.

Fig. 6-11. Vampire bat being held in a gloved hand. Photo courtesy of Bob Johnson.

PRODUCTS MENTIONED IN TEXT

Exo Terra® Faunarium: Rolf C. Hagen Inc., Montreal, Quebéc H9X 0A2 Canada; http://exo-terra.com/en/products/faunarium.php.

ACKNOWLEDGMENTS

I thank Dietrich Schaaf for his helpful comments on the manuscript.

REFERENCES

Fenton, M.B. 1992. *Bats*. Facts On File, New York, NY. 207 pp.

FoxNews.com. 2011. Saliva from vampire bats may aid stroke victims, research shows; http://www.foxnews.com/health/2011/09/04/saliva-from-vampire-bats-may-aid-stroke-victims-study-shows/?test=latestnews, accessed 14 September 2011.

Chapter 7

Exhibiting Bats in Zoological Institutions

Photo courtesy of Jessa Franck, Mesker Park Zoo and Botanic Garden

Flying Foxes

Kim Lengel and Steve Wing

INTRODUCTION

Old World fruit bats (i.e. flying foxes) are popular species for zoo exhibits. Their relative large size (e.g., genus *Pteropus*), communal nature, uncomplicated husbandry requirements, and potential to be focus animals in visitor-directed educational efforts all contribute to their persistence in zoo mammal collections. Historical archives from the Philadelphia Zoo in Philadelphia, Pennsylvania, the oldest zoo in North America, indicate that flying foxes were first exhibited in 1874. The International Species Information System (ISIS) database lists 132 institutions worldwide that exhibit flying foxes. The 15 listed species of bats in the genus *Pteropus* are *P. alecto, P. conspicillatus, P. dasymallus, P. giganteus, P. hypomelanus, P. livingstonii, P. lylei, P. poliocephalus, P. pumilus, P. rodricensis, P. rufus, P. scapulatus, P. speciosus, P. vampyrus,* and *P. voeltzkowi* (see http://www.isis.org). The methods described in this paper for exhibiting flying foxes are based on a variety of successful zoo displays.

ENCLOSURE DESIGN

One of the most important aspects of flying foxes, and one that distinguishes bats from all other mammals, is their ability to fly. Providing ample exhibit space to allow for sustained flight is critical, but is sometimes difficult to accomplish. Zoos must balance the need for large, open exhibits for their flying fox colonies against the constraints of space, funding, and guest perception. Where weather does not allow for year-round outdoor exhibition of flying foxes, conditioned indoor space must be made available, which is expensive to construct and maintain. Such space is often allocated first to species that are perceived by zoo guests and zoo collection managers to be more high-profile than flying foxes. Thus flying foxes are sometimes relegated to less optimal exhibits.

The perception that flying foxes are not charismatic enough to warrant "center-piece" exhibits has been challenged at Disney's Animal Kingdom's Cliffs of Anandapur exhibit located in Orlando, Florida. This exhibit dedicates 4,000 sq. ft. (372 sq. m) of space, with an additional 1,090 sq. ft. (101 sq. m) of off-exhibit space, to two species of flying foxes (*P. rodricensis* and *P. vampyrus*). Abundant natural and artificial perching materials allow the bats to behave as they would in the wild; they can roost together in a colony, yet have enough space to enable individual bats to establish and defend territories within the roost. The exhibit height (30 ft.; 9 m) is such that bats can roost high enough above their human caretakers (Fig. 7-1) so as not to be disturbed every time a zoo keeper enters the exhibit, and there is enough open space to allow even the large *P. vampyrus* to practice unobstructed flight.

Although Disney's Animal Kingdom has a climatic advantage that allows bats to be exhibited outside almost year-round, zoos in colder climates must develop creative ways to efficiently use their space to exhibit bats in large displays. The Oregon Zoo in Portland, Oregon displays their flying foxes in a 60-ft. (18 m) long by 10-ft. (3 m) wide

Fig. 7-1. Cliffs of Anandapur exhibit at Disney's Animal Kingdom. Photo courtesy of Disney's Animal Kingdom.

horseshoe-shaped exhibit (Mellen *et al.*, 1998) (Fig. 7-2). The rounded ends of the exhibit are wider than the center and allow bats to turn in flight and thus traverse the exhibit repeatedly in sustained flight. Both the Chester Zoo in Upton, Chester and the Wildlife Conservation Society (Bronx Zoo) in Bronx, New York have addressed the challenge of providing large spaces for their flying fox colonies by mixing the bats with other species. At Jungle World in the Bronx Zoo and Twilight Zone at the Chester Zoo bats are exhibited with a variety of mammal, bird, and reptile species. In so doing, these zoos have not only created exhibits that are engaging for guests, but have provided large, naturalistic spaces for their flying foxes (Fig. 7-3). Although a few of these exhibits contain large, open water features that have not been problematic, the authors do not recommend inclusion of pools, waterfalls, or other water features in flying fox exhibits unless they are very shallow (e.g., 3 in. [8 cm] or less in depth).

Most flying fox species maintained in zoos are not nimble flyers and need adequate exhibit space that is free of clutter to take off and maneuver in flight. Furthermore, flight is an energetically expensive mode of locomotion, and bats often choose to crawl along exhibit elements instead of flying. Placement of furnishings, exhibit shape, feeding frequency, and the physical condition of bats influence how much they fly. It is reported in the Association of Zoos and Aquariums (AZA) Bat Taxon Advisory Group (TAG) *Fruit Bat Husbandry Manual* (1995) that the minimum exhibit length needed to support sustained flight is eight times a bat's wingspan. As referenced above, rectangular, oval, u-shaped, circular, or doughnut-shaped exhibits (e.g., Lubee Bat Conservancy in Gainesville, Florida; Fig. 7-4) may be more conducive to flight than a traditional square exhibit (see LeBlanc, Chapter 9, Volume 3).

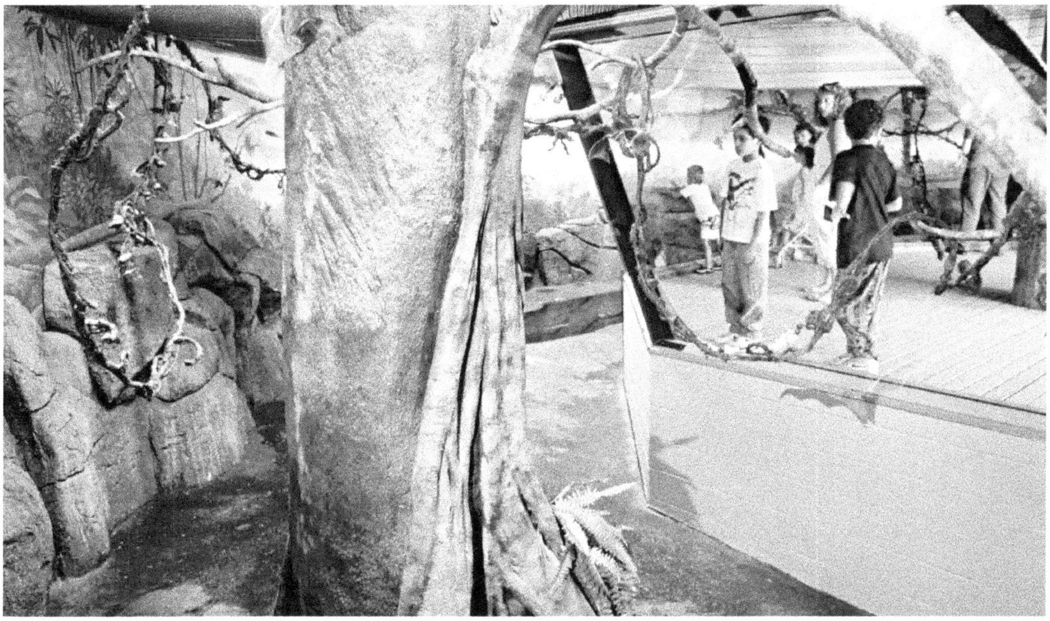

Fig. 7-2. Bat exhibit at the Oregon Zoo. This horseshoe shaped exhibit, with the rounded ends, allows bats sustained flight. The elevated visitor viewing platform provides guests a "bat's eye view." Photo by Michael Durham, courtesy of the Oregon Zoo.

Fig. 7-3. Twilight Zone at the Chester Zoo. The Twilight Zone is a multi-species, walk-through indoor enclosure featuring bats. Photo courtesy of the Chester Zoo.

Fig. 7-4. One of several bat enclosures at the Lubee Bat Conservancy. Bat enclosures are doughnut-shaped to allow bats sustained flight, and when necessary, the bats can be captured easily because of the low ceilings (i.e., 6½ ft. [2 m]). Photo courtesy of John Seyjagat, © Lubee Bat Conservancy.

CONTAINMENT

Flying fox exhibits can be constructed from a variety of materials. Wire mesh is a popular exhibit barrier, with 1-in. x 1-in. (2.5-cm x 2.5-cm) mesh being a preferred size for most flying fox species. Because of the nature of fruit bat fecal matter, recommended mesh types are vinyl-coated and stainless steel, which clean more easily than other types of wire mesh and are the best choices for frequently soiled areas. Galvanized mesh should not be used because of the potential for zinc toxicity in bats (Pye, 1967). Thin wire, such as chicken wire, is also inappropriate because it can cause toe and thumb injuries. Piano wire has been used successfully at several institutions, but wires must be placed closely enough together to prevent wing entanglement. Exhibit surfaces should be nonporous for easy cleaning, and products like gunnite and other rough materials should be avoided because of the potential to cause abrasive injuries. Glass exhibit barriers permit visitors to get very close to bats, but do not allow transmission of smells and sounds as do wire meshes (Maunder and Goodwin, 2001). In some cases, zoos make a conscious choice to eliminate transmission of smell because flying foxes are very odoriferous. Glass is a more expensive exhibit barrier than wire mesh and also may be more difficult to clean, especially in inaccessible areas. Exhibit design can eliminate most of this problem by preventing bats from roosting where their fecal material can soil the glass, and by blocking flight paths that are directly adjacent to the glass.

In addition to nonporous, easily cleanable surfaces, exhibits must have good drainage to facilitate frequent cleaning. Cracks, crevices, and holes should be eliminated where cockroaches and other pests can shelter. Because bats may frequent the floor of an exhibit, pest control devices or substances must be protected from bat contact. In outside exhibits, the use of ½-in. (1.3-cm) x 1-in. (2.5-cm) mesh should exclude most snakes, or raccoons that may reach through mesh to grab bats. Moreover, wild bats must also be excluded.

FURNITURE AND PROPS

In the wild, many species of flying foxes choose roosting sites in the upper branches of emergent forest trees. In captivity, these roosting sites are easily replicated using natural and artificial perching materials, including vines, branches (Fig. 7-5), and large-diameter, non-fraying natural ropes. Avoid the use of fabric in flying fox enclosures due to the risk of entanglement. Captive bats are generally very adaptable and often choose the wire mesh within an enclosure as their favored roosting substrate, despite the availability of more natural roosting materials. This factor should be considered when designing flying fox enclosures, and care should be taken to avoid using wire mesh or other materials in areas where bat roosting is undesirable. Wire mesh can also be covered with a smooth material such as rigid plastic to exclude roosting bats. Conversely, layering wire mesh panels on ceilings and walls gives bats substrate that is easy to traverse and attractive for roosting (Fig. 7-6).

Fig. 7-5. Rodigues fruit bat (*Pteropus rodricensis*) exhibit in the Australia House at the Brookfield Zoo in Brookfield, Illinois. The abundant natural perching is positioned to allow open flight paths for bats. Photo courtesy of the Brookfield Zoo.

Fig. 7-6. Bats roosting on the wire mesh ceiling of the bat exhibit in the Australia House at the Brookfield Zoo in Brookfield, Illinois. Photo courtesy of the Brookfield Zoo.

EXHIBIT ENVIRONMENT

LIGHTING

Although most flying fox species are nocturnal or crepuscular, they often exhibit bouts of diurnal activity in the wild and in captivity. This characteristic gives zoo managers the flexibility to display bats successfully in both nocturnal and diurnal exhibits (Carroll, 1978). Flying foxes do not require any specific type of lighting, but dimming features that gradually switch between light and dark cycles are recommended. If housing flying foxes in a nocturnal setting, blue filters can be used on lights to simulate moonlight. If displaying flying foxes in a diurnal exhibit, bats can be encouraged to be active using timed feedings, enrichment, and training. Advertising feeding times to guests or coupling interpretive talks with bat activities can greatly enhance guests' opportunities to observe active bats. Enabling guests to clearly observe the bats may help to address many of the negative images associated with them. Observing features of flying foxes such as their large eyes, fox-like faces, and lack of tail helps dispel myths that all bats are blind, ugly "flying mice." Regardless of whether they are displayed in a nocturnal or diurnal setting, bats need exposure to both day and night cycles and do well in cycles ranging from 10/12 to 12/10 hours light/dark; maintaining them in a completely nocturnal setting is inappropriate.

TEMPERATURE AND HUMIDITY

Flying foxes are tropical species and require exhibit conditions that approximate those they would experience in the wild. The AZA Bat TAG *Fruit Bat Husbandry Manual* (1995) provides a detailed overview of husbandry parameters that are highlighted here. Most tropical bats require relatively high humidity, between 60-90%. These needs can typically be met through daily exhibit washing or using mister hoses in exhibits. Exhibit temperatures range between 70-90 °F (21-32 °C) with the optimal temperatures being 80 °F (27 °C), but flying foxes can tolerate a range of temperatures as long as they have access to warm areas in cool climates and shaded areas in outdoor exhibits in warm climates. In enclosures where temperatures routinely drop below 75 °F (24 °C), heat lamps or Brood-all Heaters® (see *Products Mentioned in Text*) should be provided to give bats supplemental heating spots. In climates where temperatures drop below 50 °F (10 °C), bats must have access to heated indoor space. Due to competition and aggression between bats, exhibits must have more than one warming area, and these areas should be accessible from a variety of pathways to prevent dominant bats from completely excluding sub-dominant conspecifics. Moreover, some bats may roost lower in an exhibit due to social issues, and therefore it is important to ensure that the ambient temperature in lower exhibit areas meets the same standards as the upper areas. Because bats may choose to roost directly beneath heat sources, heating elements should be placed in a way that bats can not contact them. Also, supplemental heating should not cause surrounding enclosure barriers to become too hot.

CAPTURE

One of the challenges of housing bats in large, open exhibits is the difficulty of capturing them for routine veterinary procedures, or for other reasons. This problem can be addressed by training bats to enter enclosed areas within the exhibit where they can be captured easily. In off-exhibit areas where appearance is not a crucial factor, cages can be designed so that bats roost within reach. The bat enclosures at the Lubee Bat Conservancy have ceilings that are 6½ ft. (2 m) high (see Fig. 7-4). This ceiling height enables easy access to the bats while allowing them to maintain a comfortable distance from their human caretakers.

SOCIAL ENVIRONMENT

Most flying fox species are highly social. In the wild, colonies can number in the thousands. Because of their social nature, fruit bats should not be housed alone for extended periods. However, these animals are territorial and haremic, and within a colony males set up and defend both feeding and roosting areas. Squabbles over territory are common, but seldom result in injury beyond occasional wing tears that heal quickly. Provision of numerous feeding and watering locations in an enclosure allow all bats to have adequate access regardless of territoriality. In heavily male-biased exhibits, the ratio of feeding to watering stations to bat may need to be as high as 2 to 1 to ensure all bats have adequate access. Overcrowding can result in escalated social tensions and lead to detrimental health impacts. Determining the appropriate density of bats for an exhibit is dependent on many variables including species, sex ratio, exhibit design, and management. When setting up a new exhibit, consult species managers or AZA Bat TAG members for assistance.

Exhibit design should allow bats to separate from each other, if they choose to do so. This can be accomplished by using exhibit features such as large branches, bamboo screens, and natural foliage that subdivide the exhibit and create visual barriers, yet do not prevent bats from moving freely around an exhibit when they desire. As mentioned above, because of social conflicts, some bats may choose to roost lower in an exhibit. It is important to provide roosting, feeding, watering, and warming sites for these bats, as well as easily accessible paths from the floor and lower levels to the higher levels of an exhibit. One way to accomplish this is to include vinyl-coated wire mesh or rope ladders that span the exhibit from the ceiling to the floor (see Fig. 9-26, Volume 3). In off-exhibit areas, bats that prefer to roost in enclosed spaces can be accommodated by using common materials. For example, a plastic container lined with wire mesh and suspended from the ceiling can provide an easy, inexpensive, and effective site for bat roosting (Fig. 7-7).

MIXED-SPECIES EXHIBITS

Displaying flying foxes in mixed-species exhibits can provide excitement for guests, challenges for animals, and conserve exhibit space for zoos. However, care must be taken to choose appropriate species and to monitor them carefully so that there are no

Fig. 7-7. A simple way to provide an enclosed roosting space preferred by some species of flying foxes. Here the top of a plastic container is lined with wire mesh and suspended from the enclosure ceiling. Photo courtesy of Dana LeBlanc, © Lubee Bat Conservancy.

negative interactions. Flying foxes frequent the ground to eat dropped food; therefore, species that could injure or prey on grounded bats should not be housed with them. In indoor flying fox enclosures, it is best to avoid placing animals with them that require heavy bedding; such substrates can become a medium for fungal growth such as *Histoplasma capsulatum.* Moreover, concerns for disease transmission suggest that hoofstock, primates, and swine should not be exhibited with flying foxes (see Trupkiewicz *et al.,* Volume 1). Riger *et al.* (2001) compiled a list of species that have been housed successfully with bats in zoo collections (see also Table 7-1).

REGULATORY CONSIDERATIONS

In addition to aesthetic concerns, legal concerns influence the design of bat exhibits and the maintenance of captive bats. For information on the regulation of bats in captivity in Australia, Canada, Europe, and the United States, see *Legislation* (Chapter 1).

ADDITIONAL INFORMATION

The AZA Bat TAG *Fruit Bat Husbandry Manual* (1995) is an excellent source of additional information on maintaining and displaying flying foxes. For a copy, contact Bat TAG Chair, Steve Wing (steven.wing@louisvilleky.gov) or the Association of Zoos and Aquariums (http://www.aza.org/).

Flying foxes are under-appreciated yet potentially spectacular exhibit animals. Displayed in appropriate settings that allow for a full range of natural behaviors, flying foxes can rival any of the charismatic megavertebrates in popularity and educational opportunities.

PRODUCTS MENTIONED IN TEXT

Brood-all Heater® (Product #C14577N): Nasco®, 901 Janesville Avenue, Fort Atkinson, WI 53538 USA; http://www.enasco.com/product/C14577N.

REFERENCES

AZA Bat TAG *Fruit Bat Husbandry Manual.* 1995. Association of Zoos and Aquariums Bat Taxon Advisory Group. 39 pp; *for a copy, contact Bat TAG Chair, Steve Wing (steven.wing@louisvilleky.gov), or the Association of Zoos and Aquariums (http://www.aza.org/).*

Carroll, J.B. 1978. Behavioural observations on the Rodrigues fruit bat *Pteropus rodricensis* following a move to new accommodation and reversal of the light/dark regime. *Dodo* (Journal of the Jersey Wildlife Preservation Trust), 15:52-60.

Maunder, L. and Goodwin, I. 2001. Husbandry of Rodrigues fruit bats (*Pteropus rodricensis*): A new exhibit to Marwell Zoological Park. *Animal Keepers' Forum*, 28:448-449.

Mellen, J., Bratton, D., and Houck, B. 1998. Introduction of Rodrigues fruit bats (*Pteropus rodricensis*) into a mixed bat species exhibit. Metro Washington Park Zoo (now known as the Oregon Zoo). Portland, Oregon, USA; pdf report – 44 pp.

Pye, J.D. 1967. Bats. In: *Universities Federation for Animal Welfare Handbook on the Care and Management of Laboratory Animals*, 3rd ed. (Lane-Petter, W., Worden, A.N., and Hill, B.F., eds.). E. & S. Livingston, Edinburgh, Scotland. Pp. 491-501.

Riger, P., Bear-Hull, D., and Harmon, L. 2001. Use of chiroptera in mulit-species exhibits. *Animal Keepers' Forum*, 28:434-438.

Mixed Bat Species at the Memphis Zoo

Julie Batty, Sandi Shoemaker, Mary Manson, Emily Schultz, Janet Cooper, Doug Dalton, Ashley Averill, and Lisa Hailey

INTRODUCTION

In 1995 the Memphis Zoo in Memphis, Tennessee opened a new exhibit called "Animals of the Night," featuring small nocturnal mammals. At this writing, the exhibit displays 27 species of small mammals around a centrally located mixed-species bat exhibit that houses 2 male Jamaican fruit bats (*Artibeus jamaicensis*; Fig. 7-8), a breeding colony of 40 greater bulldog bats (*Noctilio leporinus*; Fig. 7-9), a primarily single-sex female colony (with a few hand-reared neutered males) of 186 Egyptian fruit bats (*Rousettus aegyptiacus*; Fig. 7-10), and a breeding colony of 100 Seba's short-tailed bats (*Carollia perspicillata*; Fig. 7-11). For information on the natural history of these bats, see Albayrak *et al.* (2008), Bárquez *et al.* (1999), Cloutier and Thomas (1992), Fleming (1988), Handley *et al.* (1991), Hood and Jones (1984), Kwiecinski and Griffiths (1999), and Ortega and Castro-Arellano (2001).

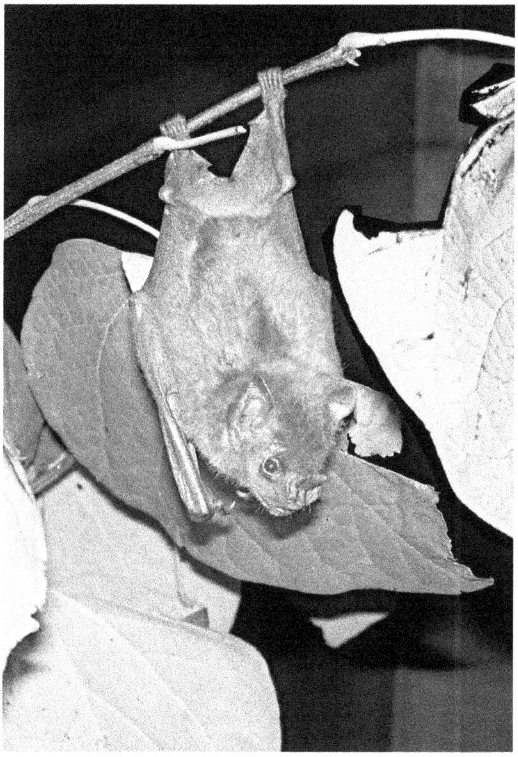

Fig. 7-8. Jamaican fruit bat (*Artibeus jamaicensis*) at Memphis Zoo. Photo courtesy of Angela Price.

Fig. 7-9. Greater bulldog bats (*Noctilio leporinus*) at Memphis Zoo. Photo courtesy of Angela Price.

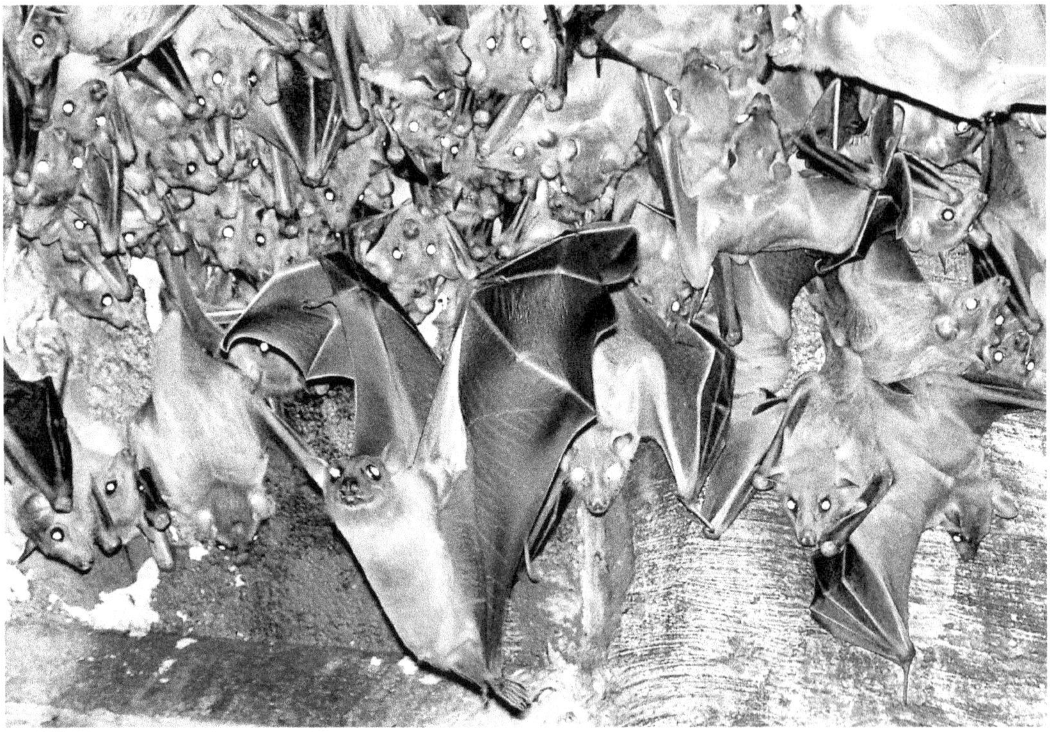

Fig. 7-10. Egyptian fruit bats (*Rousettus aegyptiacus*) at Memphis Zoo. Photo courtesy of Angela Price.

Fig. 7-11. Seba's short-tailed bats (*Carollia perspicillata*) at Memphis Zoo. Photo courtesy of Angela Price.

ABOUT THE EXHIBIT

The bat exhibit (Fig. 7-12 top photo) features two sections, a flight area, and a simulated cave (Fig. 7-13) where the Egyptian fruit bats, bulldog bats, and Seba's short-tailed bats prefer to roost. However, the three species do not roost in the same area of the cave, and the short-tailed bats have chosen a spot where they are rarely seen by the public. The Jamaican fruit bats can be found roosting throughout the exhibit flight area (Fig. 7-12 bottom photo), where they are viewed easily by zoo visitors.

Fig. 7-12. Mixed bat exhibit at Memphis Zoo. Top) exhibit viewed from public concourse; arrow points to keeper entering exhibit. Bottom) exhibit flight area. Photos courtesy of Susan M. Barnard.

Fig. 7-13. Cave portion of mixed bat exhibit. Photo courtesy of Angela Price.

The flight area is 48 ft. (14.6 m) long by 7 ft. (2.1 m) wide by 8 ft. (2.4 m) high. The cave is 18 ft. 6 in. (5.6 m) long by 16 ft. 8 in. (5.1 m) wide by 8 ft. (2.4 m) high. Natural grapevines and 1-in. (2.54-cm) square wire mesh are provided for climbing and roosting areas for the bats. The cement walls and artificial trees within the exhibit are also used by the bats for climbing and roosting. The bulldog bats are provided with wooden bat houses (Fig. 7-14). Plastic-coated wire inside and outside the boxes assists them in climbing.

EXHIBIT ENVIRONMENT

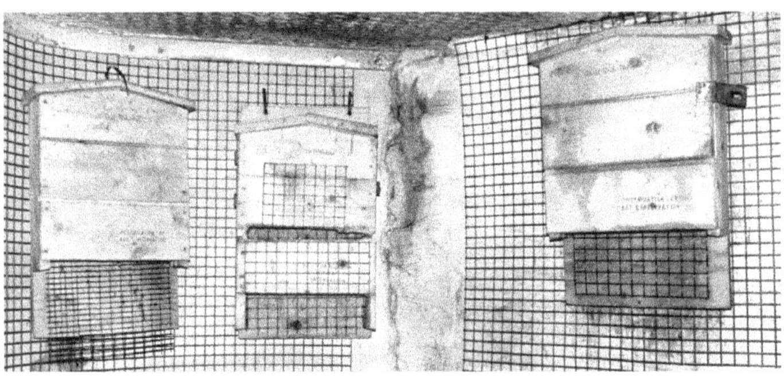

Fig. 7-14. Bat boxes provided for greater bulldog bats (*Noctilio leporinus*). Plastic-coated wire mesh is attached inside and outside the boxes to assist the bats in climbing. Photo courtesy of Angela Price.

The climate is maintained at an average temperature of 80 °F (26.7 °C) and 60% relative humidity, both of which vary seasonally. Heat and humidity are more difficult to maintain during winter. Exhibit lighting is on a reverse light cycle. The lights are set on a timer, which are turned off at approximately 10:30 A.M. [1030 hrs.] and on at 10:30 P.M. [2230 hrs.]). The lighting is adjusted for the seasons. Night time is created with heavily filtered red and blue lights, allowing just enough light for the bats to be viewed by zoo visitors. Standard fluorescent lights are used to create daylight, but the bats are always provided areas of darkness which give them choices to light exposure.

EXHIBIT MAINTENANCE

The exhibit can be entered from either the cave or flight end, but keepers tend to enter through the flight area (see Fig. 7-12 top photo). The exhibit walls and floor are rinsed with water, and the windows are cleaned daily. No chemicals are used for cleaning when bats are present in the exhibit. However, once a year all the bats are removed so the exhibit can be steam cleaned, and the entire exhibit is disinfected at this time. The animals have become accustomed to the daily cleaning routine and show no signs of stress during this activity.

ADDITIONAL NOTES

The bats are fed twice daily. The first feeding is around 10:00 A.M. (1000 hrs.) and the second occurs around 3:00 P.M. (1500 hrs.). Because the Animals of the Night building is on a reverse light cycle, both feedings are "at night" for the bats. For detailed information on bat diets, including the species exhibited at the Memphis Zoo, see Chapter 2, Volume 3.

Seba's short-tailed bats breed throughout the year, with a birth peak in late spring and early summer. The bulldog bats have exhibited aggression among themselves and toward the Seba's short-tailed bats. This aggression by the bulldog bats appears to be limited mainly to male juveniles of both species. High-ranking male bulldog bats occasionally force juvenile males out of the bat houses. Most of them roost outside the bat houses where their colony mates reside, but a male juvenile bulldog bat occasionally roosts with other species. There has never been aggression observed between bulldog bats and Egyptian or Jamaican fruit bats.

ACKNOWLEDGMENTS

We thank Matt Thompson, Angela Price, Steve Reichling, and Dietrich Schaaf for their helpful comments on the manuscript.

REFERENCES

Albayrak, I., Aşan, N., and Yorulmaz, T. 2008. The natural history of the Egyptian fruit bat, *Rousettus aegyptiacus*, in Turkey (Mammalia: Chiroptera). *Turkish Journal of Zoology*, 32:11-18.

Bárquez, R.M., Mares, M.A., and Braun, J.K. 1999. *The Bats of Argentina*. Special Publications, Museum of Texas Tech University, No. 42. 275 pp.

Cloutier, D. and Thomas, D.W. 1992. *Carollia perspicillata. Mammalian Species*, 417:1-9.

Fleming, T.H. 1988. *The Short-tailed Fruit Bat: A Study in Plant-Animal Interactions*. The University of Chicago Press, Chicago, IL. 365 pp.

Handley, Jr., C.O., Wilson, D.E., and Gardner, A.L. 1991. *Demography and Natural History of the Common Fruit Bat, Artibeus jamaicensis, on Barro Colorado Island, Panamá*. Smithsonian Institution Press, Washington, D.C. 173 pp.

Hood, C.S. and Jones, Jr., J.K. 1984. *Noctilio leporinus. Mammalian Species*, 216:1-7.

Kwiecinski, G.G. and Griffiths, T.A. 1999 .*Rousettus egptiacus. Mammalian Species*, 611:1-9.

Ortega, J. and Castro-Arellano, I. 2001. *Artibeus jamaicensis. Mammalian Species*, 662:1-9.

Mixed Species at the Biôdome de Montréal

Chantal Routhier

INTRODUCTION

Over the past few decades, bats have become very popular zoo exhibit animals. Also during this time, the philosophy behind exhibiting animals in zoological institutions has evolved. The choice of animals for display is no longer a matter of a curator's or director's personal preference (Fascione, 1996). Today, the emphasis is placed on public education, research, and conservation. The majority of zoos attempt to display animals in simulated natural habitats, as single or mixed species exhibits.

The Biôdome de Montréal in Québec, Canada comprises four ecosystems featuring the Gulf of St. Laurence islands, North and South Poles (Labrador coast and sub-Antarctic islands), Laurentian maple forest, and tropical forest (Fig. 7-15). In the latter, Seba's short-tailed bats (*Carollia perspicillata*), Jamaican fruit bats (*Artibeus jamaicensis*) and Pallas's long-tongued bats (*Glossophaga soricina*), are featured in a cave and in the open area of the forest.

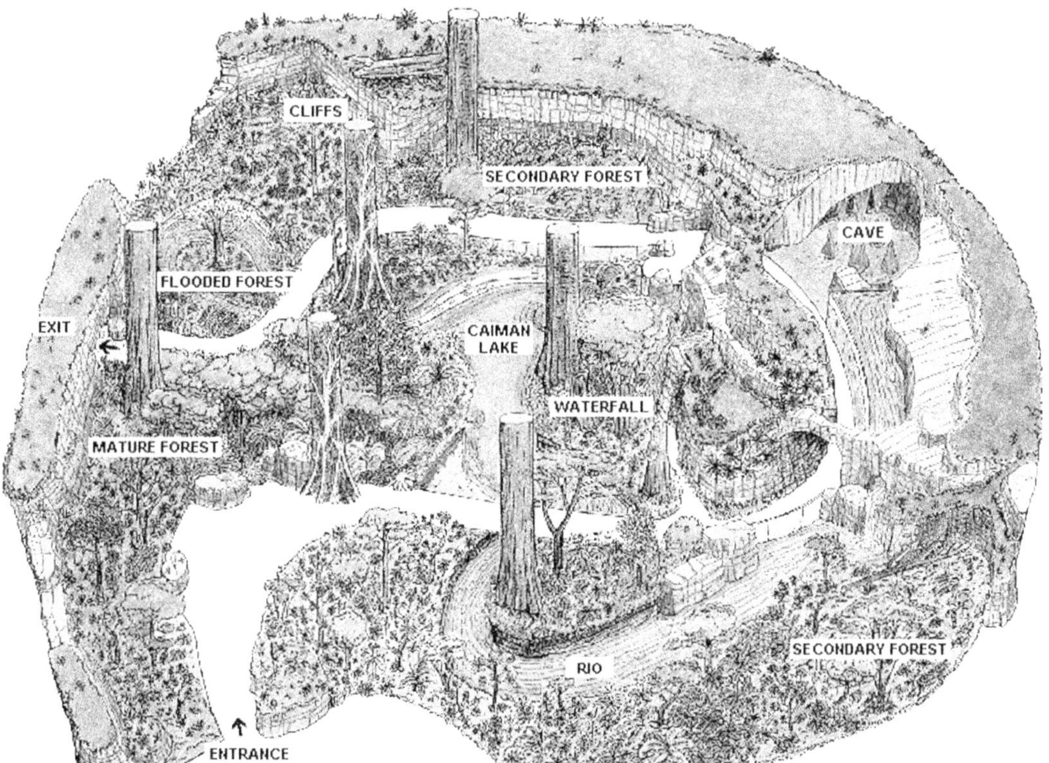

Fig. 7-15. Schematic illustration of the tropical forest exhibit at Biôdome de Montréal. Illustration courtesy of Francine Montor.

ABOUT THE EXHIBITS

The open part of the tropical forest (Fig. 7-16) houses mixed taxa (Table 7-1) with 20 male *C. perspicillata* and 10 male *G. soricina*. The forest is 27,976 ft.^{2} (2,600 m^2) in size, and the bats have access to the entire area.

When feeders are placed near visitors, they can be as close as 4ft. (1.2 m) to the bats. Because the forest is filled with live plants, visitors may sometimes be pleasantly entertained as the bats feed on the many tropical flowers (see also Appendix 2-4, Volume 3 for the diets of free-ranging *C. perspicillata* and *G. soricina*).

A simulation of a calcareous cave in Costa Rica is traversed by a glassed-in corridor (Fig. 7-17), dividing it into two exhibits that are referred to as the large and small caves. The surface area of the large cave is 323 ft.^{2} (30 m^2) and displays 64 male *C. perspicillata* and 51 male *A. jamaicensis*. The small cave is 194 ft.^{2} (18 m^2) and features 30 female *G. soricina*. The public can view bats in one or in both caves, and because the bats roost near the viewing glass, visitors can see them easily.

EXHIBIT ENVIRONMENT

TEMPERATURE AND HUMIDITY

Temperature and humidity are computer controlled throughout the Biôdome. In the tropical forest, temperatures range between 70-77 °F (21-25 °C) in winter and between 73-82 °F (23-28 °C) in summer. The cave temperature is maintained between 70 °F (21 °C) and 77 °F (25 °C) year-round. Because the tropical plants growing in the forest require high humidity, it ranges from 60% to 70% in winter, and between 65% and 85% in summer. The caves are at 80% to 100% humidity year-round. Cave humidity is at 100% between 6 A.M. (0600 hrs.) and 9 A.M. (0900 hrs.) while they are being cleaned.

Fig. 7-16. Visitors walking in the mature forest area within the tropical forest. Photo by the author.

Table 7-1. Animals at Biôdome de Montréal sharing the tropical forest with Seba's short-tailed bats (*Carollia perspicillata*) and Pallas's long-tongued bats (*Glossophaga soricina*)

Mammals	
Capybara	*Hydrochoerus hydrochaeris*
Cotton-top Tamarin	*Saguinus oedipus*
Goeldi's Monkey	*Callimico goeldii*
Golden Lion Tamarin	*Leontopithecus rosalia*
Southern Two-toed Sloth	*Choloepus didactylus*
Birds	
American White Ibis	*Eudocimus albus*
Blue Dacnis	*Dacnis cayana*
Blue-necked Tanager	*Angara Mexicana*
Brazilian Tanager	*Ramphocelus bresilius*
Brazilian Teal	*Amazonetta brasiliensis*
Green Honeycreeper	*Chlorophanes spiza*
Grey-winged Trumpeter	*Psophia crepitans*
Hyacinth Macaw	*Anodorhynchus hyacinthinus*
Military Macaw	*Ara militaris*
Northern Helmeted Curassow	*Pauxi pauxi*
Roseate Spoonbill	*Platalea ajaja*
Rufous-brested Hermit	*Glaucis hirsutus*
Saffron Finch	*Sicalis flaveola*
Scarlet Ibis	*Eudocimus ruber*
Spectacled Parrolet	*Forpus conspicillatus*
Sunbittern	*Europyga helias*
Wattled Jacana	*Jacana jacana*
White-necked Jacobin	*Florisuga mellivora*
Reptiles	
Green Iguana	*Iguana iguana*
Amphibians	
Cane Toad	*Bufo marinus*
Green and Black Poison Dart Frog	*Dendrobates auratus*

Fig. 7-17. Visitors walking between the tropical forest's small and large bat caves. Photo by the author.

LIGHTING

Fig. 7-18. LED lights used to illuminate the small and large caves in the tropical forest. Photo by the author.

The tropical forest sits under a glass dome and receives natural light and supplemented artificial light. In the cave exhibit, the light cycle has been reversed to encourage intense bat activity during the day when visitors are present. The cave lights are computer controlled LED lights (Fig. 7-18) that can be programmed to obtain lighting that ranges from daylight to moonlight. The lighting in the tropical forest and caves are on a 12:12 light:dark cycle with a choice of dimly lit roosting areas for use during the light cycles.

EXHIBIT MAINTENANCE

Bats in the tropical forest roost in the same areas of the exhibit most of the time. A broom and shovel are used to sweep up feces on a daily basis at roost sites. Once a week all walls and other surfaces in the forest are washed with a hose.

The surfaces of the cave are constructed of polyurethane, making daily cleaning easy. All surfaces, including floors and glass, are rinsed with a hose and sprayed with Nu-Bioscrub™ foam using a HydroFoamer® sprayer (Fig. 7-19a,b). Surfaces are then scrubbed (Figs. 7-19c,d) with a Plastimo boat brush to remove urine and feces, and rinsed a second time.

ADDITIONAL NOTES

One feeding a day is presented to the bats at approximately 9 A.M (0900 hrs). Clean water is also presented once daily. For information on the diets fed to the bats, see Delorme *et al.*, Chapter 2, Volume 3).

Bats are inventoried once a year; they are captured in the caves with a small bird net and in the forest with a mist net (Fig. 7-20).

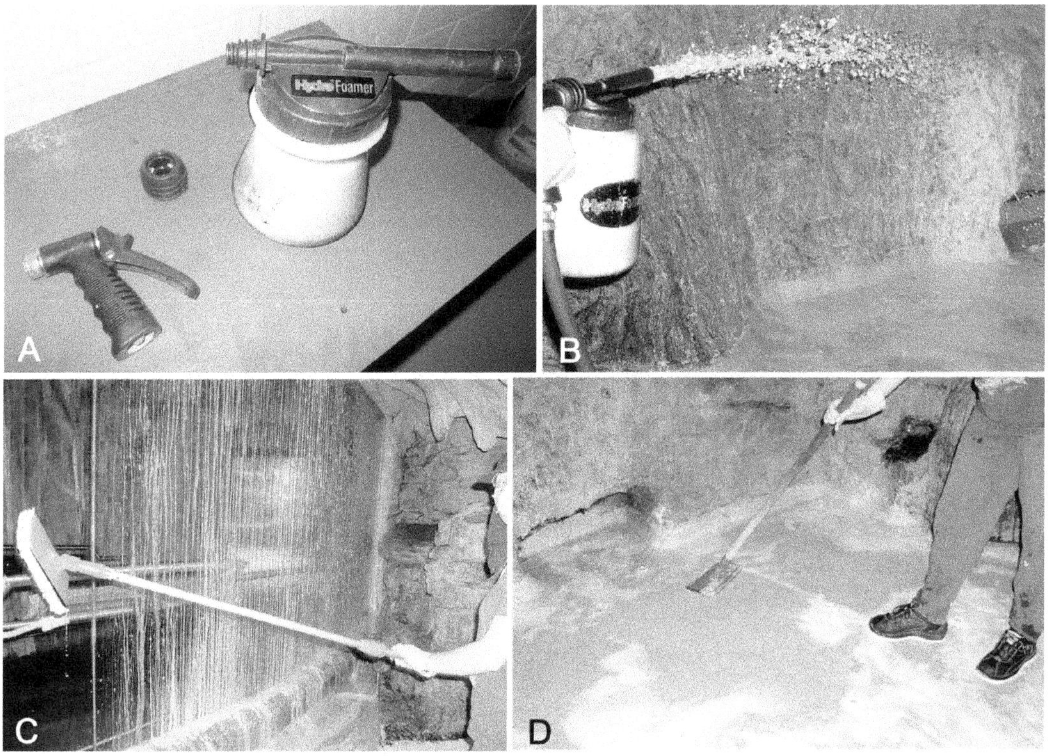

Fig. 7-19. A) HydroFoamer® sprayer; B) Nu-Bioscrub™ sprayed from the HydroFoamer® sprayer; C) exhibit glass in cave exhibit being scrubbed with a Plastimo boat brush; D) cave floor being scrubbed with a Plastimo boat brush. Photos by the author.

Fig. 7-20. Bats being captured and removed from a mist net in the tropical forest tunnel entrance leading to the cave exhibit. Photo by the author.

PRODUCTS MENTIONED IN TEXT

HydroFoamer® Sprayer: Hydro Systems Company, 3798 Round Bottom Road, Cincinnati, OH 45244 U.S.A.; http://www.hydrosystemsco.com/brands/hydro/item/hydrofoamer-sprayer.html.

LED lights: Philips Color Kinetics Headquarters, 3 Burlington Woods Drive, 4th Floor, Burlington, MA 01803 U.S.A.; http://www.colorkinetics.com/ls/rgb/colorblast12/.

Nu-Bioscrub™: Innu-Science Company, 1777 Nobel Street, bureau F, Sainte-Julie Québec, J3E 1Z6 Canada; http://www.innu-science.com/eng/Products/NU-BIOSCRUB.aspx

Plastimo Boat Brush: Marine Mega Store, Ltd., Capstan House, Hamble High Street, Hamble le Rice, Hampshire, SO31 4HA United Kingdom; http://www.marinemegastore.com/product.asp?pf_id=PLA_10574&jump=8&cat_id=MAIBRU003.

ACKNOWLEDGMENTS

I thank my colleagues in both the live collection and maintenance departments for their assistance with the manuscript. I also thank Dietrich Schaaf for his helpful manuscript edits.

REFERENCES

Fascione, N. 1996. The evolving role of American zoos in bat conservation. *Bats,* 14(1):8-14.

Pallas's Long-tongued Bats (*Glossophaga soricina*) at Zoo Atlanta

Susan M. Barnard

INTRODUCTION

Zoo reptile departments occasionally display bats, in particular the common vampire bat (*Desmodus rotundus*). A small bat species like the vampire bat is usually chosen because most enclosures built originally for reptiles are relatively small. The staff at Zoo Atlanta in Atlanta, Georgia wanted a bat exhibit that would offer more activity than is typical of most reptiles, although as is the case for reptiles, they also wanted to feature bats in a natural setting.

In September, 2001, Biôdome de Montréal in Québec, Canada shipped 15 male Pallas's long-tongued bats (*Glossophaga soricina*) to Zoo Atlanta. Pallas's long-tongued bats (Fig. 7-21) are found in forests, as well as the rural and urban areas of Central and South America (Barquez *et al.*, 2010). They are nectarivorous, weigh approximately 10 g, and roost in caves, tunnels, and houses. The small body size of this species and its natural, close proximity to human habitation made it a perfect animal to display in Zoo Atlanta's reptile facility.

Fig. 7-21. Pallas's long-tongued bats (*Glossophaga soricina*) roosting under the exhibit farm house (see also Fig. 7-23 top photo and Fig. 7-24). Photo courtesy of Joe Sebo, © Zoo Atlanta.

ABOUT THE EXHIBIT

The exhibit (Fig. 7-22) that was redesigned for the bats measured 16 ft. (4.9 m) wide by 7 ft. (2.1 m) deep by 7 ft. (2.1m) high. The exhibit theme was a small farm in Central America depicting a farm house (Fig. 7-23a, Fig. 7-24) and a variety of native tropical plants (Fig. 7-23b) growing in the farmer's yard. A diorama of crops growing on a hill-side was painted on the back wall. Simulated soil made of concrete extended from the diorama into the enclosure. The enclosure substrate itself contained approximately 2.5 ft. (0.76 m) of soil to grow tropical plants. Leaf mulch was placed on top of the soil to help maintain humidity in the exhibit and to give a natural appearance to the substrate. Two 400-W mercury vapor lamps and several two 4-ft. (1.2-m) fluorescent lamps were placed above the plants to provide illumination and promote growth. In addition, moisture was provided with very fine-spray misters, which were turned on for 15 minutes twice daily. The farm house (Fig. 7-24) was constructed of wood planks made to look old. Garden

Fig. 7-22. *Glossophaga soricina* exhibit at Zoo Atlanta. Photo courtesy of Joe Sebo, © Zoo Atlanta.

Fig. 7-23. A) view of left side of *Glossophaga soricina* exhibit showing farm house. Lower left arrow points to where visitors can view roosting bats. Upper right arrow points to where the bats roost under the house. B) view of right side of exhibit showing flight area and where the bats feed (left arrow). Right arrow points to the misters over the tropical plants. Photo courtesy of Joe Sebo, © Zoo Atlanta.

tools were propped against the side of the house to add realism to the exhibit and also for working with the live plants.

Fig. 7-24. Farm house. Photo courtesy of Joe Sebo, © Zoo Atlanta.

EXHIBIT ENVIRONMENT

TEMPERATURE AND HUMIDITY

The average temperature in the enclosure was approximately 80 °F (26.7 °C). Although humidity was not measured, it was high due to daily misting of the plants and moisture held in the natural substrate. Seemingly to enjoy misting, the bats flew back and forth through the spray while the misters were on. They were also active when fresh food was placed in their enclosure.

LIGHTING

The exhibit was maintained under a natural light cycle. Lights were turned on at 9:00 A.M. (0900 hrs.) and off when the last keeper left the building (5:30-6:00 P.M. [1750-1800 hrs.]). Fresh food was placed in the exhibit while the lights were on. In addition to the mercury vapor and fluorescent lamps, natural light entered the bat enclosure from the 60-ft. (18.1-m) wide crocodilian solarium facing it. Observations of *G. soricina* at Zoo Atlanta show that this species is active and does well under bright lights, providing they have a dimly lit retreat for roosting.

EXHIBIT MAINTENANCE

Bat feces that fell on the plants were watered into the soil over time by the misters and served as fertilizer. Feces dropped under the farm house, where the bats typically roosted, were raked up once a week. The viewing glass was cleaned every other day, and fruit that had fallen to the ground was also raked up and removed from the enclosure.

Twice a day, food and water were placed on hooks that could be moved with a pulley system. This system allowed the bats to be fed without keepers entering the exhibit, which greatly reduced stress on the animals. Although the bats never attempted to escape when the door to their enclosure was open (they flew under the house), a screened-in keeper cage was built around the door for added security.

ACKNOWLEDGMENTS

I thank Dietrich Schaaf for his helpful comments on the manuscript.

REFERENCES

Barquez, R., Perez, S., Miller, B. and Diaz, M. 2008. *Glossophaga soricina*. In: IUCN 2010. IUCN Red List of Threatened Species. Version 2010.4; http://www.iucnredlist.org, accessed June 2011.

Australian Ghost Bats (*Macroderma gigas*) at the Territory Wildlife Park

Steve Templeton and Fiona Colquhoun

INTRODUCTION

The ghost bat (*Macroderma gigas*; Fig. 7-25) is the largest microchiropteran in Australia (Nowak, 1994). Adults weigh between 130 to 170 g (Kulzer *et al.*, 1984; Pirlot and Nelson, 1980; Taylor, 1984), and their large body size makes them ideal animals for public display.

Since 1989, ghost bats have been displayed in the nocturnal house at the Territory Wildlife Park (TWP), Darwin, Australia. Currently, the group comprises 3 males and 6 females. The founders were wild-caught adults. After the bats passed quarantine and were known to be feeding themselves, they were transferred to the nocturnal house exhibit.

Fig. 7-25. Wild ghost bats (*Macroderma gigas*) in the Douglas Daly region of Northern Territory, Australia. Photo courtesy of Damien Stanioch.

ABOUT THE EXHIBIT

CAVE

The exhibit (Fig. 7-26) is a simulated cave, measuring 9.8 ft. (3 m) long by 6.6 ft. (2 m) deep by 6.6 ft. (2 m) high, with artificial rock on all surfaces. There is an artificial pond on the floor, with a thermostatically controlled heating element under it. The cave temperature ranges between 86 °F (30 °C) and 90 °F (32 °C), and the humidity is generally above 60%.

Bats are viewed through a window (Fig. 7-27) that slopes to prevent the animals from hanging against it, thereby keeping the glass free of urine and feces. Visitors can be as close as 1.6 ft. (0.5 m) to the roosting bats.

Unlike other animals in the nocturnal house, which are displayed under a reverse light cycle, the ghost bats are displayed under a natural cycle. Lighting in the cave is provided by an 8-W fluorescent tube with a partial black filter. The light is out of view to the public, near the floor of the exhibit below the window (Fig. 7-28). As the bats hang from the ceiling or walls, it illuminates their faces and upper bodies (see Fig. 7-27).

The bats do not react when visitors tap on the glass, as visitors tend to do, but when keepers enter the exhibit, they try to escape through the light lock into the outside flying area (see Fig. 7-26). The exhibit is cleaned daily.

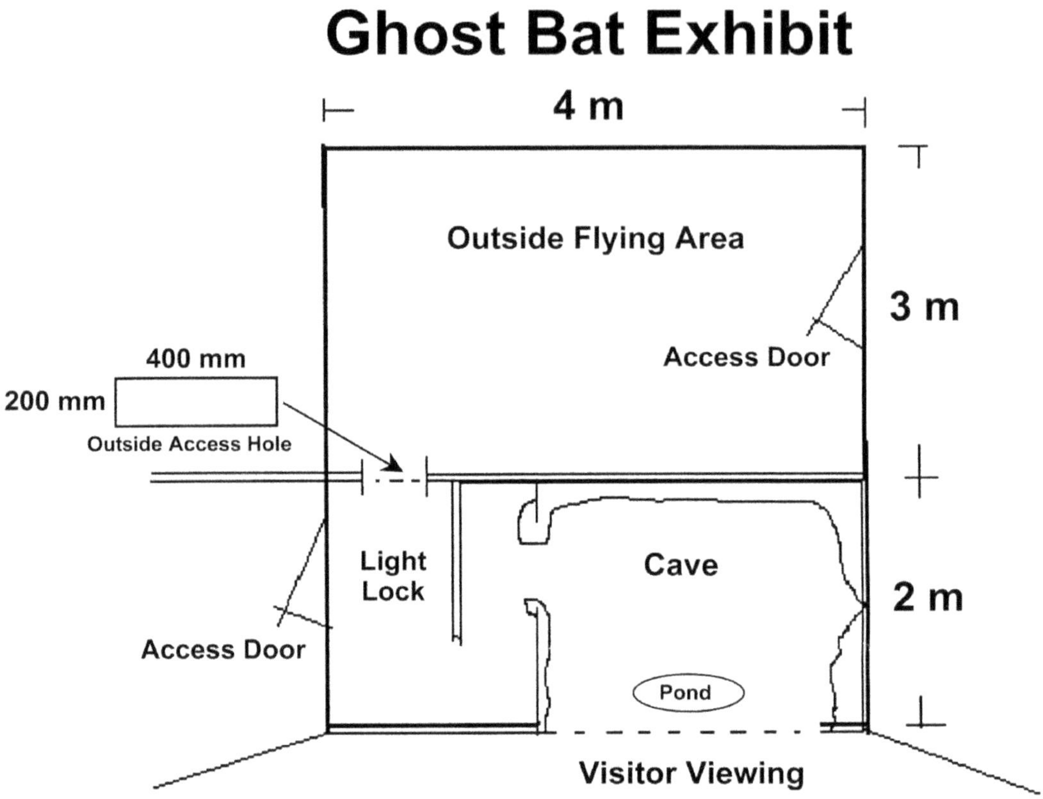

Fig. 7-26. Schematic diagram of the ghost bat exhibit at the Territory Wildlife Park. Illustration courtesy of the Territory Wildlife Park.

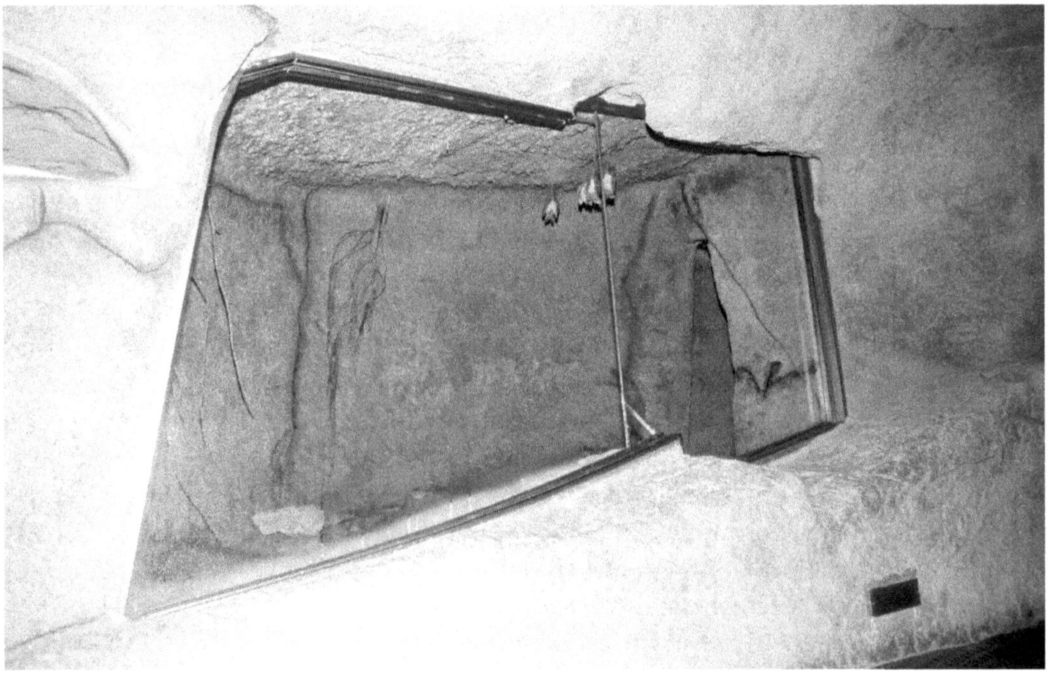

Fig. 7-27. Ghost bat cave and public viewing area. Photo courtesy of Damien Stanioch.

Fig. 7-28. Inside cave showing light (arrow). Photo courtesy of Damien Stanioch.

OUTSIDE FLYING AREA

The bats always have access to the outside flying area via the light lock, which is painted non-reflective black to prevent light entering the exhibit. The outside flying area (Fig. 7-29) is 13.1 ft. (4 m) long by 9.8 ft. (3 m) wide by 6.6 ft. (2 m) tall. The walls are constructed of compressed cement fiber sheet with wire mesh on the lower 0.98 ft. (30 cm) for ventilation. Visitors can not view bats in the flying area. The roof is composed of 0.39-in. x 0.39-in. (10 mm x 10 mm) wire mesh.

The flying area is cleaned daily. Uneaten food is removed, water is replenished, and the area is spray washed with a hose. A high-pressure cleaner is used regularly during the wet season to remove algae and mold from the ground and walls.

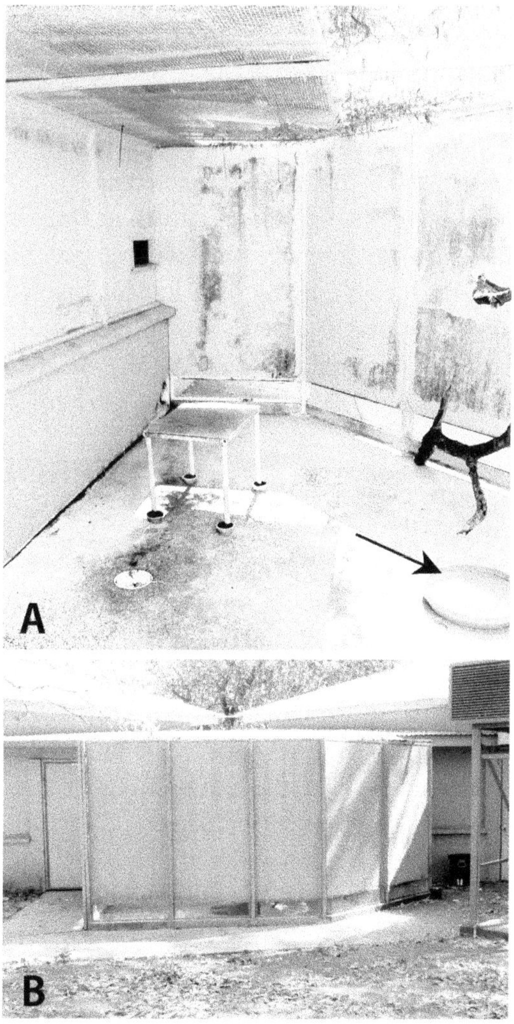

Fig. 7-29. Outside flying area where ghost bats feed. A) interior view; arrow points to water dish. B) exterior view. Photos courtesy of Damien Stanioch.

FOOD PRESENTATION

Fig. 7-30. Method of food presentation to ghost bats in outside flying area. Photo courtesy of Damien Stanioch.

Bats are fed daily late in the afternoon in the outside flying area. The diet they are offered, in order of food preference, includes frozen mice, day-old chickens, or small quail. Although the food is placed in the outside flying area at approximately 4 P.M. (1600 hrs.), the bats do not begin feeding until 7 P.M. (1900 hrs.). At this time they become active, leaving the cave to eat in the outside area. By the time they feed, the food has thawed (ambient temperature is around 82.4-86 °F [28-30 °C]) but is still fresh. Frozen food is placed on a small table (Fig. 7-30), the legs of which are in bowls of water to prevent ants from gaining access. Because of the high temperatures in Darwin, the recommended protocol for thawing food (i.e., slowly in a refrigerator) will not provide the bats with the freshest food possible.

The outside flying area has an insect-attracting lamp on the ceiling. In addition to attracting insects for the bats to catch and eat, it also attracts other prey such as geckoes, especially Asian house geckoes (*Hemidactylus frenatus*). Water is supplied in a large, shallow dish (see Fig. 7-29a) that is 19.7 in. (50 cm) by 19.7 in. (50 cm) by 0.79 in. (2 cm) deep.

REPRODUCTION

All females have bred in captivity, giving birth at the end of August and the beginning of September just prior to the monsoonal wet season. This reproductive behavior corresponds with breeding in the wild (Churchill, 1998; Richards and Hand, 2002). The fact that the captive bats breed at the same time as the wild ones may be due to their having access to ambient environmental conditions in the outside flying area. The success rate for dams raising young to weaning age varies from 50-100%. Improved success more recently is most likely due to the introduction of newly captured wild bats in 2009. Males and females share the exhibit because space is limited, and in the wild males have been observed in maternity colonies (Churchill, 1998, Richards and Hand, 2002). To reduce aggression between males and to allow one male at a time to breed, two of the three males are implanted with the contraceptive Suprelorin® (deslorelin). The hormonal implants are available in two strengths: 4.7 mg lasting 6 months, and 9.4 mg lasting 12 months respectively in dogs. In ghosts bats, the 9.4-mg implants appear to last about 9 months and are rotated so that all males eventually sire offspring.

CAPTURE TECHNIQUES

When it becomes necessary to catch the ghost bats, the least stressful and most convenient way is to herd them from the cave toward the outside access hole (see Fig. 7-26) that has been covered with a butterfly net (Fig. 7-31). As they attempt to escape to the outside flying area, they are captured in the net (Fig. 7-32). A butterfly net is used because its fine mesh is less likely to entangle the bats. As they are caught (Fig. 7-33), bats are placed in cloth bags that are suspended in a box built to hold them.

Fig. 7-31. Ghost bats are captured with a butterfly net as they are herded out of the cave to the outside flying area. Photo courtesy of Grant Husband.

Fig. 7-32. Ghost bat flying into butterfly net (arrow). Photo courtesy of Grant Husband.

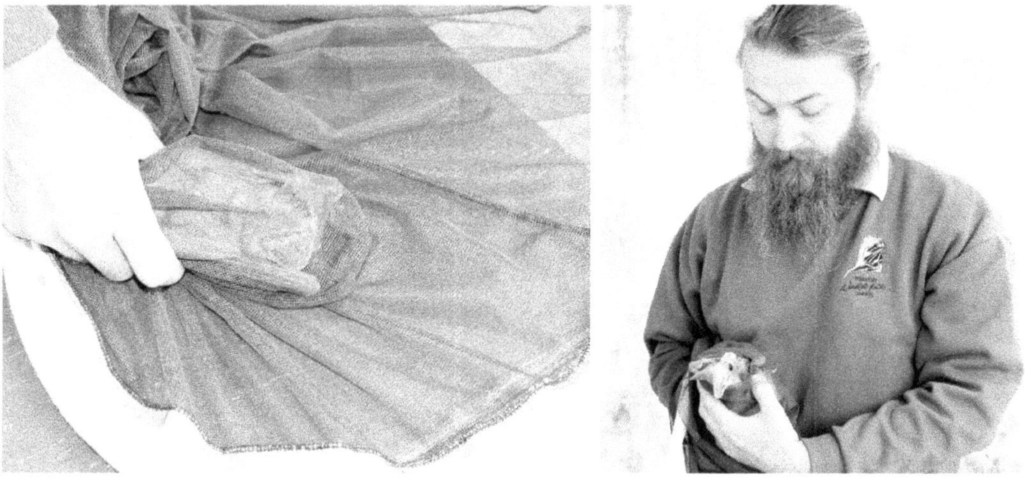

Fig. 7-33. Captured ghost bat. Photo courtesy of Grant Husband.

LONGEVITY

Life spans of ghost bats are not well known. On 17 August 1987, staff at TWP collected 2 free-ranging, female adult bats, and in 2001, they were transferred to Alice Springs Desert Park (ASDP). One died on 3 July 2003 and the other on 19 August 2004, making them over 16 and 17 years of age (Kim Branch, ASDP). Weigl (2005) reported a male and female ghost bat at the Perth Zoo living for 22 yrs, 7 mos. and 21 years, 2 mos. respectively.

AUSTRALIAN BAT LYSSAVIRUS

Staff at the Territory Wildlife Park who work with bats are vaccinated against Australian Bat Lyssavirus. Warning signs (Fig. 7-34) are placed in all bat areas to ensure that unvaccinated staff do not enter bat enclosures.

Fig. 7-34. Warning sign posted in all bat areas at the Territory Wildlife Park. Photo courtesy of Damien Stanioch.

ADDITIONAL NOTES

All bats at the Territory Wildlife Park are implanted with Passive Integrated Transponder (PIT) tags. Daily records pertaining to the husbandry and maintenance of each individual are maintained on the Animal Record Keeping System (ARKS) of the International Species Information System (ISIS).

The bats are examined at 6-month intervals for parasites. If treatment is needed, Cydectin® (5 mg/ml Moxidectin) and/or Profender® (21.4 mg/ml Emodepside; 85.8 mg/ml Praziquantel) are administered topically between the shoulder blades.

PRODUCTS MENTIONED IN TEXT

Cydectin®: Fort Dodge Australia Pty Ltd., 2152 Castlereagh Road, Penrith NSW 2750 Australia.
Passive Integrated Transponder (PIT) Tags
- AVID®: Austock Rural Pty Ltd., Unit 2/61 Gateway Drive, Noosaville, QLD 4566 Australia; http://www.austockrural.com.au/.
- LifeChip®: Digivet.com Pty Ltd., P.O. Box 6804, Baulkham Hills, NSW 2153 Australia; http://www.animal-id.com.au/contact_us.php.
- Trovan®: Microchips Australia Pty Ltd., 22 Fiveways Boulevard, Keysborough, Victoria 3173 Australia; http://www.microchips.com.au/.

Profender®: Bayer Australia Limited, 875 Pacific Highway, Pymble, NSW 2073 Australia.
Suprelorin®: Peptech Animal Health, 19-25 Khartoum Road, Macquarie Park, NSW 2113 Australia; http://www.peptech.com/HTML/Animal_Health/Superlorin_general.html.

ACKNOWLEDGMENTS

We thank the following staff at the Territory Wildlife Park for their assistance with the manuscript: Damien Stanioch, Trine Kruse, and Stephen Cutter. We also thank Dietrich Schaaf for his helpful manuscript edits.

REFERENCES

Churchill, S. 1998. *Australian Bats*. Reed New Holland, Sydney, NSW, Australia. 230 pp.
Kulzer, E., McKean, J.E., and Moehres, F.P. 1984. Prey catching behaviour and echolocation in the Australian ghost bat, *Macroderma gigas* (Chiroptera: Megadermatidae). *Australian Mammalogy*, 7:37-50.
Nowak, R.M. 1994. *Walker's Bats of the World*. The Johns Hopkins University Press, London, England. 287 pp.
Pirlot, P. and Nelson, J. 1980. Quantitative aspects of brain morphology in *Macroderma gigas* (Megadermatidae, Chiroptera). *Australian Mammalogy*, 3:105-108.
Richards, G.C. and Hand, S. 2002. Ghost bat *Macroderma gigas* (Dobson, 1880). In: *The Mammals of Australia*, Revised Edition (Strahan, R., ed.) Australian Museum/Reed New Holland, Sydney, NSW, Australia. Pp. 446–447.
Taylor, J.M. 1984. *The Oxford Guide to Mammals of Australia*. Oxford University Press, New York, NY. 148 pp.
Weigl, R. 2005. *Longevity of Mammals in Captivity; from the Living Collections of the World*. E. Schwizerbart'sche Verlagsbuchhandlung, Stuttgart, Germany. 214 pp.

Common Vampire Bats (*Desmodus rotundus*) at the Memphis Zoo

Janet Cooper, Mary Manson, Emily Schultz, Lisa Hailey, and Ashley Averill

INTRODUCTION

Although bats are not often exhibited in zoos, the common vampire bat (*Desmodus rotundus*) is an exception, primarily to satisfy public curiosity. Zoos have evolved away from merely entertaining visitors by displaying exotic creatures (Fascioni, 1996), but vampire bats continue to fascinate solely on the basis of their notoriety. The Memphis Zoo in Memphis, Tennessee has housed common vampire bats since 1990. The current colony has been featured in the "Animals of the Night" building since 1995, and at this writing, comprises 57 individuals.

ABOUT THE EXHIBIT

The vampire bat exhibit is a glass-fronted, simulated cave (Fig. 7-35) with gunite walls, floor, and ceiling. Unscented clay cat litter is used as a substrate.

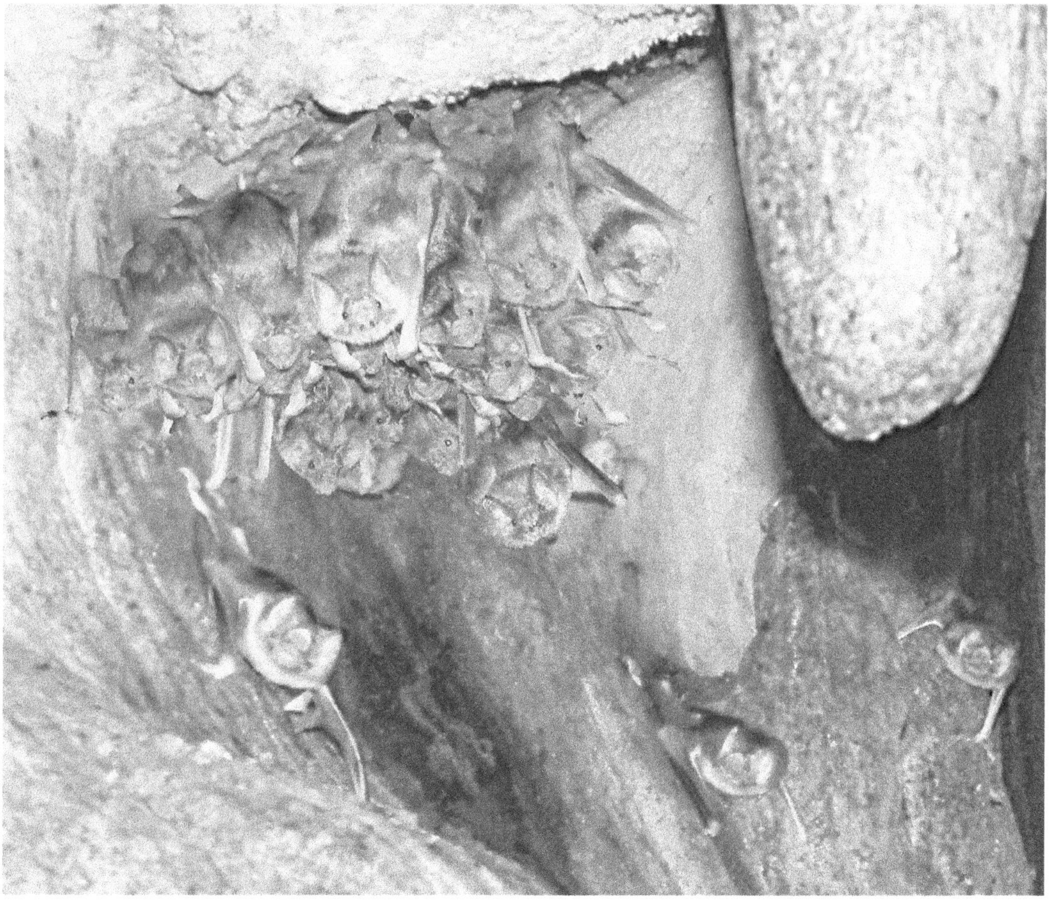

Fig. 7-35. Common vampire bats (*Desmodus rotundus*) exhibited at the Memphis Zoo. Photo courtesy of Angela Price.

The exhibit has a floor-to-ceiling public viewing window. The space in which the bats can fly is 8.9 ft. (2.7 m) high x 5.7 ft. (1.7 m) wide x 5.9 ft. (1.8 m) deep. They rest on the ceiling and walls close to the viewing glass, allowing zoo visitors a close encounter. The rear of the cave is higher than the front, forming a recessed area where the bats can find seclusion. A small pool that is constantly replenished with fresh water at the rear of the enclosure provides drinking water, although vampire bats appear to have little need for it.

EXHIBIT ENVIRONMENT

The Animals of the Night building is on a reverse light cycle. During public viewing hours the exhibit is lit by red fluorescent light fixtures at the front of the cave and a red spot bulb in the recessed area. In the evening, white fluorescent lights mounted at the front of the exhibit turn on with a timer and provide the day cycle for the bats.

EXHIBIT MAINTENANCE

Entry to the exhibit is at the rear through a 16-in. (40.6-cm) x 30-in. (76.2-cm) sliding door. Soiled cat litter is removed weekly, with spot cleaning as needed in between, and clean litter is added. The inside surface of the exhibit glass is cleaned once a week, but the public side must be cleaned daily. Glass is cleaned with Windex® and a squeegee. The bats are caught up two to three times a year so that the substrate can be removed, and the exhibit is disinfected or sometimes steam-cleaned when necessary. The exhibit viewing glass can be opened (Fig. 7-36) to allow keepers easy access to the enclosure interior.

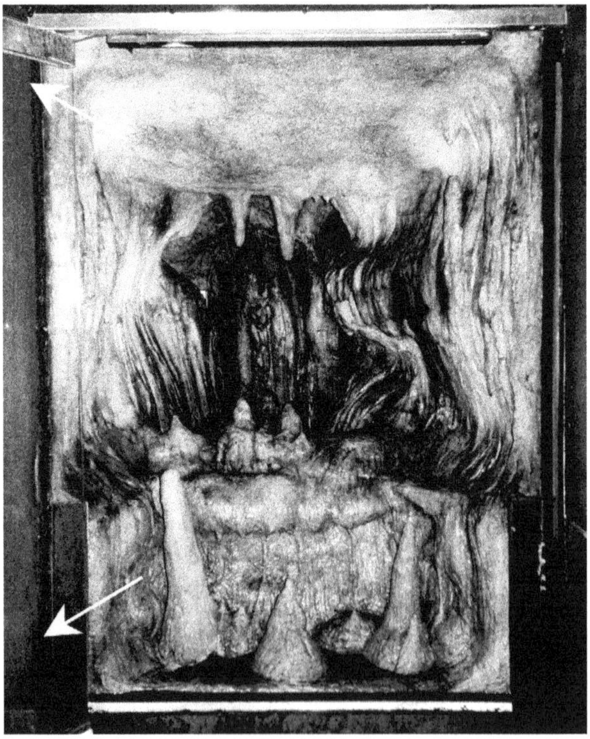

Fig. 7-36. The glass exhibit front (arrows) opens for easy access to the interior. Photo courtesy of Angela Price.

FOOD PRESENTATION

Bats are fed fresh bovine blood twice daily between 6 A.M. (0600 hrs.) and 7 A.M. (0700 hrs.) and again between 3 P.M. (1500 hrs.) and 4 P.M. (1600 hrs.). The blood is warmed and offered in four to five small, shallow, plastic dishes that are placed on the floor of the exhibit in various locations to lessen aggression and competition for food, while also lessening contamination of food with urine and feces.

Enrichment for the vampire bats has not been successful. Blood from a variety of mammals has been offered in addition to the normal bovine blood, but the bats showed little interest. Other enrichment items offered include scents, browse for climbing, and small toys, but the bats avoided them.

REPRODUCTION

The vampire bat colony breeds year-round. Young have been born in almost every month of the year, with more births in early spring and late fall. A hand rearing of an abandoned pup was achieved using methods outlined by Barnard (1995) and reintroduced successfully into the exhibit.

ADDITIONAL NOTES

Currently, the vampire bats are not marked, making individual identification impossible, and data can not be collected from individuals on reproduction or behavior. Nevertheless, individuals are monitored and weighed if a problem is observed. Weighing is also a routine procedure when bats are captured for major exhibit cleaning. Females weigh on average 31.7 g (26.4-37.3 g) and males average 26.8 g (23.4-31.6 g).

PRODUCTS MENTIONED IN TEXT

Windex®: S.C. Johnson & Son, Inc., 1525 Howe Street, Racine, WI 53403 USA.

ACKNOWLEDGMENTS

We thank Matt Thompson, Angela Price, Steve Reichling, and Dietrich Schaaf for their helpful comments on the manuscript.

REFERENCES

Barnard, S.M. 1995. *Bats in Captivity.* Wild Ones Animal Books, Springville, CA. 194 pp.
Fascione, N. 1996. The evolving role of American zoos in bat conservation. *Bats,* 14(1):8-14.

Big Brown Bats (*Eptesicus fuscus*) at Panola Mountain State Conservation Park

Susan M. Barnard

INTRODUCTION

Crevice-dwelling bats are secretive and difficult to exhibit to the public. They are also small animals, and to be seen well they must be relatively close to viewers. An exhibit for these animals must maximize viewing while minimizing stress. This can be a challenge because the public usually bangs on exhibit glass, even when "Do Not Tap On Glass" signs are conspicuously posted. With glass tapping in mind, exhibits should be designed to reduce the vibrations and sounds that are transferred to the occupants inside an enclosure.

The following exhibit design was used by Panola Mountain State Conservation Park (PMSCP) in Stockbridge, Georgia for over a decade to feature big brown bats (*Eptesicus fuscus*). This design is also suitable for other crevice-dwelling species that are known to occupy bat houses.

ABOUT THE EXHIBIT

The exhibit concept is simple and has been used in every zoological facility worldwide. An animal enclosure is placed behind a wall (see Fig. 7-40a) with a glass viewing window in it (see Fig. 7-40[7]). The space available for the big brown bat exhibit and enclosure at PMSCP was limited, but the entire structure was made as large as possible. To save space, guillotine doors were used to gain access to the exercise and feeding portion of the enclosure (Fig. 7-37[2]). Although there was not enough room for flight within the enclosure, the exercise and feeding area was large enough for the bats to stretch their wings and to flip back and forth from wall to wall, activities that kept them strong and healthy.

The enclosure comprised two features, a bat box (Fig. 7-37[1]) where the bats roosted during the day and could be viewed by park visitors, and an exercise and feeding area (Fig. 7-37[5]) that was utilized by the bats at night and closed to viewers by a door that doubled as the back wall of the bat box (Figs. 7-37[3], 7-38[3], and 7-39[3]). The entire enclosure (Figs. 7-37, 7-38) and enclosure stand (Fig. 7-37[6]) were framed in wood. The bat box walls, floor, and ceiling were constructed of ¾-in. (1.9-cm) plywood. The walls and ceiling of the exercise and feeding area were made of ¼-in. (0.64 -cm) polyethylene mesh. The plywood floor of the bat box and exercise and feeding area was covered with fiberglass resin (Figs. 7-38[4], 7-39[4]), and a drain (Fig. 7-38[4]) was installed in the exercise and feeding area to allow feces, urine, and dropped food to be flushed out with a hose.

A tree was painted on the public side of the wall (Fig. 7-40b,c), which at PMSCP was a large door (Fig. 7-40[8]) to a back room. A simulation of the front of a bat box was constructed to frame the viewing glass (Fig. 40b,c). This structure had two doors that park visitors could open (Fig. 7-40b) to look inside the bat box. To minimize stress on the bats, the doors were lightly spring-loaded so that they would close without slamming shut when viewers let go of them. A cushioning seal also dampened vibrations and sounds when the doors closed.

EXHIBIT ENVIRONMENT

TEMPERATURE AND HUMIDITY

Big brown bats at PMSCP were maintained at the same temperature and humidity as the park staff and public. The building thermostat was turned up slightly in summer and down slightly in winter to save money. However, the bats were never out of their comfort zone. In winter, a warm-mist humidifier (unknown brand) was used in the room containing the bat enclosure to combat the drying effects of the building's heat source.

LIGHTING

The exhibit was designed to take advantage of the natural behavior of bats, and because they are active at night and inactive during daylight hours, the portion of the

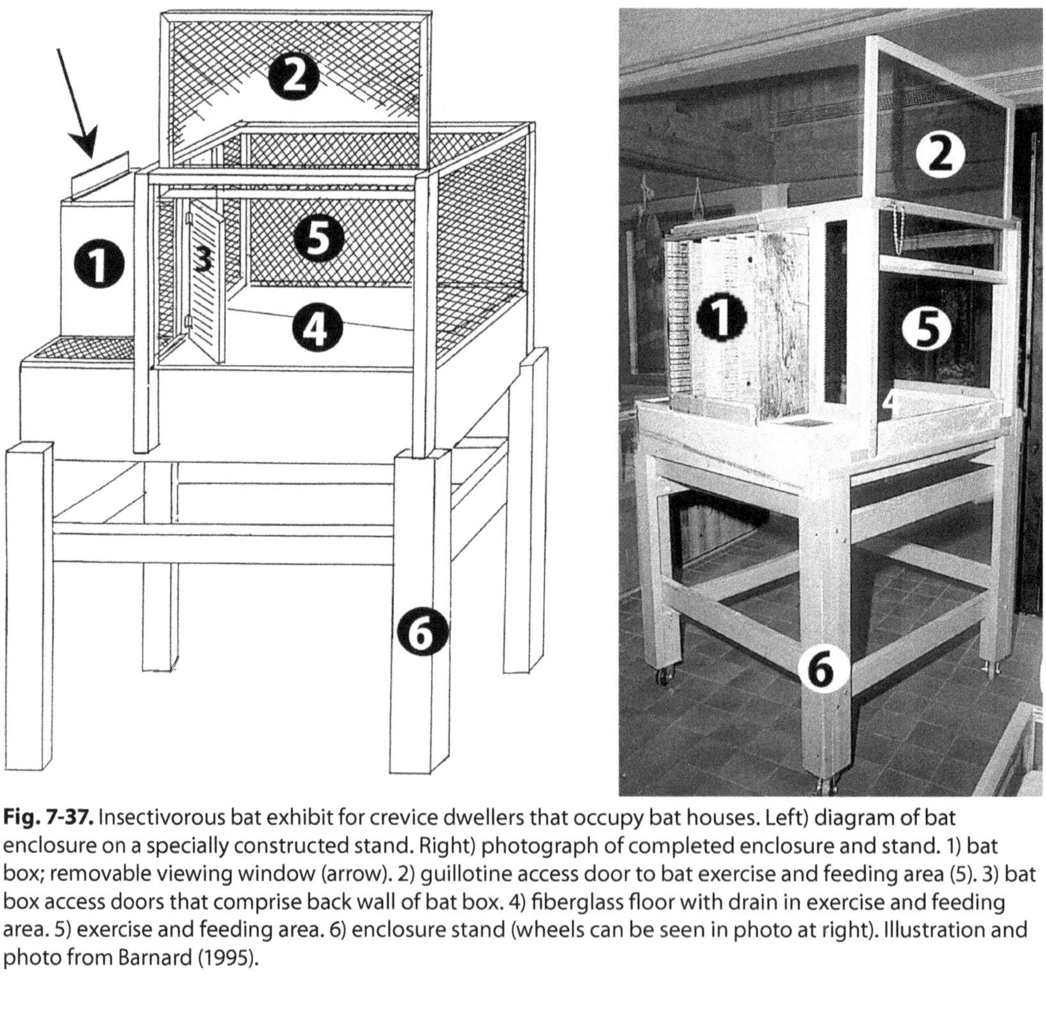

Fig. 7-37. Insectivorous bat exhibit for crevice dwellers that occupy bat houses. Left) diagram of bat enclosure on a specially constructed stand. Right) photograph of completed enclosure and stand. 1) bat box; removable viewing window (arrow). 2) guillotine access door to bat exercise and feeding area (5). 3) bat box access doors that comprise back wall of bat box. 4) fiberglass floor with drain in exercise and feeding area. 5) exercise and feeding area. 6) enclosure stand (wheels can be seen in photo at right). Illustration and photo from Barnard (1995).

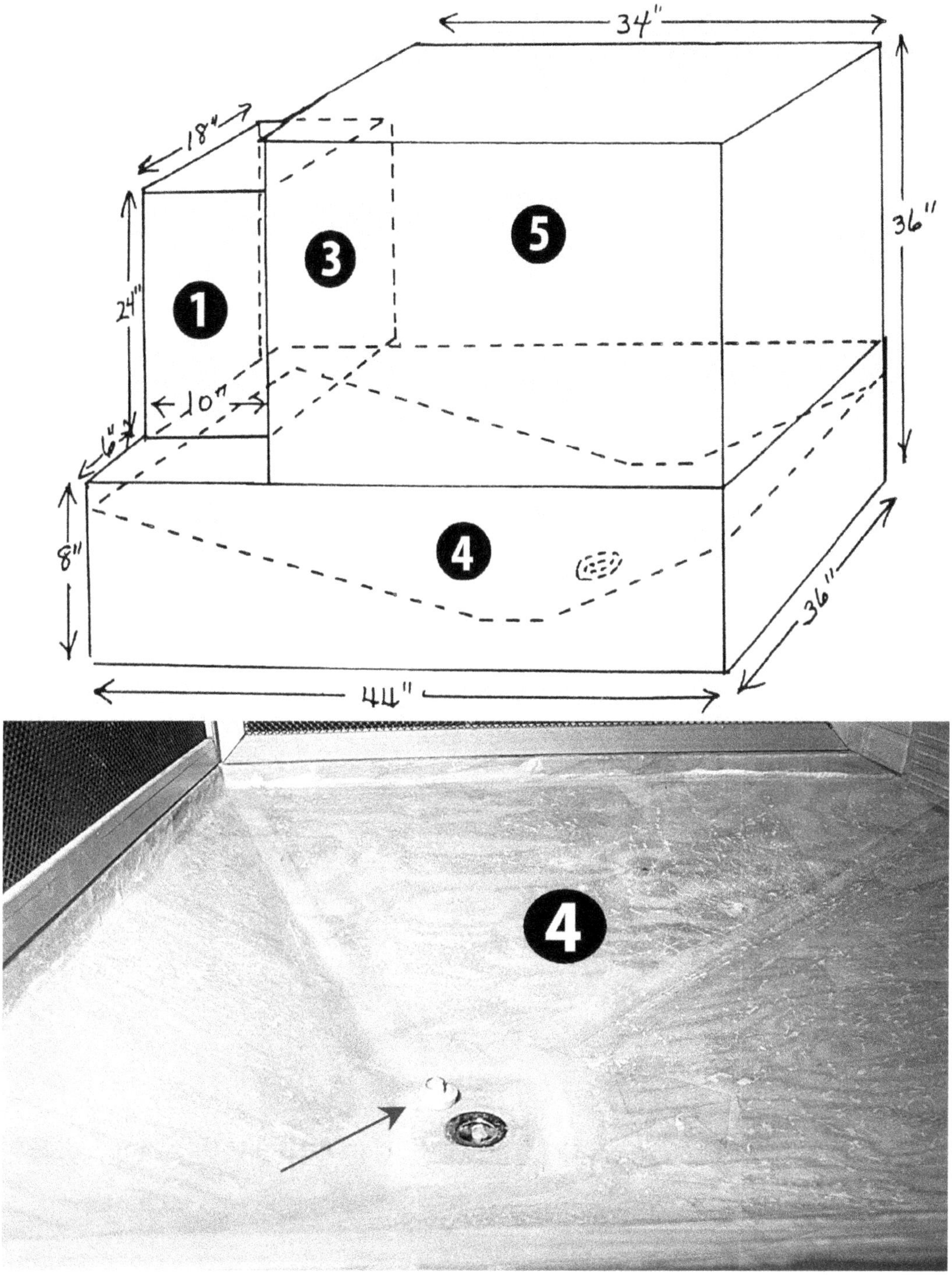

Fig. 7-38. Top) illustration showing suggested minimum dimensions for insectivorous bat exhibit. Bottom) close-up view of fiberglass floor. Drain plug (arrow) must be kept in drain when enclosure is not being cleaned to prevent bats from escaping. Illustration and photo from Barnard (1995).

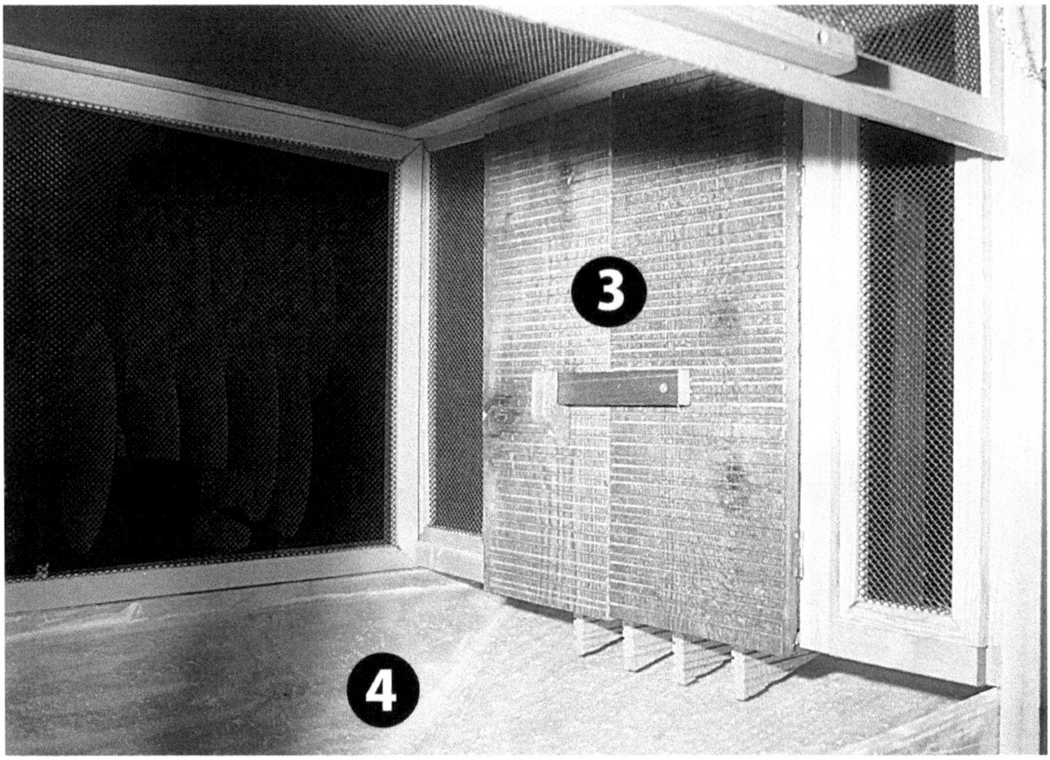

Fig. 7-39. Double doors (3) comprise back wall of the bat box. To enter the exercise and feeding area (4) from the bat box, bats must crawl through the space provided under the doors. Photo from Barnard (1995).

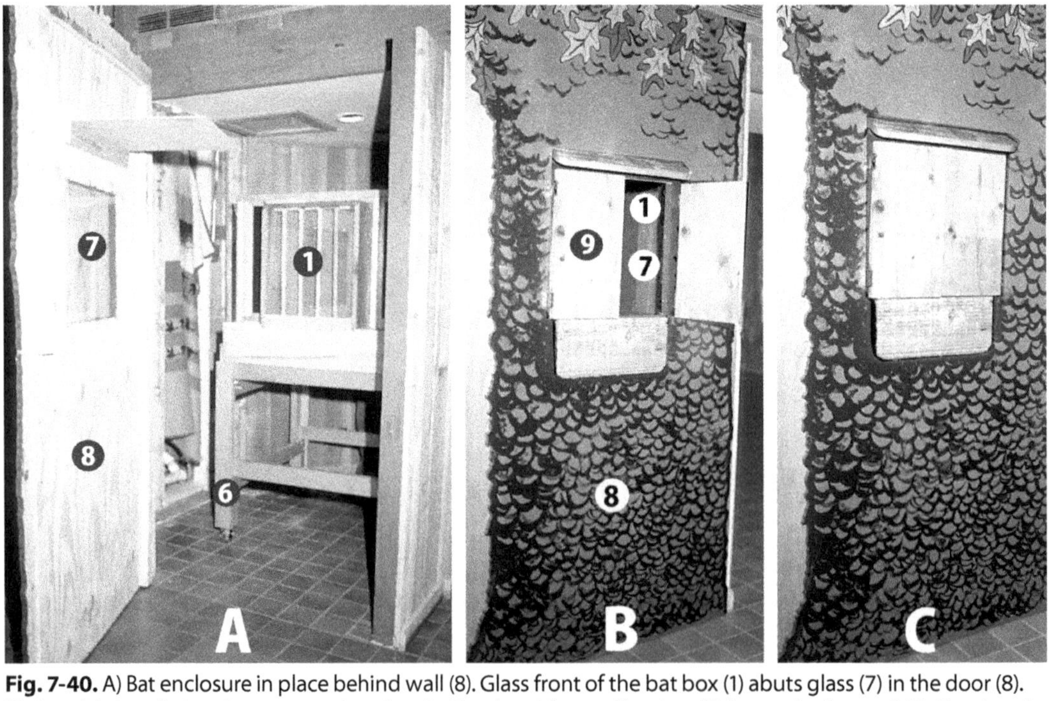

Fig. 7-40. A) Bat enclosure in place behind wall (8). Glass front of the bat box (1) abuts glass (7) in the door (8). B) door (8) closed, showing a tree painted on it. Simulated front of bat box (9) is attached to wall (8). Simulated front of bat box opens for bat viewing. C) bat box exhibit as viewed by the public. Photos from Barnard (1995).

enclosure that could not be viewed by the public was maintained under very bright lights during visiting hours, causing the animals to move to the part of the enclosure where the bat box was located. The bat box was illuminated with two red C9 7W Christmas light bulbs positioned over two 2-in. (5.1-cm) holes in the ceiling, and shed enough light for visitors to see the bats easily. The holes were covered with ¼-in. (0.64-cm) polyethylene mesh, and the low wattage of the bulbs prevented the bats from overheating. Because insectivorous bats do not see color, they perceived the interior of the bat box as a dark place, providing them a comfortable roost throughout the day. The area on the visitor's side of the wall was maintained at a low light level; when visitors opened the bat box doors to view the bats, the bats were not disturbed by light entering the bat box. Before park staff left the building for the evening, food and water were placed in the exercise and feeding area of the enclosure and all lights were turned off, with the exception of one in the building's main entrance. To reach food and water, the bats crawled under the doors (see Fig. 7-39[3]) that separated the bat box and exercise/feeding area. When park staff returned to work each morning, the bats were already on display, perhaps because a small window in the room housing the enclosure allowed light to enter at sun-up, cuing the bats to retreat to their day roost. It is also possible that the bats had been conditioned to leave the feeding area before the lights in the building were turned on.

PRODUCTS MENTIONED IN TEXT

Christmas Light Bulbs (C9, 7W, red): http://www.christmaslightsetc.com/p/C9-Christmas-Lights-Opaque-Frosted-Red-7-Watt-25-Replacement-Bulbs--15138.htm.

Polyethylene Mesh (¼-in. [0.64-cm]; item #101156-50): delstartech.thomasnet.com/item/naltex/duronet-industrial-mesh-2/1156-50-?&bc.

ACKNOWLEDGMENTS

I thank Dietrich Schaaf for his helpful comments on the manuscript.

REFERENCES

Barnard, S.M. 1995. *Bats in Captivity*. Wild Ones Animals Books, Springville, CA. 194 pp.

Chapter 8

Bat Management: Excluding Bats from Man-made Structures

Photo © Fly by Night, Inc.

Bat Management: Excluding Bats from Man-Made Structures

Laura Seckbach Finn and Thomas Gerald Finn

INTRODUCTION AND HISTORY

Bats are among the most beneficial native wildlife species in North America. Most species that roost in structures feed exclusively on insects, making them extremely valuable for their natural pest control (Tuttle and Smith 1992a). Bats are not only vital to our ecosystem, but also to our economy. Recent research has revealed that Mexican free-tailed bats (*Tadarida brasiliensis*) in south-central Texas save farmers up to $1.7 million per year due to their consumption of agricultural pests (Boyles *et al.*, 2011; Cleveland *et al.*, 2006).

Several bat species have adapted to declines in natural roost habitat by roosting in man-made structures, and some now roost almost exclusively in buildings (Greenhall and Paradiso, 1968). This adaptation initially may have allowed several species to expand their ranges (Kunz and Reynolds, 2003). However, colonies that roost in buildings can be conspicuous, and may be vulnerable to destruction. Nuisance issues, including guano accumulation, stained walls, and odor are compounded by unwarranted fear and an abundance of misleading and conflicting information about bats (Strohm, 1982). As a result, most colonies in structures are unwanted and viewed as pests that require removal or control, and it appears that the relatively new strategy of roosting in man-made structures carries with it drawbacks for bats and humans alike (Humphrey, 1982; Whitaker and Gummer, 1993).

Although there were early advocates of bat-proofing buildings (Cope 1959; Silver, 1935), chemical control was considered easier than prevention, and a variety of lethal chemical control methods were used to destroy entire colonies as recently as the late 1980s (Bat Conservation International, 1987; Marsh and Howard, 1977). Most of these control methods were also potentially lethal to the public. Over the last century, solutions for an unwanted bat colony evolved from an emphasis on bat eradication and lethal chemical pest control to those that emphasized non-chemical control methods. Although safer for the public, most of these methods remained lethal for bats. Eventually techniques using bat-proofing and non-chemical exclusion were adopted (Constantine 2009, Kunz and Reynolds, 2003), and today conservation-based bat exclusion includes high safety standards for the public as well as bats (Frantz and Laniewicz, 2000; French *et al.*, 2002).

Nonetheless, ignorance and misinformation that result in dead or injured bats are still among the greatest threats to bat populations. Mortality that is largely the result of misinformation or public ignorance is entirely preventable through education, and nuisance bat colonies can be most effectively controlled by excluding them from the structures where they roost. The objective is to get all bats safely out of a building and to keep them out permanently. This method requires simple techniques and safe, durable, high quality materials to prevent bats from re-entering a structure. However, successful exclusion is not the same as successful population management. Excluding bats without providing alternate roost sites does not solve the problem, it simply causes bats to move from one structure to another. This chapter provides solutions for the average homeowner or property manager who has bats in a structure and wants them safely outside.

BATS IN BUILDINGS

Buildings can be divided into three parts with regard to their functions. One is the interior used by humans and their pets. If there is a bat in the interior, it is unlikely to go unnoticed. The second part is the exterior, where wildlife normally resides. Part three is the voids within walls, attics and under roofs, which are the areas most likely to be utilized by bats. A bat colony living within a void may sometimes go unnoticed for years.

BATS INSIDE THE HOUSE

The most urgent problems occur when bats are found in the living area of a home, although a single bat flying inside a house does not necessarily indicate that a colony is present in the voids of the structure. Occasionally a bat may fly through an open door or window, or be carried in by a pet. Bats can usually be prevented from entering a home by keeping doors and windows screened, chimneys capped, and exterior basement or attic doors closed and in good repair.

Small bat colonies often go unnoticed until a lost bat finds its way accidentally into the living area. This can occur with lost juveniles during the summer maternity season, or when a thirsty bat awakens briefly from torpor during the winter (see Chapter 10, Volume 2). Although these events are usually rare, they may occur once or twice a year if a colony is present in the voids.

A single bat flying in the house is rarely cause for alarm and can usually be dealt with easily. In most cases, the "lost" bat is trying frantically to locate an exit and will leave on its own. The animal can be assisted by opening a widow or exterior door. All lights should be turned on, and ceiling fans turned off. Doors to adjacent rooms should be closed. It is important to remain quiet while waiting for a bat to find its way outside. If it does not, wait for it to land. Allow several minutes for it to relax and cool down, then capture it by placing a small container (i.e., a box or can) over it, and sliding a stiff piece of paper or cardboard carefully under the animal to enclose it in the container (Fig. 8-1). If a container is not available, the bat can be captured in a gloved hand, or by covering it with a small towel. To prevent injuries to the captor or bat, a bat in flight should not be chased with objects such as a net,

Fig. 8-1. Hand capture of an evening bat (*Nycticeius humeralis*). Photo courtesy of Kelli Deichmueller Gladding, © Fly By Night, Inc.

broom, or tennis racket. Once captured, the bat can be released outdoors unless someone has been bitten or scratched, in which case the local health department or a physician should be consulted immediately. In such cases, the bat should be confined in an escape-proof container until it can be delivered to animal control. Although bats may contract rabies, health hazards are generally low (Brass, 1994; Constantine, 2009). Even though the number of human rabies cases linked to bats averaged less than 1.5 deaths per year in the United States over the last 20 years (CDC data; http://www.cdc.gov), any bat that bites or scratches a person should always be tested for rabies.

STRUCTURAL VOIDS

Bats enter voids through openings on the exterior of buildings. A colony in the voids may remain unnoticed unless someone sees, hears or smells them, or as mentioned above, if a bat is found in the interior of a building. When bats are found in a void, it is best to consider options and devise a strategy to permanently exclude them.

PREVENTING BATS FROM ENTERING BUILDING INTERIORS

During maternity season, or when bats are using a structure during winter, access to the interior can be prevented without disturbing the colony. Openings that allow access to a building are typically found around attic or basement doors, or holes leading into wall spaces for plumbing, electrical wiring, or HVAC systems. Openings leading from voids to building interiors are usually found in utility closets, cabinets, behind appliances, and under sinks, and should be sealed until bats can be safely excluded from the building.

CONFIRMATION OF BAT ACTIVITY

Evidence that bats are occupying voids includes finding them in the house, seeing them entering or exiting a roost, finding them roosting in the open on an exterior wall, or seeing one or more bats on the ground near the structure. Bat guano may also be seen on the ground, or scattered on exterior walls below a roost. In many situations, audible roost chatter (high-pitched chirping), a distinct musky odor (or putrid odors if there are dead bats), or staining and rub marks at active roost entries (Fig. 8-2) may also be observed. Thus, it is usually not necessary to enter an attic, basement or other areas inside structural voids to look for bats.

Fig. 8-2. Dark stain (top center of photo) from scent glands and body oils at a bat roost entrance. Guano also peppers the walls. Photo courtesy of Kelli Deichmueller Gladding, © Fly By Night, Inc.

In some instances, a bat watch may be necessary (Greenhall and Frantz, 1994; Kern, 1995) to confirm activity and locate openings where bats are entering and leaving. During this process, several persons may be stationed around a building to watch bats exit at dusk or return at dawn.

CONFUSING EVIDENCE

Bat guano can sometimes be confused with gecko or rodent droppings. Rodent droppings are similar in color and size, but are hard and not crushed easily. Gecko droppings are soft and crushed easily, and are tipped with white uric acid deposits.

Bats are occasionally confused with chimney swifts (*Chaetura pelagica*), small migratory birds that nest inside chimneys. When the adult birds return during the day to feed their hatchlings, high pitched calls can be heard near fireplaces. Checking inside a fireplace may reveal feathers and bird droppings when swifts are present. A bat watch can also be helpful to clear up confusion between bats and chimney swifts. Swifts leave the structure in the morning and return in the evening; bats do the opposite. To prevent interior access, the flu should be kept closed and the chimney capped after the birds leave in the fall.

BAT MANAGEMENT

Successful exclusion is not the same as successful management. Excluding bats always results in roost loss and colony displacement. Alternate roosts are vital but are rarely provided, leaving bats caught in a cycle of exclusion and displacement. When large numbers of displaced bats are forced to search for suitable roosts, colony fragmentation and population decline are masked by reports of sudden "population increases" that occur when displaced bats join colonies in nearby structures. Options for mitigation range from installation of standard bat houses, full site protection, or more complex projects involving temporary exclusion and custom modifications to already occupied sites. Each project is unique, and each management plan should be project specific.

Bat houses (Chapter 9) installed prior to exclusion have an occupancy rate of over 90% (Finn, 1997a) and can be effective management tools that break the exclusion/displacement cycle (Finn, 1997b). Most bat houses installed by Fly By Night, Inc. (FBN) in Osteen, Florida were occupied within a year. The record is less than 12 hours (L. S. Finn and T. G. Finn, unpublished data 1993-2011).

ENVIRONMENTAL AND ECOLOGICAL RULES FOR TIMING EXCLUSIONS

Before excluding bats, environmental and ecological variables such as geographic location, season, bat species, and local weather conditions should be carefully evaluated. Familiarity with local species-specific variations in seasonal activity, reproductive timing, and roosting preferences are also important because bats in structures may consist of more than one species. These variables dictate project scheduling and general timing for exclusions. Timing of attachment and removal of exit tubes for final sealing (see below) can be significantly affected by weather, and local conditions must be monitored during every project. Nonspecific seasonal roosting activity is briefly summarized below.

SEASONAL ACTIVITY PATTERNS

Bat activity is generally synchronized with seasonal availability of food and local weather. In temperate regions, seasonal population fluctuations can be distinct and roosts can be described as maternity or hibernation sites. Activity fluctuations and seasonal changes are less obvious in warmer regions, where bats may be active year-round and do not hibernate.

Seasonal movement and roost switching is common among some species (e.g., *Nycticeius humeralis* and *T. b. mexicana*), whereas others may remain at the same site year-round (e.g., *Eptesicus fuscus* and *T. b. cynocephala*). Bats that are observed roosting alone or in small groups in open, easy-to-access areas are often transient occupants that are resting during spring or fall migration. These bats usually leave within in few days without intervention.

Since its discovery in 2006, the fungal disease white-nose syndrome (WNS) has resulted in bat population declines of 50% to 90% in some locations. Currently, WNS is only a risk to cave-roosting bats during winter, but many of these species form summer colonies in structures. Among affected species, the little brown myotis (*Myotis lucifugus*), one of the most common "nuisance" bats in the northeastern United States and Canada, is now in danger of extirpation by WNS (Frick *et al.*, 2010).

SUMMER ACTIVITY

Most colonial bat species give birth to one or two pups a year. This low birth rate, combined with pre-weaning mortality (i.e., predation and falling from roosts) results in slow population growth (Hermanson and Wilkins, 1986), which greatly strengthens the need for conservation-based management efforts. The annual maternity season occurs between mid-spring and late-summer for temperate species, but in subtropical and tropical regions of the world a second maternity season may occur in early fall for some species.

Females of colonial species rarely carry their pups, which remain in the roost until they are old enough to fly. Until then, with few exceptions, the adults return to the roost to nurse their young several times each night. Disturbance to a roost during maternity season may prevent adults from returning to feed their pups, resulting in significant mortality from starvation. Although adult males often roost away from females and pups (Kunz and Reynolds, 2003), they are often solitary during the maternity season, but may remain in the same structure. It should not be assumed that a small colony is simply a bachelor group that can be excluded. Population fluctuations observed before and after maternity season are generally the result of local movement between roosts, or dispersal following roost disturbance.

The time between birth and weaning can vary from 3 to 8 weeks (approximately from 15 April through 1 September in the United States). For specific information on the growth and development of selected bat species in North America and Europe, see Pages 306-308 and Pages 342-343 respectively (Chapter 13, Volume 2).

WINTER ACTIVITY

When planning an exclusion project, do not assume that all bats leave a roost in a structure during the winter. Although some bat species leave their summer roosts and migrate to other locations for the winter, other species such as *E. fuscus* and *T. b. cynocephala* may roost in buildings year-round (Henry *et al.*, 2000; Hermanson and Wilkins, 1986; Kunz and Reynolds, 2003). Bat activity is significantly reduced during periods of cold weather and torpid bats may not be noticed.

As a general rule, bats enter torpor to conserve energy during periods of reduced food availability (Altringham, 1996; see also Altringham, Chapter 10, Volume 2). In regions with distinct seasonal changes, most bats hibernate during the winter months and management efforts should be postponed until spring, when night temperatures are above 50 °F (10 °C). In the southern United States, the non-migratory *T. b. cynocephalus* is a common resident of man-made structures, and like other members of the family Molossidae, these animals do not hibernate. This species can be stunned by cold at prolonged temperatures below the mid 30s F (2s C) and individuals are often found on the ground during cold weather (L. S. Finn and T. G. Finn, pers. obs.).

ROOST MODIFICATION AND THE EXCLUSION PROCESS

Materials required for exclusions should be placed in a bucket for ready access. These include rigid plastic mesh (or galvanized mesh if rodents are an issue), adhesive caulk and caulk gun, scissors, staples and staple gun, screws and screw gun, masonry bit, and a length of rope. A bucket of water and a rag are also needed to clean up caulk residue, and for storing caulk tubes for recycling as they are emptied.

MESH

Mesh is the preferred material for bat-proofing. It does not restrict air or water flow, and can be attached with adhesive caulk, staples or screws, as needed. Only heavy-gauge, plastic or metal mesh with an opening size of ¼ in. (0.64 cm) or smaller is appropriate for bat-proofing. Rigid or extruded plastic mesh is preferred (see *Products Mentioned in Text*). It can be cut to size with a pair of sturdy scissors or fabric shears and custom fit to cover large spaces (i.e., shutters, attic and soffit vents), or folded and slipped into cracks, concrete expansion joints, or similar crevices that are too large for caulk alone (Fig. 8-3). This product is versatile, easy to work with, durable, UV resistant and does not tear easily, but it is not rodent-proof. Galvanized hardware cloth or other metal mesh should be used when squirrels or other rodents are an issue.

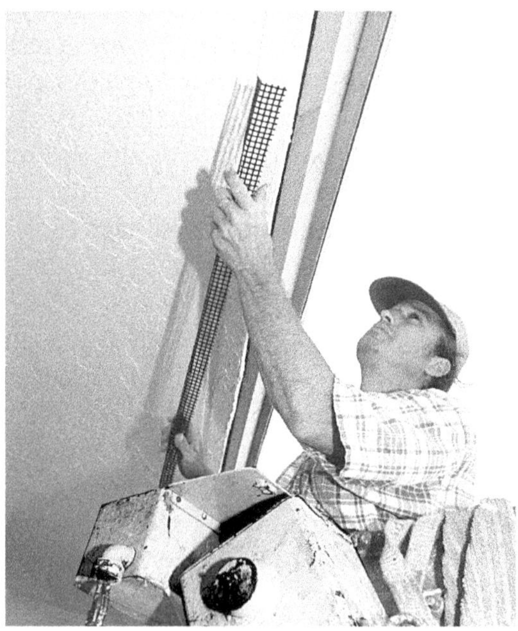

Fig. 8-3. Heavy-gauge, plastic mesh being installed during a bat exclusion. This type of mesh is versatile and easy to install. Photo courtesy of Mark Kiser, © Bat Conservation International.

ADHESIVE CAULK

High quality water-based adhesive caulk and a high quality caulk gun are the most important tools for sealing small crevices, and for the secure attachment of exclusion materials, such as mesh and modified caulk tubes (see below). Adhesive caulk should be thick, durable and aesthetically unobtrusive. It should dry quickly and have a minimal risk of failure, even on dusty surfaces. Clear PolyseamSeal® All Purpose Adhesive caulk is preferred by FBN. To prevent risk of attachment failure or potential for bats being exposed to wet caulk, it should not be applied during wet weather or less than an hour before sunset (see product label for specifications).

EXCLUSION TUBES

Exclusion tubes are one-way devises that allow bats to exit a roost easily, but not to re-enter. The use of tubes as one-way devices for excluding bats was first described by Constantine (1982). Nearly twenty years later, T. G. Finn developed a method that involves recycling empty caulk tubes used for excluding bats (French *et al.*, 2002). Empty 10-oz. plastic caulk tubes (Fig. 8-4) can be washed and modified to create effective, durable exclusion devices (Fig.8-5). The plunger from an empty caulk tube is removed by hand and the tip is cut off. After it is washed, each tube can be modified easily to facilitate attachment, using scissors to cut flair-out tabs.

Exclusion tubes can also be made from conduit, PVC pipe, hose or tubing (Greenhall and Franz, 1994). In most cases, a smooth surface without seams is preferable for making exclusion tubes, and the final product should be at least 8 in. (20 cm) long with a diameter of 2 to 3 in. (5 to 8 cm).

Fig. 8-4. Empty caulk tubes can be recycled by washing and hollowing them out to create exit tubes for bats during exclusion. Photo by Laura S. Finn, © Fly By Night, Inc.

MECHANICS OF EXCLUSION

Although the average homeowner may not be qualified to perform difficult exclusions, the following information should be useful in understanding the process. Efficient bat exclusion requires comfortable access to all exterior surfaces of a structure. A stable temporary work platform such as a ladder, scaffold, or hydraulic lift provides a safe base of operations. Workers should follow the standards set by the Occupational Safety and Health Administration's regulations and personal protective equipment guidelines for protection against falls.

It is best to begin an exclusion immediately adjacent to the most active roost entry, but to leave that opening or openings until last so bats can leave to feed while work

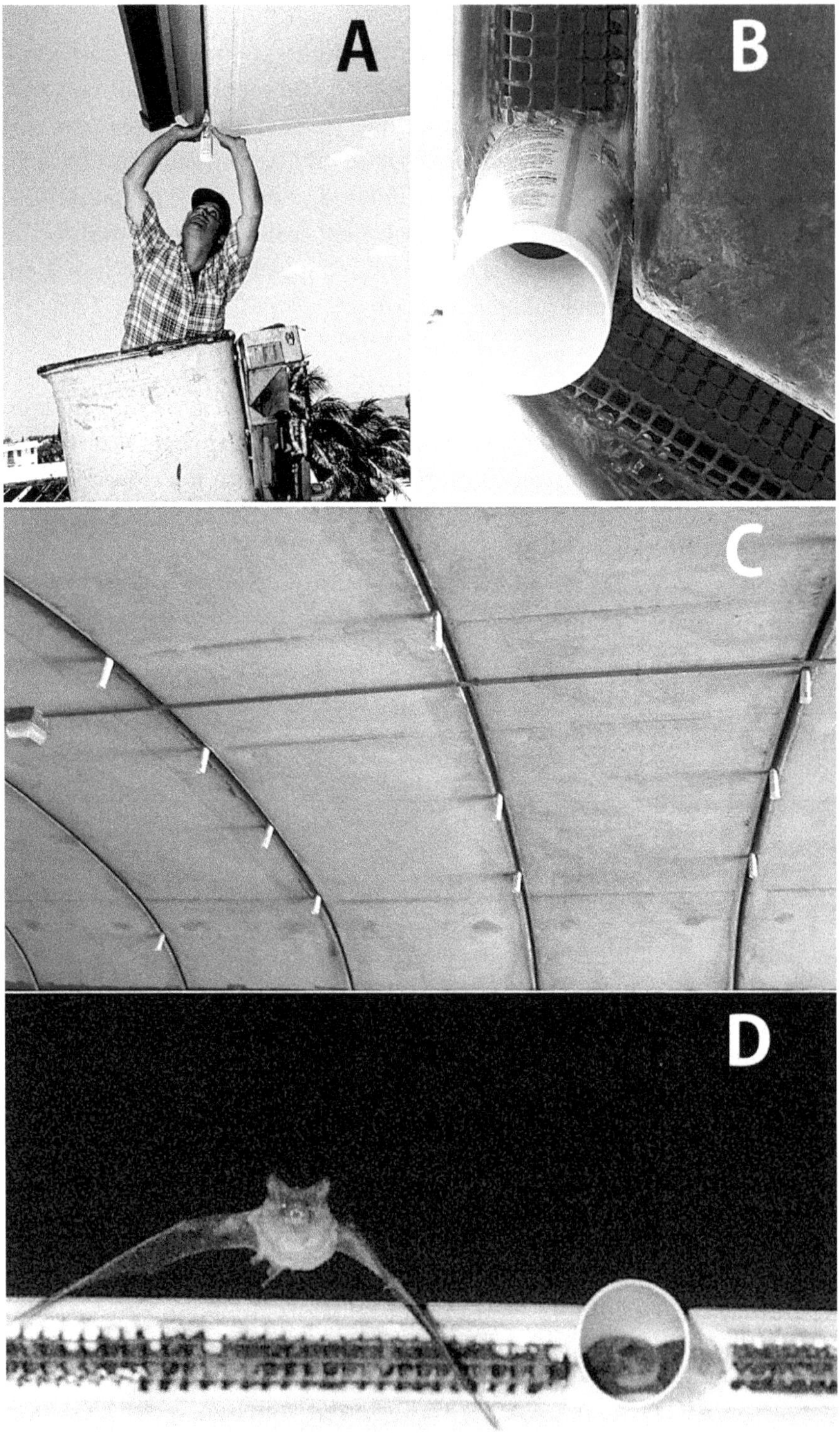

Fig. 8-5. Examples of the caulk-tube exclusion method. A) installing a recycled caulk tube at an exclusion site. B) close-up of installed caulk tube. C) heavy-gauge mesh placed between exit tubes and attached securely using adhesive caulk. Installing many exit tubes mitigates bat entrapment during emergence. D) bats exiting a roost through a modified caulk tube. A) photo courtesy of Mark Kiser, © Bat Conservation International; B-D) photos by Laura S. Finn, © Fly By Night, Inc.

is being conducted. First install an exit tube at the widest opening into the void of a building. To prevent re-entry, a near vertical placement angle is necessary. Sometimes because of difficult or unusual architecture, an exit tube extension may be needed to place the tube in an upright position (Fig. 8-6). Mesh should be attached with caulk, placing it so it flairs away from both sides of the exit tube, covering the gap completely (see Fig. 8-5a). No opening should be overlooked. As a general rule, if one can fit a pinky or little finger into a penetration, a bat can enter the void (Fig. 8-7).

Fig. 8-6. If caulk tubes can not be placed in an upright position, extensions are needed to achieve the correct angle. Photo by Thomas G. Finn, © Fly By Night, Inc.

Although exit tubes should be installed at all active locations (Fig. 8-8), if an active roost entry is found, leave it open and attach mesh at the point where the exit tube will be. Eventually mesh and exit tubes will have been applied at all openings around the perimeter of the structure, leaving only the primary high-activity areas for final lockout. It is impossible to install too many exit tubes, but installing too few can result in bottle necking or entrapment of bats.

Although exit tubes are not permanent, they should be securely attached so that they do not slip out of position. Any breach that allows bats to re-enter a roost requires repair or reinstallation, which prolongs the exclusion process.

Where minor structural repairs may be needed (e.g., rotten wood, rusted

Fig. 8-7. "Pinky test" used to identify potential access openings. If a little finger can fit through an opening, so can a bat. Photo by Thomas G. Finn, © Fly By Night, Inc.

Fig. 8-8. Exit tubes should be installed at all active locations. Photo by Laura S. Finn, © Fly By Night, Inc.

sheet metal, broken or missing roof tiles or other damage), they should be postponed until exclusion is complete. Install mesh over the damaged areas until repairs can be made.

FINAL LOCKOUT AND SEALING

After all bat-proofing is complete, for the final lockout attach exit tubes at all entry points that were left open. It is normal for bats to swarm near the roost entrance when they return before sunrise. However, when this occurs during the first morning after a final lockout and there is no available alternate roost such as a bat house, bats are at risk of predation as the displaced animals try unsuccessfully to re-enter the roost. At this time they may also be hit by cars or impaled on bird spikes (T. G. Finn pers. obs.). Displaced bats commonly roost in the open near the roost entrance, or in other suboptimal locations (Fig. 8-9). It is best to leave them alone. Bats roosting in the open typically leave at sunset and do not return the following day.

To reduce problems during and immediately after exclusion, exterior doors and windows should be closed to prevent bats from entering the living area of a home. Pet cats and dogs should be kept indoors until all bats are gone. Neighbors and their children should be advised not to disturb bats seen roosting in the open, or pick up those found on the ground.

After the final lockout, exit tubes should not be removed until after it is determined that all bats have left a roost. Most bats do not fly during inclement weather, or when temperatures are less than 50 °F (10 °C). The general rule is to leave exit tubes in place a minimum of 5-7 consecutive nights during optimal weather conditions, with night temperatures above 50 °F (10 °C). Work should be halted if it is suspected that bats are still in a roost while exclusion tubes are being removed. In some instances, there may be bats getting around an exclusion failure or breach. At this time, materials used for the exclusion should be inspected carefully to ensure that everything is secure. When exit tubes are removed, rigid mesh should be installed at all exclusion points for final sealing.

DIFFICULT ARCHITECTURE AND UNIQUE PROJECTS

Some architectural designs and roof types, such as clay tile, shake and thatch roofs, have so many access points for bats that it may seem nearly impossible to exclude them, particularly when the structure is very large or complex. Occasionally, such bat-occupied structures may require temporary exclusion of the animals prior to maintenance or renovation. Repairs and renovations can be followed by projects that include bat-proof designs.

TILE ROOFS

Clay tile roofs are often seen as problematic to bat proof (Constantine, 1979; Corrigan, 1984; Loven, 2003). Although bat proofing clay tiles is time-consuming and challenging, permanent exclusion is not impossible and can be completed successfully, using the same combination of rigid mesh, exit tubes, and adhesive caulk described earlier in this chapter (Fig. 8-10).

Fig. 8-9. After excluders are installed, bats may hang near access area(s). Photo by Laura S. Finn, © Fly By Night, Inc.

Fig. 8-10. Bats can be excluded from tile roofs using exit tubes and heavy gauge mesh. Photo by Thomas G. Finn, © Fly By Night, Inc.

NIGHT PROJECTS

When exclusion requires night work, such as projects in high-traffic public areas like theme parks, a thin layer of tissue paper, sand, leaves, or similar materials can be applied to cover exposed caulk while it is wet. This serves as a barrier, leaving a surface that is safe for bats instead of sticky glue. This method can also be used to camouflage exit tubes (Fig. 8-11) until they are removed

PORCH ROOSTS

Some species such as pallid bats (*Antrozous pallidus*) use open porches, patios, or garages as temporary night roosts for feeding or social activity (Kunz, 1982). Bats are absent from these sites during the day, and insect parts or guano may be the only evidence that bats were roosting the night before. These animals can often be discouraged from roosting by attaching strips of aluminum foil or mylar balloons to the ceiling. The movement of these items in the breeze frighten the bats (French *et al.*, 2002). A smooth surface that bats can not traverse can also be created by attaching a sheet of plastic at the roost site. Course fiberglass batting tacked to the roost surface has been suggested as well (Constantine, 1979; Loven, 2003). Chemical dog and cat repellents have also been suggested (Greenhall and Frantz, 1994) as a temporary solution, but these products are not currently legally approved bat repellents and should not be used when bats are present. Bats roosting behind shutters can be discouraged by opening the shutters at night. Rigid mesh is then attached the following day to prevent future entry.

Fig. 8-11. As mentioned in the text, various materials can be used to cover wet caulk, or to camouflage an exclusion device. Photo by Laura S. Finn, © Fly By Night, Inc.

PREVENTING PROBLEMS RELATED TO EXCLUSION

Injury or mortality as a consequence of incorrect exclusion methods or attachment failures often results in bats becoming entangled or trapped. Entrapment occurs after an active exit is improperly blocked by sealing a structure at the wrong time of year or by not allowing enough time for bats to exit before sealing a roost. Bats can also become trapped inside a building after an exit is closed inadvertently by construction workers such as roofers or painters.

Evidence of entrapment includes numerous bats found alive or dead and dying in the living space of a building. Most trapped bats die of dehydration and starvation, and mortality is often hidden inside the walls and attic. Decaying bats may cause dark, moist stains on walls or ceilings, seepage near blocked roost entries (Fig. 8-12), and putrid odors. These conditions may constitute a health hazard. Although mortality in such cases may be significant, it is rarely documented. Cleanup following such an event can be costly and is totally preventable.

Fig. 8-12. When bats are trapped and die after a roost exit is blocked, decaying carcasses produce fluids that may cause seepage. In this example, seepage occurred after a roost was blocked using expandable foam. Note the difference in appearance between seepage and staining from body oils and glands shown in Figure 8-2. Photo by Laura S. Finn, © Fly By Night, Inc.

Bats are also known to enter openings in metal frames around windows, support beams for screen patio enclosures (Fig. 8-13), and light fixtures (Fig. 8-14). These can become problematic "bat traps."

Any event that significantly alters the microclimate may cause bats to move within a roost, become trapped in a void, or to leave a structure and roost in exposed areas outside, if no alternative is available.

MAINTENANCE AND PREVENTION OF FUTURE PROBLEMS

Although a structure may be bat proofed and free of bats after a project is completed, bats are generally faithful to a roost site, and if new access areas are created by squirrels or woodpeckers, lack of general maintenance, storm damage, or natural deterioration of building materials, they may attempt re-entry. To prevent re-occupation, property owners should maintain structures and make repairs as needed. Maintenance workers like pressure washers, painters, and handymen should be instructed not to disturb or remove any bat-proofing materials.

CLEANING UP AFTER BATS ARE GONE

After bats are excluded, insects and other arthropods may be noticed near guano accumulations. These are bat ectoparasites such as bat bugs, mites, and bat flies that usually die quickly in the absence of their hosts and are rarely a problem (Greenhall and Frantz,

Fig. 8-13. Hundreds of bats died accidentally at this central Florida home. Although they roosted below the roof tiles outside the pool enclosure, they sometimes exited from the tiles inside the enclosure. When they were unable to re-enter the roof tiles, they tried to leave the screened-in pool through small openings at the top of the hollow aluminum frame, where they were trapped inside the vertical support beams (top arrow). The stain on the pool deck was seepage from decomposing bats (bottom arrow), mistaken for rust by the homeowner, who also complained of a persistent odor. Photo by Laura S. Finn, © Fly By Night, Inc.

Fig. 8-14. Mexican free-tailed bats (*Tadarida brasiliensis cynocephala*) roosting in a light fixture on a porch in Jacksonville, FL. Light fixtures are often overlooked as roost sites during bat-proofing. Photo by Laura S. Finn, © Fly By Night, Inc.

1994). If parasites are a concern, application of boric acid or diatomaceous earth can provide safe long-term control.

Direct contact with bat guano or inhalation of guano dust should be avoided. However, removal of guano may be necessary if the histoplasmosis-causing fungus (*Histoplasma capsulatum*) has been confirmed or when guano is found in the living area of a house (e.g., around ceiling tiles, light fixtures, electrical outlets, or heating and air systems). Respirator masks should always be worn when working around dried guano, and both respirators and vacuums should be equipped with high efficiency particulate absorbing (HEPA) filters, effective for particles as small as 2 microns, to avoid breathing dust that may be laden with fungal spores. Before cleaning up bat guano, it is best to consult with the Centers for Disease Control and Prevention, Atlanta, Georgia, local health department, and a hazardous waste disposal company concerning safety precautions and disposal regulations.

REGULATIONS GOVERNING BATS

It is important to be familiar with local, state, and federal wildlife and environmental regulations pertaining to bats before beginning an exclusion project (see Chapter 1). Over 50% of bat species in the United States and Canada are included on official status lists, or are otherwise considered of management concern. In the United States, any chemical use by professionals is regulated by the Federal Insecticide, Fungicide and Rodenticide Act (FIFRA), and there are currently no pesticides registered for bat control. Most states prohibit lethal bat control methods, and many require special permits or certification to work with these animals.

In the United Kingdom (UK), bat exclusion without prior notification to the proper authorities is a punishable offense. UK residents should seek guidance from one of the following: English Nature, Scottish Natural Heritage, The Countryside Council for Wales, or the Countryside and Wildlife Branch of the Department of the Environment in Northern Ireland. For additional information on bat legislation throughout Europe, see Mitchell-Jones (Chapter 1). Chapter 1 also covers bat legislation in Australia, Canada, and the United States.

DO-IT-YOURSELF OR HIRE A PROFESSIONAL?

If the services of a professional are required, homeowners should seek the following information:

- Is the company approved by Bat Conservation International's Bats in Buildings Program (see http://www.batcon.org/index.php/bats-a-people/bats-in-buildings/professional-excluders.html). In countries outside the United States, local or regional wildlife organizations or bat groups should be consulted (see also *Legislation*, Chapter 1).
- Does the professional understand the facts and provide truthful information regarding bats in buildings? Are scare and high-pressure tactics used to sell add-on services? If so, do not use the service.
- Can the professional provide proof of rabies immunization?
- Is the professional willing to provide at least three recent client references and a guarantee of two or more years?
- What rating does the professional have with the Better Business Bureau (United States) or similar agencies in other countries, and can the professional provide proof of appropriate insurance and permits?
- Does the professional plan to use the approved methods outlined in this chapter?
- Is the professional planning to trap and relocate bats? If this is the case, do not use the service.

EXCLUSION MATERIALS AND TECHNIQUES TO AVOID

Bat-proofing materials should be durable, permanently attached, preserve the aesthetics of a structure, and should not alter natural ventilation or water flow. The following is a checklist of common exclusion materials and techniques that do not work.

- Avoid using temporary materials that deteriorate quickly to close holes, such as paper towels, steel wool, or soft rags.
- Avoid using products or making structural modifications that may block natural ventilation, like hanging plastic sheeting over an active roost entrance, thereby altering roost microclimate.
- Avoid using chemical and ultrasonic repellents. These items are ineffective and do not resolve problems for the long-term. Ultrasonic devices may even *attract* bats (Tuttle, 1988).
- Avoid the use of traps to remove bats from a structure. Traps require frequent monitoring to prevent overcrowding. Furthermore, not all bats leave a roost each night, and because these animals have excellent homing instincts, they may return to their roosts, even when released far from the trap site.
- Avoid the use of silicone, polyurethane, or similar non-water-based caulk products. They are odorous, take several hours to dry, and solvents are required for cleanup.

- Avoid the use of metal window screen and metal flashing, which have sharp edges that may injure or impale bats if not attached properly.
- Avoid the use of expandable foam products at occupied sites. Foam can block ventilation and alter the microclimate. Expandable foam products can also kill bats that come into contact with the material before it dries. Dead bats have been found entombed in foam (T. G. Finn pers. obs.; Mark Kiser, pers. comm.).
- Avoid the use of flexible netting attached with duct tape. These products may be the most widely suggested and commonly used materials for excluding bats (Corrigan, 1984; Greenhall and Frantz, 1994; Kern, 1995; Loven, 2003; Marks and Marks, 2006; Tuttle and Smith, 1992b; Williams-Whitmer and Brittingham, 1995). Duct tape or similar adhesive tapes fail when surfaces are rough, coated with dust, mold or mildew, or when used in high humidity or during rain. Failure to adhere properly can result in re-entry or entrapment when a closure is breached. Installation of bird (flexible) netting is overly complicated, and there are multiple drawbacks. The most significant flaw is that the ½-in. (1.27-cm) or even the ¼-in. (0.64-cm) mesh can entangle bats (Fig. 8-15), resulting in permanent injury or death (T. G. Finn, pers. obs.; M. Kiser, pers. comm.). There is also a high risk of attachment failure and undetected bat re-entry. Flexible netting also loses integrity over time. Holes, tears or other damage can be caused by severe weather, fallen tree limbs, maintenance

Fig. 8-15. An evening bat (*Nycticeius humeralis*) caught in flexible bird netting. Photo by Laura S. Finn, © Fly By Night, Inc.

activities and birds or rodents, all of which may allow bats to re-enter a building.

- Although technically not exclusion materials, items to avoid when controlling bats are sticky bird repellents and glue boards because these are lethal traps and not approved for use against bats in most states (Fig. 8-16; see also *Dermatological Hazards from Crude Oil Substances*, Chapter 6, Volume 1). Other lethal items to avoid are wasp sprays, car exhaust, fire extinguishers, and termite tents.

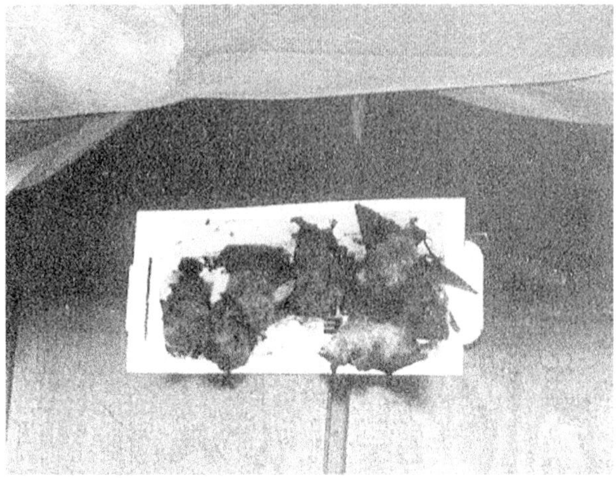

Fig. 8-16. Bats trapped on a glue board stapled below a roost entrance. The animals were left to die by the pest control operator. Photo courtesy of Elise Able, © Foxwood Wildlife Rescue, Inc.

PRODUCTS MENTIONED IN TEXT

Rigid plastic mesh (XV1170): Industrial Netting, 7681 Setzler Pkwy N. Minneapolis, MN 55445 USA; http://www.industrialnetting.com/plastic_extruded.htm.

PolyseamSeal® All Purpose Adhesive Caulk: Henkel Corporation, 1600 Executive Drive, LaGrange, GA 30240 USA; http://www.polyseamseal.com/pdf/tech_all-purpose.pdf.

ACKNOWLEDGMENTS

We thank Barbara French and Mark and Selena Kiser for reviewing and editing early versions of the manuscript, and Dietrich Schaaf for his helpful edits on the final draft.

REFERENCES

Altringham, J.D. 1996. *Bats: Biology and Behavior.* Oxford University Press, New York, NY. 292 pp.

Bat Conservation International. 1987. Profiles: Rozol use against bats banned in North Dakota. *Bats*, 5:6.

Boyles, J.G., Cryan, P.M., McCracken, G.F., and Kunz, T.H. 2011. Economic importance of bats in agriculture. *Science*, 332:41–42.

Brass, D.A. 1994. *Rabies in Bats.* Livia Press, Ridgefield, CT. 335 pp.

Cleveland, C.J., Betke, M., Federico, P., Frank, J.D., Hallam, T.G., Horn, J., López, J.D., Jr, McCracken, G.F., Medellín, R.A., Moreno-Valdez, A., Sansone, C.G., Westbrook, J.K., and Kunz, T.H. 2006. Economic value of the pest control service provided by Brazilian free-tailed bats in south-central Texas. *Frontiers in Ecology and the Environment*, 4:238–243.

Constantine, D.G. 1979. Bat rabies and bat management. *Bulletin of the Society of Vector Ecology*, 4:1-9.

Constantine, D.G. 1982. Bat proofing of buildings by installation of valve-like devices in entryways. *Journal of Wildlife Management*, 46:507-513.

Constantine, D.G. 2009. *Bat Rabies and Other Lyssavirus Infections.* Reston, VA., U.S. Geological Survey Circular 1329. 68 pp.; http://pubs.usgs.gov/circ/circ1329/pdf/circ1329.pdf.

Cope, J.B. 1959. Build bats out. *Pest Control*, 27:28-29.

Corrigan, R.L. 1984. Nuisance bats: Current technology in their management and control. In: *Proceedings of the Eleventh Vertebrate Pest Conference* (Clark, D.O., ed.). University of Nebraska, Lincoln, NE. Pp. 174-179.

Finn, L.S. 1997a. The use of bat houses in the management of central Florida bats. *The Bat House Researcher*, 5:5-6.

Finn, L.S. 1997b. *Bat House Use in Central Florida, with Emphasis on Tadarida brasiliensis cynocephala and Nycticeius humeralis.* MSc. Thesis, University of Central Florida, Orlando, FL. 194 pp.

Frantz, S.C. and Laniewicz, B.R. 2000. Comprehensive management of commensal bats. In: *The Ninth Wildlife Damage Management Conference Proceedings* (Brittingham, M.C., Kays, J., and McPeake, R., eds.). State College, PA. Pp. 172-185.

French, B., Finn, L., and Kiser, M. 2002. *Bats in Buildings: An Information and Exclusion Guide.* Bat Conservation International, Austin, TX. 5 pp.; http://www.batcon.org/pdfs/education/fof_ug.pdf and http://www.flybynightinc.org/bibguidelines.htm.

Frick W.F., Pollock, J.F., Hicks, A.C., Langwig, K.E., Reynolds, D.S., Turner, G.G., Butchkoski, C.M., and Kunz, T.H. 2010. An emerging disease causes regional population collapse of a common North American bat species. *Science*, 329:679-682.

Greenhall, A.M. and Frantz, S.C. 1994. Bats. In: *Prevention and Control of Wildlife Damage* (Hygnstrom, S.E., Timm, R.M., and Larson, G.E., eds.). University of Nebraska Cooperative Extension, Lincoln, NE. Pp. D5-D24.

Greenhall, A.M. and Paradiso, J.L. 1968. *Bats and Bat Banding.* U.S. Fish and Wildlife Service Resource Publication No. 72., Washington, D.C. 47 pp.

Henry, T.H., Best, T.L., and Hilton, C.D. 2000. Body size, reproductive biology, and sex ratio of a year-round colony of *Eptesicus fuscus fuscus* and *Tadarida brasiliensis cynocephala* in eastern Alabama. In: *Fourth Colloquium on Conservation of Mammals in the Southeastern United States* (Chapman, B.R. and Laerm, J., eds.). Occasional Papers of the North Carolina Museum of Natural Sciences and the North Carolina Biological Survey, No. 12. Pp. 50-56.

Hermanson, J.W. and Wilkins, K.T. 1986. Pre-weaning mortality in a Florida maternity roost of *Myotis austroriparius* and *Tadarida brasiliensis. Journal of Mammalogy*, 67:751-754.

Humphrey, S.R. 1982. Bats: Vespertilionidae and Molossidae. In: *Wild Mammals of North America: Biology, Management and Economics* (Chapman, J.A. and Feldhammer, G.A., eds.). Johns Hopkins University Press, Baltimore, MD. Pp. 52-70.

Kern, W.H., Jr. 1995. Bat exclusion methods. *Proceedings of the Seventh Eastern Wildlife Damage Management Conference*, 7:139-148.

Kunz, T.H. 1982. Roosting ecology of bats. In: *Ecology of Bats* (Kunz, T.H., ed.). Plenum Press, New York, NY. Pp. 1-55.

Kunz, T.H. and D.S. Reynolds. 2003. Bat Colonies in Buildings. In: *Monitoring Trends in Bat Populations of the United States and Territories: Problems and Prospects* (O'Shea, T.J. and Bogan, M.A., eds.). U.S. Geological Survey, Biological Resources Discipline, Information and Technology Report, Fort Collins, CO. Pp. 91–102

Loven, J. 2003. *Animal Damage Management, ADM-11.* Purdue University Cooperative Extension Service, Purdue University, West Lafayette, IN. 4 pp.; http://extension.entm.purdue.edu/publications/ADM-11.pdf.

Marks, C.S. and Marks, G.E. 2006. *Bats of Florida.* University of Florida Press, Gainesville, FL. 176 pp.

Marsh, R.E. and Howard, W.E. 1977. Vertebrate pest control manual: VII. Bats. *Pest Control*, 45:24-39.

Silver, J. 1935. *Eliminating Bats from Buildings.* U.S. Department of Agriculture Leaflet No.109. 5 pp.

Strohm, B. 1982. Most facts about bats are myths. *National Wildlife Magazine,* 20:35-39.

Tuttle, M.D. (Undated/Unpublished). *Why Pesticide Use Against Bats Should Be Banned.* Milwaukee Public Museum. 12 pp

Tuttle, M.D. 1988. *America's Neighborhood Bats.* University of Texas Press, Austin, TX. 96 pp.

Tuttle, M.D. and Smith, E.C. 1992a. Bats: Natures own pest control. *Pest Management,* 11:10-13.

Tuttle, M.D. and Smith, E.C. 1992b. Evicting Unwanted Tenants. *Pest Management*, 11: 18-19, 22.

Whitaker, J.O., Jr. and Gummer, S.L. 1993. The status of the evening bat, *Nycticeius humeralis*, in Indiana. *Proceedings of the Indiana Academy of Science,* 102:283-291.

Williams-Whitmer, L.M. and Brittingham, M.C. 1995. *A Homeowners Guide to Northeastern Bats and Bat Problems.* The Pennsylvania State University, University Park, PA. 22 pp.

Chapter 9

Bat Houses and Artificial Roosts

Photo by Mark and Selena Kiser, © Bat Conservation International

Bat Houses and Other Artificial Roosts

Mark Kiser and Selena Kiser

INTRODUCTION

Because bats spend over half their lives in roosts, they depend heavily on the quality and abundance of roost sites, so much that roost availability may be a major cause in determining bat population sizes and distributions (Kunz, 1982). However, many natural roosts have disappeared with the loss of forests and disturbances of caves. As a consequence, artificial roosts are becoming increasingly important to bats, and in some areas, may provide the only remaining places for bats to raise young or hibernate. Artificial roosts are defined as human-made structures, such as buildings, bridges and culverts, as well as structures designed specifically for bats, including bat houses.

Many species of bats have adapted to the loss of natural roosts by taking advantage of a myriad of artificial roosts. Some North American species, such as big brown bats (*Eptesicus fuscus*) and little brown myotis (*Myotis lucifugus*), have adapted to human-made structures so well that there are few recent records of maternity colonies in natural roosts (Barbour and Davis, 1969). Brazilian free-tailed bats (*Tadarida brasiliensis*) even have expanded their historical range northward by using artificial roosts for maternity sites (Kunz, 1982). Most bat species in the United Kingdom use buildings or other structures for at least part of the year (Stebbings and Walsh, 1991); some are highly dependent on buildings throughout the year (Shore *et al.*, 1991). These examples illustrate the potential that artificial roosts have for bat management and conservation. At least 17 species of bats in the eastern United States and Canada have been found to use human-made structures as maternity and hibernation sites, night roosts, or migratory stopovers (Barbour and Davis, 1969; Keeley and Tuttle, 1999), and at least 25 species in the United States and Canada have been documented to use either bridges, road culverts, or bat houses (Keeley and Tuttle, 1999; Kiser and Kiser, 2001a).

Artificial roosts can be created for numerous species of colonial bats, some quite easily, nearly any place where suitable habitat occurs. The main factors that determine use of a particular structure are proximity to water and foraging sites, roost design, thermal buffering/microclimate, in addition to protection from predators, adverse weather, and sunlight (Kunz, 1982; Pierson *et al.*, 1996; see also Appendix 9-1). The presence of existing bat colonies in nearby human-made structures and the physical eviction of bats from structures also appear to be significant factors in bat house occupancy (Kiser and Kiser, 2000; Tuttle *et al.*, 2004; see also Finn and Finn, Chapter 8). Certain structures appear to be better suited to a given region than others; for example, bridges are more successful as roost sites in the southern United States, where ambient temperatures are warmer than those in northern latitudes (Keeley and Tuttle, 1999). Use of human-made structures also depends heavily upon species' microclimate and other roost preferences.

Bats that use artificial roosts select either narrow crevices or open areas. Crevice-dwellers, like little brown myotis, prefer narrow spaces behind siding on buildings (Barbour and Davis, 1969), box-beam bridges (Keeley and Tuttle, 1999), and crevice-style bat houses (Tuttle and Hensley, 1993), whereas open space-dwellers such as Rafinesque's big-eared bats (*Corynorhinus rafinesquii*) prefer spacious sites in buildings, large hollow trees, and open areas between beams in bridges (Barbour and Davis, 1969; Keeley and Tuttle, 1999). By understanding roost preferences, one can create the type of artificial roost that a certain species might select.

Bat houses and other artificial roosts designed specifically for bats have successfully attracted at least 23 species in Europe, 20 in the United States and Canada, 12 in Australia, 10 in Costa Rica, 10 in South Africa, 9 in Japan, and at least 1 in India, Cayman Islands, New Zealand, and the Virgin Islands (Appendix 9-2). The vast majority of bat house occupants around the world are insectivorous species. However, frugivorous, nectivorous, carnivorous, and even one sanguivorous bat species have also used specially-designed bat houses in tropical forest settings (Kelm, 2008; Kelm *et al.*, 2008). Insect pest control, forest regeneration, and conservation of endangered species are some of the primary reasons that farmers, scientists, conservationists, and the general public install and experiment with artificial roosts. This chapter explores the history of artificial roosts for bats and some of the lessons learned along the way.

EARLY BAT HOUSE RESEARCH

The idea of creating artificial roosts specifically for bats dates back to at least the early 20th century. In 1902, a physician named Dr. Charles Campbell began experimenting with small, wooden bat houses attached to buildings, bridges, and trees in central Texas (Campbell, 1925). Although only two of the bat houses were successful, he continued to study bats' natural roost preferences and later developed large, free-standing towers designed to hold thousands of insectivorous bats. From 1909 to the 1920s, at least 16 of Campbell's 30-ft. (9.1-m) bat towers were constructed in the southeastern United States and Italy (Campbell, 1925). Although Dr. Campbell's dreams did not materialize that the bat towers would aid in the suppression of malaria-spreading mosquitoes, at least five of the towers in the United States attracted colonies of bats (Campbell 1925; Kiser, 2002; Murphy, 1989). Today, small numbers of *T. brasiliensis* and cave myotis (*Myotis velifer*) continue to use one of the original Campbell towers in Comfort, Texas (http://templeterracebattower.com/about.html), and *T. brasiliensis* uses a renovated tower in Orange, Texas (Michael Hoke, pers. comm.).

Bat houses and other types of artificial roosts designed for bats have evolved and improved steadily since Dr. Campbell's first experiments in Texas. Whereas little research was conducted on bat houses in the United States from the 1930s to the 1980s (in part due to hysteria following the discovery of rabies in bats in the 1950s), bat house experimentation in Europe has continued unabated for nearly a century. The first small, wooden bat boxes in Europe were tested in forested areas of France following the discovery that bats sometimes roosted inside bird houses (Jolyet, 1918). Jolyet' s publication describing the construction of small bat boxes inspired others, and by the late 1930s, bat houses were widely used in central and eastern Europe (Mayle, 1990; Stebbings and Walsh, 1991). In 1968, a pilot project involving 26 boxes in Dorset, United Kingdom, had promising results, with the first bats moving in within five months (Stebbings and Walsh,

1991). This success led to a larger experiment in the 1980s involving 3,000 bat houses installed in six forests across the United Kingdom. This experiment was coordinated by the Institute of Terrestrial Ecology, the Forestry Commission, BBC TV, and the World Wildlife Fund (Mayle, 1990; Stebbings and Walsh, 1991). By 1987, nearly 5,000 bats representing 7 species had moved into the boxes (Stebbings and Walsh, 1988). Results from these two experiments led to popular publications by the Bat Conservation Trust, which chronicled lessons learned regarding construction and installation of boxes (Stebbings and Walsh, 1988, 1991). By 1990, at least 12 species were known to have regularly used bat houses in Britain and Europe (Mayle, 1990).

Thanks in part to these accomplishments in Europe, bat houses in North America underwent a resurgence in popularity in the 1980s (Tuttle, 1988), and a number of backyard enthusiasts began erecting birdhouse-style models. When these did not achieve the same success rates as their European counterparts (see Nielson and Fenton, 1994), other experimenters began testing larger, less conventional models of their own design (Tuttle and Hensley, 1993). To better understand the factors that influenced successful occupancy in North America, to encourage additional, widespread experimentation, and to share results with a large audience, Bat Conservation International (BCI) launched a citizen-science project in 1993, known as the North American Bat House Research Project (Tuttle and Hensley, 1993). From 1993 to 2003, hundreds of volunteers (Fig. 9-1) across North America submitted data on several thousand bat houses (Kiser and Kiser, 2004). Following several successful long-term projects in the 1990s in Australia (see Bender and Irvine, 2001; Irvine and Bender, 1995 for review), the Bat Roost Box Network debuted in 2001 to foster experimentation and share knowledge in that country. In addition, via newsletters, listservs, forums and websites, numerous bat working groups, bat clubs, and backyard enthusiasts across the globe also share results gained from local testing of numerous bat house designs, materials, and placement techniques. Local testing is a key element of bat house success regardless of target species or location. The good news is that homemade and commercially produced bat houses of various sizes have already proven to be a popular means of conserving bats around the world.

Fig. 9-1. Kent Borcherding, one of hundreds of volunteer researchers for Bat Conservation International's Bat House Project, inspects a series of bat houses he constructed in Wisconsin. Photo by Merlin D. Tuttle, © Bat Conservation International, Austin, Texas.

BAT HOUSE DESIGNS

Taking into account decades of experimentation by researchers from around the globe, it is apparent that there is no one bat house design that works equally well for all species, locations, or times of the year. A tremendous diversity of models of different sizes, shapes, materials, and features has proven effective in attracting colonial bats (Appendix 9-3). Regardless of the actual design, all bat houses must posses certain key features if they are to function well, including a suitable range of temperatures; protection from rain, drafts, and predators, and rough texture enabling bats' claws to grip the interior surfaces (Krzanowski, 1955; Tuttle *et al.*, 2004). Given that most bats are small and have a high surface area-to-volume ratio compared to large mammals, bats are highly susceptible to heat and water loss to their environment (Bassett and Studier, 1988; Feldhamer *et al.*, 2004). Energy conservation is therefore of utmost importance to bats, and roost choice plays a significant role in that regard. Bats often adjust their position along a temperature gradient according to their thermoregulatory needs; therefore, bat houses must provide a wide temperature gradient, as well as high daily temperatures suitable for maternity use (Williams and Brittingham, 1997).

As a rule of thumb, the species of bats that use bat boxes prefer darkened interiors. No matter what shape or size, bat houses invariably have some type of roof and an outer shell, which retains some heat absorbed during the day, and which prevents most sunlight and precipitation from entering the roost. Bats generally enter from (or near) the bottom of a bat house and crawl upwards into the interior to roost. A rough-surfaced board (landing pad) extending underneath the entrance at least 3 to 6 in. long (7.6 to 15.2 cm) facilitates entry and exit by bats (Marks and Marks, 2006; Tuttle *et al.*, 2004). Some house exteriors have sufficiently rough texture that bats can land on the outside and crawl in.

Most bat houses fall into one of two basic categories: those with a central, hollow chamber, and those with one or more narrow roost chambers (crevices).

Birdhouse-style bat houses used widely in Europe and Australia traditionally have one, hollow interior space (sometimes two) that permits bats to move freely about inside (Fig. 9-2). Many of those designs, though not all, also have closed bottoms. In general, bats enter birdhouse-style boxes through a narrow slot at or near the bottom. In contrast, North American crevice-style models typically are much taller than birdhouse-style models, have open bottoms and narrow crevices that run all the way up to the roof (or nearly so) and the roost chambers themselves are just

Fig. 9-2. A birdhouse-style bat box installed at Organ Pipes National Park in Victoria, Australia. Photo courtesy of Robert Bender.

large enough to accommodate the target species (Fig. 9-3). Of course, there are exceptions to the trends listed above, and there are multiple advantages and disadvantages to both types of houses.

Fig. 9-3. Little brown myotis (*Myotis lucifugus*) roosting in bat house crevices at Mowrer's Farm in Petersburg, PA. Photo by the authors, © Bat Conservation International, Austin, Texas.

Birdhouse-style designs with one or two chambers are generally small, lightweight, inexpensive, and simple to construct and install. Large numbers of birdhouse-style units may be deployed over a wide area relatively cheaply (e.g., in forest settings). Given their small size, however, they accommodate fewer bats per house and are best deployed in groups to provide bats with greater roosting options. Crevice-style bat houses on the other hand, are typically larger, heavier, more expensive, and more difficult to build and install. This type of artificial roost restricts bats' movements within the interior to some degree (some bat species practically wedge themselves into the chambers), but the provision of one or more partitions (dividers) together with tall chamber heights creates multiple temperature zones within a bat house, offering bats more microclimate choices. Their larger size and taller height also permits the formation of larger colonies; therefore, fewer houses need be erected at a given location. Birdhouse-style houses tend to hold anywhere from a few individuals to a few dozen bats on average, although 90 to 100 individuals have been found in some European and Australian boxes (Bender, 2010a; Mayle, 1990). Crevice-style houses on the other hand often shelter hundreds of bats, depending on the number of chambers and the overall width of the house (Kiser and Kiser, 2004).

Stebbings and Walsh (1991) reported that the minimum effective internal size for birdhouse-style houses in Europe is 3.9 x 3.9 x 3.9 in. (100 x 100 x 100 mm). This interior volume was determined sufficient to allow colonies to conserve heat by clustering together. They also concluded that larger boxes with internal dimensions of 5.9 in. (150 mm) wide x 11.8 in. (300 mm) tall or taller would accommodate colonies of 40 to 100 medium-sized bats, such as brown long-eared bats (*Plecotus auritus*). Stebbings and Walsh (1991) recommend keeping the front-to-back internal dimension at 3.9 in. (100 mm), as bat house occupants prefer to cluster in confined spaces. Gerell (1985) tested boxes in Swedish forests and found that bats preferred small boxes with internal volumes of 97.6 $in.^3$ (1600 cm^3) versus large boxes with 494 $in.^3$ (8100 cm^3) volumes. For most North American bats, the evidence to date suggests that large bat houses are superior to small ones (Kiser and Kiser, 2004; Marks and Marks, 2006; Tuttle and Hensley, 1993). However, forest-dwelling species, such as Northern myotis (*Myotis septentrionalis*) and long-eared myotis (*M. evotis*) that do not form large colonies, may not require bat houses as large as those preferred by such species as *M. lucifugus*, *E. fuscus* and *T. brasiliensis*. In a study outside the Indianapolis International Airport, Indiana, *M. septentrionalis* routinely occupied small, birdhouse-style boxes on trees, and even used them as maternity sites (3D/ESI, 1995, 1996).

Open-bottomed designs have fewer problems with guano build-up and parasites than close-bottomed designs and are easy to inspect for occupants (Krzanowski, 1955; Tuttle *et al.*, 2004). Bats may also be able to enter and exit more easily. Compared with open-bottomed models, closed-bottomed houses need more frequent cleaning to remove guano, but they offer better protection against certain types of predators such as raccoons. They are also more efficient at trapping heat inside and therefore buffer nighttime temperature drops better than open-bottomed houses. To reduce air circulation inside open-bottomed bat houses in cold, northern climates, they can be modified by attaching a hinged-bottom cover at a downward sloping angle, leaving one narrow opening for bat access (Kiser and Kiser, 2004). The hinged cover can be opened when needed for inspection or to facilitate cleaning; the cover can be secured in the closed position with a j-hook and eye screw combination, or with a sliding barrel bolt.

The entrance slot on birdhouse-style boxes is typically just wide enough to accommodate a bat's body (see Fig. 9-2), usually less than 0.8 in. (20 mm). Gerell (1985) recommended 0.7 in. (17 mm) for most European bats. Any larger, and birds and other non-target animals may take over the boxes. The vast majority of North American bat house occupants seem to prefer ¾- to 1-in. (19- to 25-mm) roost chambers, although a few small species accept chambers between ½ and ¾ in. (13 and 19 mm). Large species, such as pallid bats (*Antrozous pallidus*) and Florida bonneted bats (*Eumops floridanus*), appear to prefer 1½- to 1¾-in. (38- to 44-mm) crevices. A restriction plate, or partial bottom that leaves a ¾-in. (19-mm) gap across the entire width of the bat house, is recommended for houses with 1½-in. (38-mm) or larger chambers to deter non-target wildlife and predators (Tuttle *et al.*, 2004).

At least 10 species in the United States, Canada, and the Caribbean are consistently using crevice-style bat houses with multiple roost chambers and open bottoms, and other species have also been documented using them (Lisa Godinho, pers. comm.; Kiser and Kiser, 2001a; Korsten, 2006; Tuttle and Hensley 1993). Based on 1,553 bat houses reported to BCI's North American Bat House Research Project, the most successful houses in the United States and Canada are tall and/or wide with multiple crevices (Kiser and Kiser, 2004; Tuttle *et al.*, 2004). For example, nursery-style (Appendix 9-3) and rocket-style houses (Appendix 9-3), with roost crevices at least 25 in. (64 cm) tall are highly successful (Dourson and MacGregor, 1997; Kiser and Kiser, 2001a; Tuttle and Hensley, 1993). Of 1,163 North American bat houses surveyed, those with internal crevice widths of 15 to 24 in. (38 to 61 cm) performed better than those with smaller widths; houses with 25-in. (63.5-cm) or greater crevice widths had the highest occupancy rate (Tuttle *et al.*, 2004).

Ventilation holes or slots are features found on many modern North American bat houses; they are highly recommended for bat houses installed in warm climates (Tuttle *et al.*, 2004). Horizontal or vertical ½-in. (13-mm) slots on the front and sometimes on the back are generally located in the lower third of the house to provide cooling for the lower portion of the chambers (Marks and Marks, 2006; Tuttle and Hensley, 1993; Tuttle *et al.*, 2004). As bats adjust their positions inside a bat house throughout the day to find the most optimal temperatures, they tend to cluster at or near the top of the house during the coolest hours, and at or near the bottom during the warmest portion of the day. Ventilation slots are thought to enhance the overall temperature gradient from top to bottom, permitting bats to remain comfortable during periods of heat build-up in summer. Nevertheless, the successful rocket boxes designed and tested by Dourson and MacGregor (1997) in Kentucky have ½- to ¾-in. (13- to 19-mm) circular vents covered with screen on two opposing sides at the tops of the boxes, which reduce heat build-up at the top of the roost.

BAT HOUSE CONSTRUCTION MATERIALS

Bat houses have been constructed from a wide variety of materials such as lumber, plywood, PVC plastic, clay, concrete, concrete-sawdust blends, and rigid foam insulation (Gerell, 1985; Kiser and Kiser, 2001b; Mayle, 1990; Stebbings and Walsh, 1991). Some commercially built bat houses with metal exteriors affixed over wood or rigid foam insulation have also attracted bats (Kiser, 2002a). Regardless of the design, the vast majority of bat houses around the world are constructed from some type of wood or wood product, owing to its relatively low cost, widespread availability, and ease of cutting and assembly.

Many successful bat houses in North America today are constructed using a combination of solid lumber and exterior-grade plywood (Tuttle *et al.*, 2004). Plywood should have a minimum of four plies and have exterior-grade glue to withstand the elements. Virtually any type of wood species can be used for most designs, although some wood species resist deterioration, splitting and warping better than others. Stebbings and Walsh (1991) suggested that elm should be avoided due to problems with warping, and that some species of cedar might have objectionable odors to bats. In North America, western red cedar and tulip poplar are favored by many builders; softwoods like pine, spruce and fir are often used in European and Australian boxes (Bender, 2010b; Mayle, 1990; Stebbings and Walsh, 1991). Some researchers have suggested that bats prefer aged wood and may object to odors in newly-constructed boxes (Krzanowski, 1955), although there are numerous examples of bats using newly constructed boxes within one month of installation (Kiser and Kiser, 2004; Kiser and Kiser, 2005a; Marks and Marks, 2006; Tuttle *et al.*, 2004). Where roost surface texture is not adequately rough, horizontal grooves in wooden houses can be cut ½ in. (13 mm) apart, by approximately 1/16 in. (1.6 mm) deep. Grooves cut into plywood should never go through the first ply, or the wood will delaminate. For longevity, exterior-grade screws are always preferable to nails when assembling wooden boxes, as they hold wood pieces together more securely (Marks and Marks, 2006; Tuttle *et al.*, 2004). Because wood expands and contracts as temperatures fluctuate, over time gaps may form between the boards, which may force bats to abandon a box. All exterior joints should be caulked or sealed in some fashion to keep out rain and drafts.

Lumber used to construct most small boxes has a thickness of ¾ to 1 in. (19 to 25 mm). This size is a good compromise between weight and cost; the heat retention and insulative properties for this thickness are sufficient for occupancy most months of the year (Stebbings and Walsh, 1991). In Australia, researchers experimented with 1.8-in. (45-mm) thick lumber to develop boxes suitable for hibernation (Bender, 2005; Bender, 2010b). For the first seven years, all but two of the thick-walled boxes were used sparingly in winter, if at all, and received more spring and fall use. However, by 2010, the boxes were receiving considerable use as winter roosts, which surprisingly led to more winter use in the thin-walled boxes at the study site (Bender, 2010b).

Fig. 9-4. Schwegler bat box 2FN. A) box mounted on a tree with front panel closed; B) box interior with roosting brown long-eared bats (*Plecotus auritus*). Photo courtesy of Schwegler Vogel-und Naturschutzprodukte GmbH.

In Europe, bats regularly use small boxes made from a mixture of concrete and sawdust (Mayle, 1990; Stebbings and Walsh, 1991; Tuttle and Hensley, 1993). Advantages of this material include long life spans, good heat retention properties, and a good surface texture for bat claws to grip (Fig. 9-4). Gerrell's (1985) comparison of box designs in Sweden found that bats preferred porous concrete or cement-sawdust boxes to wooden models. Graczyk (1974) found that bats in Poland preferred cement-sawdust boxes as well. A simple box with one hollow space can be fashioned in a mold at home by the do-it-yourselfer, but owing to the time-consuming process, today most are commercially produced (Stebbings and Walsh, 1991). Schwegler in Germany is one such manufacturer with numerous ready-made models available (Appendix 9-3) that are used extensively by European bats, including greater and common noctules (*Nyctalus lasiopterus* and *N. noctula*), mouse-eared myotis (*Myotis myotis*), Bechstein's myotis (*M. bechsteinii*), *Plecotus auritus* and common pipistrelle (*Pipistrellus pipistrellus*) (Gianna Dondini, pers. comm.; Schwegler, 2011). These concrete-sawdust houses have not been tested widely in North America, but at least one has attracted southwestern myotis (*Myotis auriculus*) near Flagstaff, Arizona (Horncastle and Bayless, 2005), a species not previously reported in bat houses.

A number of bat houses constructed of polyvinyl chloride (PVC) have attracted bats. In Australia, in 1995, Terry Reardon (pers. comm.) reported success with boxes made of 5.9-in. (15-cm) diameter PVC sewer pipe painted black. This box design was 23.6 in. (60 cm) long, had a removable cap on top, and was left open at the bottom. Coarse fabric (hessian bagging) was used on the interior to provide a rough surface for bats to grip. Lesser long-eared bats (*Nyctophilus geoffroyi*) moved in within three weeks of placement in 1995. Texas-based bat house builder Marvin Maberry designed numerous PVC-shelled models with welded seams, both nursery- and rocket-style designs, with and without insulation, that are attractive to bats (Bat Conservation International, 1995a, 1995b, 1996; Kiser and Kiser, 1999; Kiser and Kiser, 2001b). Maberry's first non-insulated, nursery models had exterior wall thicknesses of only ⅛-in. (3.2 mm) and thus heated up and cooled down extremely quickly (Kiser and Kiser, 2001b). Later models were coated inside and out with a commercial-grade stucco mixture to provide a rough surface and to reduce temperature fluctuations. Insulated models made with either plastic or galvanized steel shells and ½-in. (13-mm) thick foam board performed well; one 4-ft. (1.2-m) wide version attracted a maternity colony of 332 *E. fuscus* in northeast Texas (Kiser, 2002a). A rocket-style house made from PVC pipe with an internal reservoir in the upper half capable of holding 12 pounds (5.4 kg) of dry sand was also successful (Kiser and Kiser, 2005a); several of these sand-filled models in northeast Texas are used by up to 60 *E. fuscus* (Marvin Maberry, pers. comm.) and others around the United States were used by *M. evotis* and *M. lucifugus*. These species, along with *T. brasiliensis*, have adopted Maberry's insulated and thin-shelled, non-insulated designs, in both the United States and Canada.

Because bats often select roost sites that are warmed by the sun, painting or staining exteriors of bat houses can help promote a suitable range of temperatures inside. Paint or stain can also help protect wood and other materials against the elements. The actual paint or stain color is usually not important, but the shade of the color (dark, medium or light) is. Depending on the shade of color, warmer or cooler temperatures can be offered versus untreated boxes. Krzanowski (1969) was one of the first to recommend painting exteriors dark colors to promote heat gain to foster maternity use. However, Tuttle (1988) and Stebbings and Walsh (1991) cautioned against painting houses due to the possibility that toxic odors may repel bats. Over time, as wide-spread experimentation demonstrated the benefits of painting and staining bat houses, BCI and others began recommending exterior-grade, water-based products to protect wood, avoid toxicity problems and to influence temperatures. BCI suggests colors for North American houses based upon average daily high temperatures in July: black or dark colors for areas 85 °F (29.4 °C) or less; dark or medium shades for areas 85 °F to 95 °F (29.4 °C to 35 °C); medium or light shades for areas 95 °F to 100 °F (35 °C to 37.8 °C); and light shades or white where temperatures exceed 100 °F (37.8 °C).

BAT HOUSE MOUNTING AND PLACEMENT STRATEGIES

Proper placement is a crucial factor in any bat house' s ability to attract bats. Even the most tried-and-tested models will likely fail if installed in an unsuitable or unsafe location. Tree branches, other vegetation and similar obstacles should be kept at least 15 ft. (4.6 m) away from bat houses (Stebbings and Walsh, 1991; Tuttle *et al.*, 2004) to enable bats to find the bat house easily, and to reduce problems with shading and predators. Mounting height is a particularly important variable to consider for both the "landlord" maintaining the house and for the bats themselves. Placement height should be convenient for conducting inspections, performing occasional cleaning tasks, and making necessary repairs from time to time. From a bat' s perspective, however, roost height is important for other reasons. One primary concern relates to protection against climbing predators like snakes, raccoons and cats. Because a bat's wing structure dictates its flight and foraging characteristics, and because wing morphology and foraging heights vary from species to species, it follows that different bat species have different roost height preferences. Long-eared bats (*Plecotus* spp.) and a few additional species in Britain accept boxes mounted 4.9 ft. (1.5 m) above the ground, whereas noctules (*Nyctalus* spp.) prefer houses at least 16.4 ft. (5 m) high (Stebbings and Walsh, 1991). Molossid bats such as *Tadarida* spp., with their long, narrow wings, also seem to prefer high roosts: These species are not as maneuverable as other bats and "free-fall" out of a roost before becoming fully airborne. Kiser and Kiser (2002) recommended that all boxes be installed at least 12 ft. (3.7 m) high for vespertilionid bats and 15 ft. (4.6 m) high for molossids; Krzanowski (1969) recommended heights of at least 13 ft. (4 m) for all boxes in Europe.

Thermal conditions within a bat house are ever-changing and are influenced by multiple factors, including mounting substrate, amount of direct sun exposure, type and thickness of the construction materials, roof color, amount of roof overhang and exterior color. Mounting substrate is an important variable for bat house landlords to consider. Most occupied houses in North America are situated on wooden or masonry buildings or on sturdy, wooden or metal poles (Kiser and Kiser, 2001a; Tuttle and Hensley, 1993). Buildings and other large structures typically provide the strongest attachment points and even more important, a measure of thermal stability not possible with pole- or tree-mounted bat houses. Regardless of any other considerations, bat houses mounted on buildings in BCI' s survey were occupied 64% of the time (Tuttle *et al.*, 2004). Masonry structures excel at absorbing and retaining heat compared to structures made of wood or metal. In general, the exteriors of metal buildings seem less attractive to bats, owing to great temperature fluctuations and a high likelihood of overheating. However, the authors have witnessed occupied bat houses in central California that were suspended from the ceilings of metal barns and sheds. Flat, single-chambered houses like those in *The Bat House Builder's Handbook* (Tuttle and Hensley, 1993; Tuttle *et al.*, 2004) should be mounted on building exteriors for best results.

Wooden and steel poles allow greater flexibility and freedom for bat house landlords in terms of placement choice. Poles must be sturdy enough to withstand high winds. Steel poles with a 2-in. (5.1-cm) inside diameter work well for many types of bat houses. Wooden 4 x 4-in. (10.2 x 10.2-cm) posts are less expensive than steel pipe, but sometimes they warp, and the portion buried in the ground eventually rots. Four x 6-in. (10.2 x 15.2-cm) or larger wooden posts are a better choice because they are stronger than 4 x 4s and less prone to warping. Of 960 bat houses mounted on poles in BCI' s North American survey, 52% were used by bats, regardless of other factors (Tuttle *et al*, 2004). Pole-mounted houses appear to work best in areas where temperatures fluctuate less than 28 °F (15.6 °C) from day to night (Kiser and Kiser, 2002).

Tree-mounted bat houses seem to work well for a few forest-dwelling species in North America, such as *M. septentrionalis*, but not for most species (Ritzi *et al.*, 2005; Tuttle *et al.*, 2004). For example, of 974 tree-mounted houses installed as mitigation for Indiana myotis (*Myotis sodalis*) near the Indianapolis International Airport, less than 10 were ever used by *M. sodalis* (Ritzi *et al.*, 2005). According to BCI' s research, tree-mounted bat houses in the United States and Canada are occupied only 20% of the time (Tuttle *et al.*, 2004). In contrast, European and Australian bat species readily accept bat houses mounted on trees, especially when multiple houses are installed on each tree (Bender, 2010a, 2010b; Krzanowski, 1955; Stebbings and Walsh, 1991). Whether North American occupancy rates for tree-mounted houses could be improved, and additional species accommodated on a regular basis with better designs, remains to be seen. Rocket-style boxes mounted on poles, which simulate exfoliating bark and tree crevices, are already being used by a number of forest-dwelling bats in the United States and Canada, including *M. sodalis* (Carter, 2002; Carter *et al.*, 2001; Dourson and MacGregor, 1997).

Installing several bat houses in groups to provide greater options is highly recommended whenever possible, especially when the houses are small because bats often move from one bat house to another. Advantages to multiple roosts include access to a variety of microclimates, ectoparasite avoidance, supplemental room for colony expansion, and alternative roosts if one roost site is disturbed, destroyed or predated (Whitaker, 1998; Williams-Whitmer, 1994). Data from BCI indicate that when three or more bat houses were installed in a group, 80% were successful, versus 49% compared to single bat houses (Kiser and Kiser, 2004). Whenever large colonies are to be evicted from buildings or other large structures, multiple houses are recommended to provide sufficient room for the entire colony.

Compass orientation plays a large role in the amount of sun exposure a typical bat house receives, and therefore directly affects the microclimate inside the roost. Rocket boxes with continuous, 360° roost chambers do not have a set orientation, although sunny and shaded sides are still present; bats are free to move about inside and to select warmer or cooler chambers as needed (Dourson and MacGregor, 1997). Other bat house designs may be successful no matter what mounting orientation is chosen; however, orientation preferences appear to vary with the seasons. Of 43 boxes mounted in 8 directions at Organ Pipes National Park in Victoria, Australia, those oriented to the east, northeast, and southeast (receiving morning sun exposure) were most favored by bats. Boxes facing west, northwest and southwest received afternoon sun and were favored slightly less (Bender, 2010b).

European and Australian bat houses are often mounted in groups of two (90° or 180° apart) or three (120° apart in a triangular pattern) on tree trunks such that some houses receive sun and some are shaded (Fig. 9-5). The placement of boxes on the north, southeast and southwest-facing sides of a tree has proven to be a particularly successful strategy for European houses, as the southeast-facing box is warmed by the morning sun, the southwest-facing box receives afternoon sun, and the north-facing box is largely shaded and therefore cooler (Stebbings and Walsh, 1991). Bender (2010b) found that boxes in Australia mounted side by side on tree trunks, 90° apart, often had complementary use patterns; when one box was occupied, the adjacent box typically was not.

Fig. 9-5. Mounting multiple bat boxes on trees, with both sunny and shaded orientations, is a highly successful strategy for attracting European and Australian bats. Photo courtesy of Robert Bender.

In North America, research indicates that bat houses located within ¼ mi. (400 m) of an open freshwater source are more likely to be occupied by vespertilionid bats than houses installed farther away (Kiser and Kiser, 2004; Tuttle *et al.*, 2004). However, molossid bats such as *T. brasiliensis* are able to travel longer distances given their wing morphology, and can occupy bat houses more than 1 mi. (1.6 km) from water (Tuttle *et al.*, 2004). In the Daniel Boone National Forest in Kentucky, rocket-style bat houses were most successful in upland forests with open canopies or in small openings along edge habitat near ponds or in riparian zones (Dourson and MacGregor, 1997). In one BCI study, bat houses were more successful when the nearest freshwater source was a lake greater than 25 acres (10.1 hectares) or a river wider than 50 ft. (15.2 m), compared to marshes, streams and ponds (Tuttle *et al.*, 2004).

BCI' s North American research also indicated that bat houses installed in wetlands and desert areas had lower occupancy rates (28% and 31%, respectively) than houses installed in forests (54%), grasslands (55%) and overgrown fields/shrubs (61%) (Tuttle *et al.*, 2004). The authors speculate that wetlands such as freshwater marshes may have a paucity of natural roost sites, and therefore may have lower bat populations, which lowers the odds of box use. Bat houses have not been tested as extensively in desert areas of the United States and as such, accommodating bats there has proven more difficult. The major challenge for bat house builders is that roosts in extremely hot climates are prone to overheating. Tatarian (2001) designed a nursery house with roost chambers under a large overhanging roof (http://www.wildliferesearchassoc.com/index_files/Page1095.htm) that proved attractive to *A. pallidus* in California. The house resembles an upside-down "L" and contains both vertical and angled roost chambers. Entry slots near the top of the box where the top of the box and overhang join also serve as ventilation slots, reducing the possibility of overheating. Pallid bats prefer roost temperatures in the 80° F to 90° F (27° C to 32° C) range, and usually abandon roosts if temperatures exceed 100° F (38° C). Tatarian modified his original design after overheating problems were discovered; his newer, taller model accommodated maternity colonies of up to 300 *A. pallidus* and 400 *T. brasiliensis.*

Some farmers install bat houses on their farms and in their orchards for biological pest control. According to BCI' s data, those boxes installed on farms with a combination of pastures, row crops, and orchards performed better than boxes on farms with row crops only (68% vs. 24% occupancy, respectively) (Tuttle *et al.*, 2004). Where the dominant land use within a 1-mi. (1.6-km) radius was either natural vegetation or a combination of farms and natural vegetation, occupancy rates were higher (59% and 57%, respectively) than for residential areas and cattle ranches (48% occupancy). North American bat houses installed in rural settings were more likely to attract bats than those in urban or suburban areas (Tuttle *et al.*, 2004).

BAT HOUSE USAGE PATTERNS

Bats utilize bat houses for a variety of reasons, as bachelor roosts, migratory stopovers between summer and winter sites, maternity roosts, overwintering/hibernation roosts, and as night roosts between foraging bouts. A single bat house may function as more than one type of roost at different times of the year. Of the colonial species that use them, males and females typically form separate colonies during the breeding season. Males may be solitary, or they may form small bachelor groups of several dozen individuals; males are generally less demanding about their roost requirements and select cooler sites than are used by females for rearing young.

Bat houses may be used in any month of the year, though occupancy is generally highest in the warmest months of the year, during the maternity season. In one study involving 20 boxes at Wilson Reserve, Ivanhoe, Australia, the greatest usage reported was from December through February, during the breeding season; the largest groups of bats in each box were found during these months, when juveniles are dependent on their mothers. No groups larger than 20 bats were found during Australia's winter months of June, July and August (Bender, 2010a). A similar pattern was evident at Organ Pipes National Park in Victoria, Australia; of three designs tested (37 boxes total), monthly usage was greatest from October through February (Bender, 2010b). In one study conducted in England, Ireland and Wales, bat house usage was year-round but was highest in August and September, and lowest in February and March (Poulton, 2006). Stebbings and Walsh (1991) reported that in lowland areas of Great Britain, boxes are used mostly from April to November, but at higher elevation locations, boxes may be used only from late May to September. Boxes made from 0.8 to 1.2-in. (20 to 30-mm) thick lumber were acceptable to bats year-round, except when winter temperatures fell below 24.8° F (4° C).

In North America, bats may begin returning to bat houses from hibernation or overwintering sites such as caves or mines as early as February or March, or as late as July, depending on species, latitude, and elevation. Most North American maternity colonies in bat houses typically stay no later than August or September, but some individuals may linger until November or December (Kiser and Kiser, 2000). Non-migratory populations of *T. brasiliensis* in the southeastern United States and the west coast, and Pallas's mastiff bats (*Molossus molossus*) in the Cayman Islands occupy bat houses year-round.

Bat houses are used as hibernation or overwintering sites by numerous species, though in North America, fewer bats have been reported using them in winter than in Europe. Concrete-sawdust houses, such as those built by Schwegler in Germany, are commonly used for hibernation by European bats (Poulton, 2006; Schwegler, 2011). Schwegler's "Woodcrete" models 2FN and 1FF, and 1FW in particular, are used by many species in winter (Schwegler, 2011); Henry Schofield (pers. comm.) observed up to 60 noctules (*Nyctalus noctula*) in the large 1FW model. A simple wooden design with three narrow crevices known as the CJM box has been used as winter roosts by at least 7 European species (Vincent Wildlife Trust, 2011).

There is no proven method of attracting bats to a bat house (Marks and Marks, 2006). Bats have to locate, investigate, and accept a new roost site on their own. Using echolocation, bats are able to detect the presence of potential roost cavities. Various methods of coaxing bats into bat houses have been tried, from applying guano to using aged wood from an actual bat roost to broadcasting recordings of captured bats (Marks and Marks, 2006). Although no conclusive evidence exists to suggest that these strategies are effective, one experimental trial in Australia showed that Gould's wattled bats (*Chalinolobus gouldii*) released into an outdoor flight cage preferred a roost box treated with guano collected from their own species versus an untreated box (Payne *et. al.*, 2011). If there is a local need for roost space, and the bat house meets bats' basic requirements for shelter, then bats will likely move in. How long it takes for a bat house to first become occupied varies widely. In one North American study, time until first occupancy for 735 bat houses ranged from 0 to 158 months; average time was 9 months. Sixty-one percent of occupied houses in that study were used within 6 months; 89% of occupied houses had bats within 24 months of installation (Tuttle *et al.*, 2004). In Dorset, England, where the first bat houses in Great Britain were tested in 1968, bats moved in within 5 months (Stebbings and Walsh, 1991). At Organ Pipes National Park in Victoria, Australia boxes were not used by bats for the first two years (Bender, 2010). Bat houses not occupied after several years can be repositioned to provide different conditions and to improve odds of inhabitation.

OTHER TYPES OF ARTIFICIAL ROOSTS DESIGNED FOR BATS

RETROFITTING STRUCTURES

Existing bridges and culverts can be made to attract bats with retrofits (http://environment.transportation.org/environmental_issues/construct_maint_prac/compendium/manual/7_3.aspx). A variety of artificial roost types like the "Texas Bat Abode" composed of vertical plywood panels can be adapted to fit between steel I-beams and cast concrete beams under bridges, and single plywood panels such as the "Oregon Wedge" can be secured to vertical surfaces (e.g., pillars and bent caps) (Keeley and Tuttle, 1997, 1999). Biologists in Oregon successfully attracted bats to flat-bottomed bridges with simple wooden boxes anchored to the undersides of bridge decks (Arnett and Hayes, 2000). A manufacturer of commercial bat houses, Maberry Centre Bat Homes of Daingerfield, Texas developed a large, rectangular model with 16 roost chambers that fits between bridge beams. The roost was 24.5 in. (62.3 cm) tall x 25 in. (63.5 cm) wide x 54.5 in. (138.4 cm) long and was composed of a galvanized steel shell and 1/2-in. foam board insulation. The authors inspected one of these bridge roosts installed in an interstate overpass in Hays County, Texas, which had attracted approximately 1,000 *T. brasiliensis* in less than two years.

Today, several state transportation departments have taken the next step by constructing bat-friendly bridges and culverts with built-in bat roosts (Keeley and Tuttle, 1999). At the Mokelumne River in Sacramento County, California the California Department of Transportation removed a wooden trestle bridge used by 40,000 bats of several species and replaced it with a new concrete, bat-friendly bridge (http://www.dera.saccounty.net/portals/0/Docs/ProtectedResources/SpecialStudies/BatReportProjectReport.pdf). Roost chambers made from the old wooden bridge timbers were cast in the concrete on the underside of the new bridge deck (Johnston, 2005). The chambers are 13 in. (33 cm) deep, and there are 13,600 ft. of linear roost space (about 2.5 mi. [4.1 km]) of bat habitat that could potentially house up to 680,000 bats. By summer 2004, approximately 10,000 bats had moved into the bridge chambers, despite the fact that the bridge would not be completed until 2005 (Kiser and Kiser, 2005b).

ARTIFICIAL BARK

Fig. 9-6. Tom Finn with Fly By Night, Inc. installs a piece of Bat Bark on a pine tree in Volusia County, Florida. Photo by the authors, © Bat Conservation International, Austin, Texas.

Other kinds of artificial structures designed for bats have also been successful. One of these, Bat Bark (Fig. 9-6), mimics loose tree bark found on snags (dead or dying trees). Because numerous species roost underneath loose bark, which often is in short supply and falls off after several years, synthetic Bat Bark was designed as a substitute to last for decades (Noteman, 1998). In one study in the Kaibab National Forest, Arizona, two types of Bat Bark were tested on trees; one group was made from polyurethane foam and another was made from fiberglass. Bark pieces were curved, flexible, and patterned to match natural tree bark. Individual Bat Bark pieces were 24 in. (61 cm) tall x 20 in. (50.8 cm) wide and were secured to trees with screws along the top and sides. A narrow gap was left at the bottom between the tree and the bark pieces for bats to enter and exit. Bats used 90% of the Bat Bark units by the second year of installation (Siders, 1999). Southeast, southwest, and west orientations worked best; mounting heights of 23 ft. (7 m) or higher were preferred.

Another similar but simpler (and less expensive) type of artificial bark can be fashioned from a piece of corrugated metal wrapped around a tree trunk. Because Bat Bark must be handmade, the average homeowner would probably not want to attempt it. It would be best to use pieces of corrugated metal or corrugated fiberglass cut to appropriate size. At one location in Missouri, metal sheets were used as predator guards for wood duck boxes; however, several nursery colonies of *M. lucifugus* were found underneath them within three years of installation (Clawson and Gardner, 1993). On live trees, artificial bark pieces must be monitored often to ensure that trees are not girdled and that proper spacing between the tree trunk and bark piece is maintained as the tree grows. For additional information on artificial (bark) roosts, see http://library.eri.nau.edu/gsdl/collect/erilibra/index/assoc/HASH0eda.dir/doc.pdf.

ARTIFICIAL TREES AND OTHER HOLLOW STRUCTURES

In the United States, at least 18 artificial hollow trees made from concrete culverts, manhole shaft sections, concrete/wood fiber pipes (HardiePipe™, Fig. 9-7), cinder blocks (Fig. 9-8), and metal culverts have successfully attracted 5 species of bats in 6 southeastern states, including colonies of *C. rafinesquii* in Georgia, Texas, Mississippi and Kentucky (Kiser and Kiser, 2003; Bayless, 2006). Concrete has several advantages as a construction material, including thermal buffering, heat retention, and long life span; disadvantages of concrete tree roosts include cost and difficulty in transporting materials. Heavy equipment is required for installation and adequate roads to the construction site must be available, which may preclude installation in bottomland forest locations where *C. rafinesquii* resides. Two similar, but less expensive, tree roosts made from metal culverts at the Lower Suwannee National Wildlife Refuge in Dixie County, Florida were covered with cypress boards on the exterior (for thermal buffering) and had an elastomeric compound applied to the roost interior to provide surface texture for bats to grip (Steve Barlow, pers. comm.); the authors have observed individual *C. rafinesquii* in both metal roosts. Artificial tree hollows range in size from 2 to 4 ft. (0.6 to 1.2 m) in diameter and from 12 to 16 ft. (3.7 to 4.9 m) in height. Most have a large opening at or near the bottom to simulate the irregular, often triangular, openings found in tupelo (*Nyssa* spp.) and cypress (*Taxodium distichum*) trees used by bats in swamps and floodplain forests. One or more small entrance slots or holes are created near the top (simulating knotholes in natural trees), which allow bats to escape in the event floodwaters or predators enter the roost. Artificial tree hollows are usually installed in groups of two or three in sunny and shady locations to provide options for bats because *C. rafinesquii* frequently switches tree roosts; having a variety of microclimates to choose from appears to be important for forest-dwelling bats (Bayless, 2006). Roost-switching may also fool predators or reduce ectoparasites.

In Costa Rica, researchers have attracted at least 10 species of insectivorous, frugivorous, nectivorous, carnivorous and sanguivorous bats to bat boxes and artificial tree hollows constructed of concrete/wood fiber materials (Kelm, 2008; Kelm *et al.*, 2008a; Reid *et al.*, 2009). The roosts constructed for one study at La Selva Biological Station in Costa Rica were successful

Fig. 9-7. An artificial hollow tree constructed from concrete/fiber pipe, St. Catherine Creek National Wildlife Refuge, Adams County, Mississippi. Photo by the authors, © Bat Conservation International, Austin, Texas.

Fig. 9-8. An artificial hollow tree constructed from concrete blocks, Trinity River National Wildlife Refuge, Liberty County, Texas. Photo by the authors, © Bat Conservation International, Austin, Texas.

in shaded locations within both primary tropical forest and deforested (agricultural) areas. Roosts were constructed of 9 concrete-sawdust slabs that formed a square, free-standing, hollow box with a single, 1.3 x 1.3-ft. (0.4 x 0.4-m) square opening in the lower half. Two roost sizes were built; internal measurements were 1.8 x 1.8 x 6.4 ft. (0.5 x 0.5 x 1.9 m) and 2.4 x 2.4 x 5 ft. (0.7 x 0.7 x 1.5 m). Concrete-sawdust slabs were selected for their ability to withstand the tropical wet climate, termite resistance, their weight (ease of transport) and cost-effectiveness. Plastic netting was attached to the ceilings to provide a textured surface for bats to grip (Kelm, 2008). All 43 structures were occupied (100% colonization); 10 species of bats used the artificial tree hollows, including 3 species of *Carollia* and 3 species of *Micronycteris*; 2 roosts in forested habitat also attracted 1 to 2 common vampire bats (*Desmodus rotundus*). Average colonization time for all roosts was 3 weeks in both forested and agricultural locations; fastest time to first colonization was 3 days (Kelm *et al.*, 2008a). Numbers of bats ranged from 1 to approximately 100; 4 species formed colonies of 19 or more individuals. Numbers of bats were comparable to those in natural roosts, except for *D. rotundus*, which forms large colonies and is considered undesirable in agricultural areas. *D. rotundus* would likely prefer larger artificial roosts of the same design, but keeping roosts at the current size may prevent large numbers of these bats from using them (Kelm *et al.*, 2008a). Nine bat species used the roosts in the forested habitat and 6 species used the roosts in disturbed habitat; 5 species used the boxes in both habitat types (Kelm *et al.*, 2008a). Given the success of these artificial roosts and rapid colonization rate, Kelm *et al.* (2008b) concluded that installing the roosts in forest patches, or under shade trees in cleared areas of the tropics, would increase the likelihood that seed-dispersing bats would colonize the area, thus aiding reforestation efforts.

In another similar study conducted at Coto Brus, Puntarenas in Costa Rica, researchers installed 30 concrete/wood fiber (fibrolit) boxes on poles or live tree branches (stakes) inserted into the ground. Fibrolit boards are lightweight, inexpensive, and readily available in Central American hardware stores. The boxes were smaller (1/5 to ⅛ the size) than those in Kelm *et al.* (2008a); they also had no bottoms to better facilitate seed and macronutrient deposition below the roosts (Reid, 2009; Reid *et al.*, 2009). Six of the boxes were used as night roosts by striped hairy-nosed bats (*Mimon crenulatum*) and possibly other species in both forested and disturbed habitats.

LARGE-SCALE/COMMUNITY BAT HOUSES

Large, free-standing bat houses (often called "community bat houses") on poles are sometimes successful, but owing to their extensive costs, they are recommended only when large colonies are facing eviction/exclusion from a nearby roost. Perhaps the most successful large-scale bat house in the United States is an 18 x 18-ft. (5.5 x 5.5-m) roost constructed in 1991 at the University of Florida in Gainesville, Alachua County. This experimental design (Fig. 9-9) featured a large, open attic, a gabled roof, extensive insulation in the floor, walls and roof, as well as 180 vertical roosting chambers 18 in. (45.7 cm) tall on the underside of the roost. Three chamber sizes were provided for bats: ¾ in. (19 mm), 1 in. (25.4 mm), and 1½ in. (38 mm). The bat house is supported by five, 18-ft.

(5.5-m) utility poles and was built to provide alternative accommodations for a colony of *T. brasiliensis* displaced from multiple campus buildings (Glover, 1994; Kiser and Glover, 1997). Located in an open field 267 ft. (81 m) from Lake Alice, it attracted 20,000 free-tailed bats and southeastern myotis (*Myotis austroriparius*) by 1996. Bats used the ¾- and 1-in. (19- and 25.4-mm) chambers first, then later moved into the widest chambers. As of 2010, the number of bats had increased to more than 100,000 (Ken Glover, pers. comm.). A second, similar roost with improved features was built adjacent to the first one in spring 2010 to provide additional space for the growing colony.

Fig. 9-9. The University of Florida bat house, situated across from Lake Alice on the university campus, Gainesville, Florida, attracted approximately 100,000 bats. Photo courtesy of Susan M. Barnard.

The Pennsylvania Game Commission (PGC) developed an 8 x 8 x 8-ft. (2.4 x 2.4 x 2.4-m) Bat Condo with a closed bottom that proved highly effective in accommodating large numbers of *M. lucifugus* at multiple locations across the United States (Cal Butchkoski, pers. comm.; Kennedy and Tuttle, 1997; Kiser and Kiser, 2000). The Bat Condo sits atop four, 10-ft. (3-m) posts and features an enclosed flight space and 80 plywood roosting partitions inside. The enclosed flight space provides a protected place for young bats to develop their flight skills. Construction plans are available from the PGC (see *Products Mentioned in Text* for address).

There are numerous examples of large-scale bat houses failing to attract bat colonies, either because of design flaws, unsuitable locations, or other reasons. The majority of Dr. Campbell' s bat towers were not successful (Goldman, 1926; Mohr, 1972). In 1992, Auburn University in Lee County, Alabama constructed a large wooden roost with a spacious attic similar to the University of Florida' s design to shelter *T. brasiliensis* displaced from campus buildings. Unfortunately, this experimental bat house was situated amongst tall trees where it received partial shade much of the day. The roost also lacked insulation and suitable roosting crevices underneath the structure (Kiser, 1996). No bats ever adopted the structure. The authors have observed numerous, unsuccessful large houses of various experimental designs throughout central Texas and central California; experimental community bat houses in Oldsmar, Florida and three in Tallahassee, Florida have likewise failed to attract colonies (Marks and Marks, 2006). It is recommended that prior to erecting any large bat roost that smaller, proven nursery-style houses be installed on site as a first step to gauge whether bats can be attracted to that locale. If the small-sized houses become occupied, then there is a good chance that an appropriately designed, large-scale roost will eventually be accepted.

The large bat house at the University of Florida was not used as a year-round roost until 1995, four years after its completion, despite excluding bats from multiple buildings and translocating or "seeding" captured bats into the bat house early on. Translocating bats into the Auburn bat house was likewise attempted, but those efforts did not result in successful occupancy (Kiser, 1996). The practice of relocating bats to small bat houses installed less than 1 mi. (1.6 km) from a roost site is sometimes successful when performing a bat exclusion from a structure (Frantz, 1989), although the average person should not attempt this due to the possibility of being bitten. Only trained professionals (with appropriate immunizations) should ever attempt to handle bats. Bats forcibly relocated to a new roost tend to return to their original roost within a short time period.

SHEDS AND BUILDING MODIFICATIONS FOR BATS

For species that do not use crevices or standard types of bat houses, such as Townsend's and Rafinesque's big-eared bats (*Corynorhinus townsendii* and *C. rafinesquii*), other types of hollow roosts can be created. Bat Sheds are free-standing structures resembling buildings that are set on the ground. A 4 x 8-ft. (1.2 x 2.4-m) wooden shed built as a replacement roost for *C. rafinesquii* using a dilapidated trailer attracted up to 11 individuals at the Disney Wilderness Preserve in central Florida (Laura Finn, pers. comm.). In the Aomori Prefecture in Japan, a shed-like structure was constructed in 1975 as a replacement home for 10,000 Asian particolored bats (*Vespertilio sinensis*) that were to be evicted from a shrine. After 2 years, approximately 5,000 bats moved in and later used the new roost as a maternity site for 30 years (Kuniko Kawai, pers. comm.). Nigel Fernsby (pers. comm.) in South Africa renovated an unused, two-room shed for bats that attracted a maternity colony of 100 Egyptian slit-faced bats (*Nycteris thebaica*). The corrugated metal roof of the 9.8 x 21.3-ft. (3 x 6.5-m) shed was replaced with a thatched roof; the windows were covered, and the interior and exterior walls were plastered. Bats leave the shed in summer (February) and return in winter (July). The shed is also used as a night roost. A pair of Mauritian tomb bats (*Taphozous mauritianus*) later took up residence on the outside of the shed under the roof overhang.

In west Dorset, England, lesser horseshoe bats (*Rhinolophus hipposideros*) that had been evicted from a large building moved to the roof area of a small, 8.2-ft. (2.5-m) tall outbuilding with a reduced range of temperature options. Researchers constructed a hollow, cinder block cool tower inside the building to provide the bats with a roosting area that would not overheat. Thirty bats began using the cinder block tower within days. The voids in the cinder blocks can be filled with earth to provide additional insulation (Vincent Wildlife Trust, 2010). In Canada, a colony of Townsend's big-eared bats moved into a special room created for them by subdividing the attic in a large hotel (Dixie Pierson, pers. comm.). Crevice-dwelling species such as *M. lucifugus* that use traditional bat houses can be accommodated in attics of buildings by providing a series of full sheets of roughened plywood secured to vertical posts (Kennedy, 1996).

ARTIFICIAL CAVES

Where suitable, natural caves are scarce, artificial caves may be acceptable for either maternity use or as hibernacula, as long as the structures provide appropriate microclimates for the target species. Concrete military bunkers, both above- and below-ground, have attracted bats in North America, Europe and Asia (Voute, 1989; Levin, 2009). However, few real-world examples of purpose-built, artificial caves for bats are available. A three-domed, artificial cave made of steel lath, steel rebar and concrete (gunite) set into a hillside in Blanco County, Texas (Sander, 1997), attracted *M. velifer* and *T. brasiliensis* within several years of completion (Locke, 2003). Not until the cave was buried underneath the soil did internal temperatures stabilize enough for bats to show interest. Initially, bats used the cave only during migration. However, in recent years the artificial cave has supported a maternity colony of approximately 60,000 *T. brasiliensis* and 7,000 *M. velifer* (David Bamberger, pers. comm.). Up to 200,000 bats use the cave at certain times of the year.

CONCERNS REGARDING BAT HOUSES

Some researchers have reservations about the use of bat houses and other purpose-built artificial roosts. One concern is the possibility that a species may adapt so well to bat houses that it might gain a competitive advantage over other species, thereby altering the composition of local bat communities. For example, Robert Bender and other researchers captured few *C. gouldii* prior to the installation of 34 bat houses at Organ Pipes National Park in Victoria, Australia, whereas little forest bats (*Vespadelus vulturnus*) were the most common species captured at the time. After the bat houses became occupied (Fig. 9-10), the population of *C. gouldii* increased so dramatically that they later comprised 91% of captures; little forest bats, which did not use the bat houses, were rarely detected afterwards (Bender, 2005).

In South Africa, Wendy White commented that some researchers do not recommend installing bat houses, "except on monoculture farms where no natural vegetation exists, or when a colony is evicted from a structure ... bat houses encourage crevice-dwelling species at the expense of less common foliage- and cave-dwelling bats."

Fig. 9-10. Gould's wattled bat (*Chalinolobus gouldii*) is the most frequent user of bat boxes in Victoria, Australia. This species roosts in bat boxes all year long and uses them as breeding sites. Photo courtesy of Robert Bender.

It is not uncommon for multiple species to share a natural roost or a bat house.

For example, *T. brasiliensis* often shares roosts with either *M. austroriparius* or *M. velifer* in the southeastern United States, and little free-tailed bats (*Chaerephon pumilus*) have been documented sharing bat houses with Angolan free-tailed bats (*Mops condylurus*) and African yellow bats (*Scotophilus dinganii*) in KwaZulu-Natal province in South Africa. The latter occurrence is unusual because African yellow bats are extremely territorial and have been known to kill other bat species attempting to roost in the same bat house (Wendy White, pers. comm.). At one location in northeast Texas, *E. fuscus* roosting in multiple bat houses since the early 1990s (Kiser and Kiser, 2002) were largely displaced by *T. brasiliensis* beginning around 2005 (Marvin Maberry, pers. comm.). As these 2 species commonly share building roosts across the southern United States, the disappearance of *E. fuscus* is unusual in this case. *E. fuscus* rarely moves more than 48 mi. (77 km) between their summer and winter roosts, and they are extremely loyal to the roosts where they were born (Kiser and Kiser, 2002). It may be possible that the preferred roost microclimates changed due to the presence of the additional bats. *E. fuscus* prefers lower roost temperatures than does *T. brasiliensis*, and if temperatures exceed 95 °F (35 °C) they will abandon a roost (Davis *et al.*, 1968; Nagorsen and Brigham, 1993).

Other researchers have speculated that bat houses could become population sinks if the roosts are not maintained properly. In this case, local bat numbers could actually decrease because of unsuitable conditions with the bat houses. Wooden bat houses are prone to deterioration, and may need to be repainted or re-caulked every three to five years, depending on climate, quality and type of materials used, size of roof overhang, mounting site, amount of sun exposure and other factors. Wooden bat houses often need to be replaced within 10 years (Stebbings and Walsh, 1991). Failure to maintain or replace bat houses could result in abandonment, or in the death of pups or adults. Bat houses mounted on trees or poles occasionally fall to the ground, sometimes due to violent storms or high winds, or simply from the deterioration of wood or mounting hardware over time (Bender, 2010b). Tree growth may also push the bat house away from the trunk, eventually causing the mounting board or hardware to fail (Stebbings and Walsh, 1991). These unfortunate events may result in the complete destruction of the bat house and sometimes in the death of bats (Laura Finn, pers. comm.; Tom Finn, pers. comm.). In Arizona, a bat house mounted on a pole was reportedly struck by lightning, killing 75 *T. brasiliensis* inside. Lightning rods with copper grounding wires could be installed on pole-mounted and large, free-standing houses in lightning-prone areas.

Problems with insects often arise with bat houses, which may result in the displacement of bats. Bat houses were avoided when infested by ant colonies at Wilson Reserve Ivanhoe near Melbourne, Australia (Bender, 2010a). Only when control measures succeeded in keeping ants out of the houses did bats return. Stebbings and Walsh (1991) also reported problems with ants in European bat houses. In the United States, wasps known as mud daubers (family Sphecidae) may fill roost chambers with tubular nests constructed of mud. The authors have witnessed a pair of occupied houses in central Texas that were so full of mud nests they had been rendered nearly uninhabitable by bats. Within 12 hours of the nests' removal, more than 50 *T. brasiliensis* returned, and more than 100 bats had returned by the following day (Kiser, 1998). Paper wasps, red wasps, and hornets (family Vespidae), all of which have painful stings, also build nests inside roost chambers and on the exterior of bat houses (Stebbings and Walsh, 1991; Tuttle *et al.*, 2004). Wasp stings have been known to kill bats (Barbara French, pers. comm.), and bats apparently avoid individual roost chambers or entire houses if wasps are present. Roost chambers should be inspected at least once a year, and any wasp nests found should be carefully removed with a long pole (make sure bats are not present when attempting to remove nests inside the chambers).

Other animals may take up residence inside bat houses (especially those mounted on trees), including spiders, mice, squirrels, woodpeckers, songbirds, lizards, and tree frogs. Australian bat boxes have even accommodated sugar gliders, rats, ringtail possums and brushtail possums (Bender, 2010b). Spider webs should be removed from unoccupied roost chambers whenever they accumulate because bats are unlikely to use these houses. In fact, the presence of spider webs in an unoccupied bat box may indicate that conditions are unsuitable for bats, and the box may need to be repaired or repositioned. However, bats sometimes coexist with spiders such as huntsmen (Sparassidae) and sac spiders (Clubionidae) (Robert Bender, pers. comm.). Woodpeckers (family Picidae) sometimes destroy the interior roosting partitions to make the bat house more suitable for their own use. The authors have witnessed both wooden and rigid foam partitions heavily damaged by woodpeckers. Woodpeckers and squirrels also cause damage to the exteriors of wooden houses. Metal plates attached around the entrance slots of birdhouse-style houses can deter chewing by squirrels (Stebbings and Walsh, 1991).

Predators such as raccoons (*Procyon lotor*), house cats (*Felis sylvestris*), and rat snakes (*Elaphe* spp.) have been known to climb trees and wooden poles to prey on bats inside bat houses. Because rat snakes are constrictors, they are fully capable of climbing metal poles. Bats may abandon bat houses, sometimes for long periods, following successful predation attempts. Therefore, predator guards (cylindrical baffles or cones; http://www.noble.org/Ag/Wildlife/PredatorGuards/#) constructed of sheet metal should be attached to each pole. Predator guards need to be installed high enough above the ground that mammalian predators can not reach above them, but not so high that they would enable perched hawks or owls (which also prey on bats) to easily reach the bats as they enter or exit the bat house. One researcher in California witnessed a Swainson's hawk (*Buteo swainsonii*) eating bats while perched on the roof of a relatively short, 20-in. (50.8-cm) bat house. On this occasion, high temperatures during the day had apparently driven the bats to roost on the landing pad below the bat house, where they were exposed. A taller bat house than the one being used would have offered better protection. An owl in Wareham Forest, England apparently learned to open the removable lids of birdhouse-style boxes and thus preyed on more than 100 *P. auritus*. The problem was solved by permanently securing the roofs shut. When a red-shouldered hawk (*Buteo lineatus*) was disturbing Wayne Bryan's occupied bat houses in northeast Texas, he devised a series of anti-predator spikes made from ⅛-in. (3-mm) galvanized steel wire. The 20-in. (51-cm) long spikes were set into a piece of lumber anchored on the roof of the houses, which eliminated the problem (Kiser, 2001). Tree branches, electrical wires, or other obstacles near bat houses should always be removed (or the houses repositioned elsewhere) to eliminate potential perches for hawks and owls.

CONCLUSION

Bats are very adaptable and opportunistic in their selection of roost sites (Kunz, 1982), which makes bat houses and artificial roosts such an important and feasible aspect of bat conservation. Human-made structures can not entirely replace existing natural roosts, but they can supplement them; for example, suitable tree roosts are often in short supply in many habitats. Because roost selection is highly dependent on proximity of foraging sites, protection of high quality natural habitat is important for foraging opportunities. Artificial roosts can be excellent research tools because they are often more easily protected than natural roosts and are accessible to researchers. Of course, visitation to bat roosts should be limited, even for research, because bats may leave roosts when disturbed by humans (Waldien, 1996). In addition, bats are quite loyal to a specific roost when roost sites are long-lived, as are many human-made structures. One example is an adult male *E. fuscus* that was captured four times between 1992 and 1995 at the same bridge (Pierson *et al.*, 1996). This has significant implications for management because roosts can be created and used by bats for many years.

Just like natural roosts, artificial roosts must be monitored, protected, and maintained over time, in order for them to become reliable long-term sites for bats. Whether or not artificial roosts are capable of aiding in the long-term recovery of endangered or declining species (including North American bat populations decimated by white-nose syndrome) remains to be seen. To date, few artificial caves have ever been tested. Further experimentation regarding artificial roost designs for maternity and hibernation use is highly encouraged, and will lead to improved models and strategies to help a growing number of species. As natural roost availability and bat populations continue to decline, investigations into the roost preferences for additional species become all the more important for the conservation of our beneficial bats.

PRODUCTS MENTIONED IN TEXT

Bat Condo (Construction Plans): Pennsylvania Game Commission, 2001 Elmerton Avenue, Harrisburg, PA 17110 USA; see also http://www.depweb.state.pa.us/portal/server.pt/document/714602/bat_condo_pdf.

Fibrolit Boards: Plycem®, Paraíso de Cartago, 5 Km este de la basílica de Cartago, contiguo al restaurante Casa Vieja, Costa Rica; http://www.plycem.com/index.php?option=com_content&view=article&id=48&Itemid=55&lang=en.

HardiePipe™ Concrete Pipes: James Hardie Concrete Pipes, 46 Randle Road, Meeandah, QLD 4008 Australia; http://www.jameshardiepipes.com.au/.

Plastic Mesh or Bat House Netting (Product XV1672): Industrial Netting, 7681 Setzler Pkwy N., Minneapolis, MN 55445 USA; http://www.industrialnetting.com/.

Schwegler 'Woodcrete' Bat Boxes: Schwegler Vogel- und Naturschutzprodukte GmbH, Heinkelstrasse 35, D-73614 Schorndorf, Germany; http://www.schwegler-nature.com.

ACKNOWLEDGMENTS

We thank Meera Banta (Bat Conservation International), Silke Weidler (Schwegler Vogel- und Naturschutzprodukte GmbH), and all individuals (see photo credits) who graciously allowed us to use their photographs in this chapter. We also thank Susan M. Barnard for her assistance with Appendices 9-2, 9-3, and 9-4, and Drs. Henry Schofield (The Vincent Wildlife Trust) and Gianna Dondini (Nature and Archaeological Center of Pistoiese Apennine) for their helpful reviews of Appendix 9-4.

REFERENCES

3D/ESI. 1995. 1995 field studies for interim Indiana bat habitat mitigation at the Indianapolis International Airport. *Environmental Technical Report*. 23 pp.
3D/ESI. 1996. 1996 field studies for interim Indiana bat habitat mitigation at the Indianapolis International Airport. *Environmental Technical Report*. 87 pp.
Arnett, E.B. and Hayes, J.P. 2000. Bat use of roosting boxes installed under flat-bottom bridges in western Oregon. *Wildlife Society Bulletin*, 28:890-894.
Baranauskas, K. 2010. Diversity and abundance of bats (Chiroptera) found in bat boxes in east Lithuania. *Acta Zoologica Lituanica*, 20:39-44.
Barbour, R.W. and Davis, W.H. 1969. *Bats of America*. The University Press of Kentucky, Lexington, KY. 286 pp.
Bartonička, T. and Řehák, Z. 2007. Influence of the microclimate of bat boxes on their occupation by the soprano pipistrelle *Pipistrellus pigmaeus*: Possible cause of roost switching. *Acta Chiropterologica*, 9:517-526.
Bassett, J.E. and Studier, E.H. 1988. Methods for determining water balance in bats. In: *Ecological and Behavioral Methods for the Study of Bats* (Kunz, T.H., ed.). Smithsonian Institution Press, Washington, D.C. Pp. 373-386.
Bat Conservation International. 1995a. Results from 1994 season. *The Bat House Researcher*, 3(1):1-2.
Bat Conservation International. 1995b. Advances in bat house materials. *The Bat House Researcher*, 3(2):1.
Bat Conservation International. 1996. Update on plastic bat houses. *The Bat House Researcher*, 4(2):5.
Bayless, M. 2006. Designing Homes for Forest Bats. *Bats*, 26:9-11.
Bender, R. 2005. Bat boxing update from Organ Pipes National Park and Wilson Reserve. *The Australasian Bat Society Newsletter*, 24:11-17.
Bender, R. 2010a. Bat roost box project: Wilson Reserve Ivanhoe, 2005 to 2010. *Report to Banyule City Council*. 22 pp.
Bender, R. 2010b. Bat roost boxes at Organ Pipes National Park, Victoria: Seasonal and annual use patterns. (In Review). 30 pp.
Bender, R. and Irvine, R. 2001. Bat boxes at Organ Pipes National Park. *Australasian Bat Society Newsletter*, 16:19-23.
Campbell, C.A.R. 1925. *Bats, Mosquitoes, and Dollars*. The Stratford Company, Boston, MA. 262 pp.
Carter, T. 2002. Bat houses for conservation of endangered Indiana myotis. *The Bat House Researcher*, 10:1-3.
Carter, T., Feldhamer, G., and Kath, J. 2001. Notes on the summer roosting of Indiana bats. *Bat Research News*, 42:197-198.
Clawson, R.L. and Gardner, G. 1993. Creating new bat roosting habitat. *Bats*, 11: 20.
Davis, W.H., Barbour, R.W., and Hassell, M.D. 1968. Colonial behavior of *Eptesicus fuscus*. *Journal of Mammalogy*, 49:44-50.
Dourson, D. and MacGregor, J. 1997. A post bat house design. *Bat Research News*, 38:21-22.
Feldhamer, G.A., Drickamer, L.C., Vessey, S.H., and Merritt, J.F. 2004. Chiroptera. In: *Mammalogy: Adaptation, Diversity, Ecology*, 2nd ed. McGraw Hill, Boston, Massachussetts. 356 pp.
Fowler, J. 1996. Bat Abatement at Chehaw Park. *ZOOMagazine* (A publication for the members and supporters of Zoo Atlanta), Summer/Fall 1996.
Frantz, S.C. 1989. Bat houses in state parks: An experiment in New York. *Bats*, 7:14.
Fukui, D., Okazaki, K., Miyazaki, M., and Maeda, K. 2010. The effect of roost environment on roost selection by non-reproductive and dispersing Asian parti-coloured bats *Vespertilio sinensis*. *Mammal Study*, 35:99-109.
Gerell, R. 1985. Tests of boxes for bats. *Nyctalus*, 2:181-185.
Glover, K.V. 1994. *Bats at the University of Florida*. Unpublished report, Environmental Health and Safety Division. 5 pp.
Goldman, E.A. 1926. Recent literature. *Journal of Mammalogy*, 7:136-138.
Graczyk, R. 1974. Study on population of bats (Chiroptera) settled in sawdust concrete hatching boxes in forests. *Prace Komisji Nauk Rolniczych i Leśnych (Poznanskie)*, 37:89-94.
Griffin, T. and Stuart, J. 2010. New Mexico. *Western Bat Working Group Newsletter*, 6:11-12.
Horncastle, V. and Bayless, M. 2005. Forest treatment study. *Arizona Bat Resource Group Newsletter*, Winter:3.
Irvine, R. and Bender, R. 1995. Initial Results from Bat Roosting Boxes at Organ Pipes National Park. *Victorian Naturalist*, 112:212-217.
Johnston, D. 2005. Recreating battered bat roosts. *Bats*, 23:1-4.
Jolyet, A. 1918. Cabanes a chauvres-souris. *Revue Eaux et Forets*, 56:121-126.
Juškaitis, R. 1999. Mammals occupying nestboxes for birds in Lithuania. *Acta Zoologica Lituanica*, 9:19-23.
Keeley, B. and Tuttle, M.D. 1997. *The Texas Bats and Bridges Project*. Texas Department of Transportation, Austin, TX. 16 pp.
Keeley, B. and Tuttle, M.D. 1999. *Bats in American Bridges*. Bat Conservation International, Inc. Resource Publication No. 4. 42 pp.
Kelm, D.H., Wiesner, K.R., and Helverson, O.v. 2008a. Effects of artificial roosts for frugivorous bats on seed dispersal in a neotropical forest pasture mosaic. *Conservation Biology*, 22:733-741.
Kelm, D.H., Wiesner, K.R., and Helverson, O.v. 2008b. Are there benefits of bat roosts for tropical forest restoration? *Conservation Biology*, 22:1090-1091.
Kelm, D.H. 2008. Restoring lost rainforests. *Bats*, 26:6-7.
Kennedy, J. 1996. What if you want bats in your attic? *The Bat House Researcher*, 4(2):3-5.

Kennedy, J. and Tuttle, M.D. 1997. Extra-large bat houses. *The Bat House Researcher,* 5(1):2.
Kerth, G. and König, B. 1999. Fission, fusion and non-random associations in female Bechstein's bats. *Behaviour,* 136:1187–1202.
Kiser, M. 1998. Bat houses and wasps. *The Bat House Researcher,* 6(2):3-4.
Kiser, M. 1999. Pallas' mastiff bats in bat houses. *The Bat House Researcher,* 7(1):6-7.
Kiser, M. 2001a. Meet *Myotis evotis. The Bat House Researcher,* 9(1):1-2.
Kiser, M. 2001b. Anti-predator device for bat houses. *The Bat House Researcher,* 9(2):6.
Kiser, M. 2002a. Big brown bats in northeast Texas. *The Bat House Researcher,* 10(1):1-3.
Kiser, M. 2002b. Dr. Charles Campbell: Bat house pioneer. *The Bat House Researcher,* 10(1):5-6.
Kiser, M. and Kiser, S. 1999. Bat houses and exclusion in British Columbia. *The Bat House Researcher,* 7(2):3-4.
Kiser, M. and Kiser, S. 2000. 1999 season results. *The Bat House Researcher,* 8(1):1-3.
Kiser, M. and Kiser, S. 2001a. Survey results from 2000. *The Bat House Researcher,* 9(1):2-4.
Kiser, M. and Kiser, S. 2001b. Temperature test results for "plastic insulated" houses. *The Bat House Researcher,* 9(2):6-7.
Kiser, M. and Kiser, S. 2002. 2001 Bat house research project survey results. *The Bat House Researcher,* 10(2):3-5.
Kiser, M. and Kiser, S. 2003. Innovative homes for bats that shun bat houses. *The Bat House Researcher,* 11(1):1-2.
Kiser, M. and Kiser, S. 2004. A decade of bat house discovery. *The Bat House Researcher,* 12(1):1-7.
Kiser, M. and Kiser, S. 2005a. Sand gives bat houses a new twist. *Bat Conservation Times e-Newsletter,* May, 3(5). Bat conservation International Inc., Austin, TX.
Kiser, M. and Kiser, S. 2005b. The bat box. *Bat Conservation Times e-Newsletter,* March, 3(3). Bat conservation International Inc., Austin, TX.
Kiser, W.M. 1996. *Conservation of LeConte's Free-tailed Bat (Tadarida brasiliensis cynocephala): Environmental Parameters of a Natural and An Artificial Roost.* M.S. Thesis, Auburn University, Auburn, AL. 245 pp.
Kiser, W.M., and Glover, K.V. 1997. Acceptance of an artificial roost by LeConte's free-tailed bats, *Tadarida brasiliensis cynocephala. Bat Research News,* 38:1-4.
Korsten, E. 2006. A bigger bat house for the Netherlands. *Bats,* 26:12-13.
Krzanowski, A. 1955. A new type of box for the forest bats. *Chronmy Przyrode Ojczysta* 11:25-27.
Krzanowski, A. 1969. The protection of bats. In: *Lynx* (Museum Nationale, Praha), Proceedings of the First International Bat Conference, Czechoslovakia, 6-9 September, 1968. Pp. 41-43.
Kunz, T.H. 1982. Roosting ecology of bats. In: *Ecology of Bats* (Kunz, T.H., ed.). Plenum Press, New York, NY. Pp. 1-55.
Lesiński, G., Skrzypiec-Nowak, P., Janiak, A., and Jagnieszczak, Z. 2009. Phenology of bat occurrence in boxes in central Poland. *Mammalia,* 73:33–37.
Levin, E. 2009. Bats along the Jordan River. *Bats,* 27:7-10.
Marais, W.C. 2009. *The Potential of Using Insectivorous Bats (Microchiroptera) as a Means of Insect Pest Control in Agricultural Areas.* Magister Scientiae Dissertation, University of Johannesburg, Johannesburg, South Africa. 120 pp.
Marks, C.S. and Marks, G.E. 2006. *Bats of Florida.* University Press of Florida, Gainesville, FL. 177 pp.
Mayle, B.A. 1990. A biological basis for bat conservation in British woodlands—A review. *Mammal Review,* 20:159-195.
McAney, K. 2006. *A Conservation Plan for Irish Vesper Bats.* Irish Wildlife Manuals, No. 20, National Parks and Wildlife Service, Department of the Environment, Heritage and Local Government, Dublin, Ireland. 57 pp.
Michaelsen, T.C. 2011. BCI bat houses pay off in Norway. *Bats,* 29(3):9-11.
Mohr, C.E. 1972. *The World of the Bat.* J.B. Lippincott Co., Philadelphia, PA. 162 pp.
Mooseman, P.R. 2005. *Roosting Behavior and Habitat Dynamics of Male Indiana Bats (Myotis sodalis) Following a Large Scale Natural Disturbance.* Ph.D. Dissertation, Auburn University, Auburn, GA. 95 pp.
Morris, C. 2010. *Cool Tower for Lesser Horseshoe Bats.* Vincent Wildlife Trust, Herefordshire, U.K. 4 pp.; http://www.vwt.org.uk/downloads/The_Cool_Tower.pdf.
Murphy, M. 1989. Dr. Campbell' s malaria-eradicating, guano-producing bat roosts. *Bats,* 7:10-13.
Nagorsen, D.W. and Brigham, R.M. 1993. *The Bats of British Columbia.* University of British Columbia Press, Vancouver, B.C., Canada. 164 pp.
Noteman, L. 1998. Hidden housing: Artificial bark for bats. *Bats,* 16:12-15; http://www.batcon.org/index.php/media-and-info/bats-archives.html?task=viewArticle&magArticleID=873.
Payne, C.J., Griffiths, S.R., Godinho, L.N., and Coulson, G.M. 2011. The influence of scent on roost selection in Gould's wattled bat *Chalinolobus gouldii.* (Unpublished). 13 pp.
Pierson, E.D., Rainey, W.E., and Miller, R.M. 1996. Night roost sampling: A window on the forest bat community in northern California. In: *Bats and Forests Symposium,* October 19-21, 1995, Victoria, British Columbia, Canada (Barclay R.M.R. and Brigham, R.M., eds.). Research Branch, British Columbia Ministry of Forests, Victoria. Pp. 151-163.
Poulton, S.M.C. 2006. *An Analysis of the Usage of Batboxes in England, Wales and Ireland for The Vincent Wildlife Trust;* http://www.jwaller.co.uk/batgroup/pdf/VWTReportBatboxesMay2006.pdf.
Reid, L. 2009. Putting bats to work for forest restoration. *Amigos Newsletter* (Las Cruces Biological Station, Costa Rica), 72:8-9.

Reid, L., Holste, E.K., and Zahawi, R.A. 2009. *The Development of Artificial Roost for Tropical Forest Restoration.* Progress Report to Bat Conservation International. 12 pp.

Ritzi, C.M., Everson, B.L., and Whitaker, J.O., Jr. 2005. Use of bat boxes by a maternity colony of Indiana myotis. *Northeastern Naturalist*, 12:217-220.

Sander, S.M. 1997. How to build a cave: A bold experiment in artificial habitat. *Bats,* 15:8-11.

Schlapp, G. 1999. *Myotis bechsteinii* (Kuhl, 1817). In: *The Atlas of European Mammals* (Mitchell-Jones, A.J., Amori, G., Bogdanowicz, W., Kryštufek, B., Reijnders, P.J.H., Spitzenberger, F., Stubbe, M., Thissen, J.B.M., Vohralík, V., and Zima, J., eds.). Academic Press, London, England. Pp. 100-101.

Schober, W. and Grimmberger, E. 1997. *The Bats of Europe and North America.* T.F.H. Publications, Neptune City, N.J. 239 pp.

Schwegler Bird and Nature Conservation Products. 2011. Catalogue No. 68; http://www.schwegler-natur.de, accessed March, 2011.

Shore, R.F., Myhill, D.G., French, M.C., Leach, D.V., and Stebbings, R.E. 1991. Toxicity and tissue distribution of pentachlorophenol and permethrin in pipistrelle bats experimentally exposed to treated timber. *Environmental Pollution*, 73:101-118.

Siders, M.S. 1999. *Artificial Bat Bark Development and Testing.* Report to Bat Conservation International and Max McGraw Wildlife Federation. 22 pp.

Stebbings, R.E. and Walsh, S.T. 1988. *Bat Boxes.* Flora and Fauna Preservation Society, London, England. 24 pp.

Stebbings, R.E. and Walsh, S.T. 1991. *Bat boxes: A Guide to the History, Function, Construction and Use in the Conservation of Bats,* 3rd ed. The Bat Conservation Trust. London, England. 24 pp.

Swaenen, D. 2010. Artificial Hibernating Roost for Geoffroy's bat (*Myotis emarginatus*). http://www.eurobats.org/news_events/news/pdf/Artificial%20hibernating%20roost.pdf, accessed August, 2011.

Tatarian, G. 2001. Successful pallid bat house design in California. *The Bat House Researcher*, 9(2):2-4.

The Timaru Herald. 2008. Bats are having a ball at Temuka. http://www.stuff.co.nz/timaru-herald/news/611564, accessed August, 2011.

Tuttle, M.D. 1988. *America's Neighborhood Bats.* The University of Texas Press, Austin, TX. 96 pp.

Tuttle, M.D. 1989. Extending an invitation to bats. *Bats*, 7:5-9.

Tuttle, M.D. and Hensley, D.L. 1993. *The Bat House Builder's Handbook.* Bat Conservation International, Inc., Austin, TX. 35 pp.

Tuttle, M.D., Kiser, M., and Kiser, S. 2004. *The Bat House Builder's Handbook*, 2nd ed. Bat Conservation International, Inc. Austin, TX. 35 pp.

Vincent Wildlife Trust. 2011. Design of the CJM bat box; http://www.vwt.org.uk/downloads/CJM%20Batbox.pdf, accessed August, 2011.

Waldien, D.L. 1996. *Bat Communities of Western Oregon: Landscape Relationships of Roost Structures and Activity Areas of Myotis evotis (Long-eared Myotis).* Report to Bat Conservation International, Austin, TX. 32 pp.

Williams, L. M. and Brittingham, M.C. 1997. Selection of maternity roosts by big brown bats. *Journal of Wildlife Management*, 61:359-368.

Williams-Whitmer, L.M. 1994. *Maternity Roost Selection and the Use of Bat Boxes by Displaced Colonies of Big Brown and Little Brown Bats.* M.S. Thesis, The Pennsylvania State University, State College, PA. 147 pp.

Whitaker, Jr., J.O. 1998. Life history and roost switching in six summer colonies of eastern pipistrelles in buildings. *Journal of Mammalogy*, 79:651-659.

Voute, A.M. 1991. Poland' s unique bat reserve: A resource in trouble. *Bats,* 9:17-18.

Appendix 9-1. Bat Conservation International's Tips for North American Bat Houses (*summarized from Tuttle et al., 2004*)

DESIGN

Bat houses should be at least 2 ft. tall (61 cm), 14 in. (36 cm) or more wide (except for rocket boxes), and have a 3- to 6-in. (8- to 15-cm) landing area extending below the entrance. Taller and wider houses are preferred. Most houses have one to four roosting chambers, the more the better. Roost partitions should be carefully spaced ¾ to 1 in. (19 to 25 mm) apart, except for boxes designed for pallid bats (*Antrozous pallidus*) and Florida bonneted bats (*Eumops floridanus*), which use 1½- to 1¾-in. (38- to 44-mm) crevices. All partitions and landing areas should be roughened. Wood surfaces can be scratched or grooved horizontally, or covered with ⅛- or ¼-in. (3- to 6-mm) durable plastic mesh (see *Products Mentioned in Text*). Include vents 6 in. (152 mm) from the bottoms of all houses to be used where average July high temperatures are 80° F (30° C), or above. Front vents are as long as a house is wide, side vents are 6 in. tall (152 mm) x ½ in. (13 mm) wide.

CONSTRUCTION

For most designs, a combination of exterior plywood and quality lumber such as cedar is best. Do not use pressure-treated wood. Any screws, hardware, or staples used must be exterior grade or galvanized. To increase longevity, use screws rather than nails. Caulk all seams, especially around the roof.

WOOD TREATMENT

For the exterior, apply three coats of exterior grade, water-based paint or stain. Available observations suggest that color should be black or dark colors where average high temperatures in July are 80 to 85° F (27 to 30° C), dark or medium colors (such as dark brown or dark gray) where they are 85 to 95° F (30 to 35° C), medium or light colors where they are 95 to 100° F (35 to 38° C), and light colors or white where they exceed 100° F (38° C). Much depends upon amount of sun exposure; adjust to darker colors for less sun. For the interior, use two coats dark, exterior grade, water-based stain. Apply stain after creating scratches or grooves (do not use paint as it will fill in grooves) or prior to stapling plastic mesh.

SUN EXPOSURE

Houses where high temperatures in July average 80° F (27° C) or less should receive at least 10 hours of sun; more is better. In most areas, at least six hours of direct daily sun are recommended where daily high temperatures in July average less than 100° F (38° C).

HABITAT

Most nursery colonies of bats choose roosts within ¼ mi. (400 m) of water, preferably a stream, river, or lake. Greatest bat house success has been achieved in areas of diverse

Appendix 9-1 continued

habitat, especially where there is a mixture of differing agricultural use and natural vegetation. Bat houses work extremely well in regions where bats are already attempting to live in buildings.

MOUNTING

Bats find houses mounted on poles or buildings more than twice as fast as on trees, which are also less preferred. Houses mounted on metal siding are rarely used. Wood or stone buildings with proper solar exposure are ideal, and locations under the eaves often have been successful. Mounting two bat houses back to back, ¾ in. (19 mm) apart on poles, both covered by a tin roof, helps protect the interior from overheating in hot climates. All bat houses should be mounted at least 12 ft. (3.7 m) above ground; 15 to 20 ft. (4.6 to 6.1 m) is better. Bat houses should not be lit by bright lights.

PROTECTION FROM PREDATORS

Houses mounted on sides of buildings or high up on poles provide the best protection from predators. This may be a key factor in determining bat choice. Locations at least 20 to 25 ft. (6.1 to 7.6 m) from the nearest tree branches, wires or other obstacles are best. However, houses may be found more quickly if located along forest or water edges where bats tend to fly.

AVOIDING UNINVITED GUESTS

Wasps can be a problem before bats fully occupy a house. Use of ¾-in. (19-mm) roosting spaces reduces wasp use. If nests accumulate, they should be removed in late winter or early spring before either wasps or bats return. Open-bottom houses greatly reduce problems with birds, mice, squirrels or parasites, and guano does not accumulate inside.

TIMING

Bat houses can be installed at any time of the year, but are more likely to be used their first summer if installed before the bats return in spring. When using bat houses in conjunction with excluding a colony from a building, install the bat houses two to six weeks before the actual eviction.

IMPORTANCE OF LOCAL EXPERIMENTATION

It is important to test for local needs before putting up more than three to six houses, especially when comparing houses of different exterior colors and sun exposure.

Appendix 9-2. Bats Reported to Occupy Bat Houses

Countries and Bat Species	References
Australia	
Chalinolobus gouldii (Gould's Wattled Bat)	Bender (2005); Lisa Godinho (pers. comm.); Glenn Hoye (pers. comm.); Ian Kitchen (pers. comm.); Monika Rhodes (pers. comm.)
Chalinolobus morio (Chocolate Wattled Bat)	Bender (2005); Lisa Godinho (pers. comm.); Glenn Hoye (pers. comm.); Ian Kitchen (pers. comm.)
Mormopterus norfolkensis (Eastern Little Mastiff Bat)	Glenn Hoye (pers. comm.); Bradley Law (pers. comm.)
Nyctophilus bifax (Bifax Long-eared Bat)	Jennefer Maclean (pers. comm.); Monika Rhodes (pers. comm.)
Nyctophilus geoffroyi (Lesser Long-eared Bat)	Glenn Hoye (pers. comm.)
Nyctophilus gouldi (Gould's Long-eared Bat)	Bender (2005); Glenn Hoye (pers. comm.); Monika Rhodes (pers. comm.)
Scoteanax rueppellii (Rüppell's Broad-nosed Bat)	Monika Rhodes (pers. comm.)
Scotorepens balstoni (Western Broad-nosed Bat)	Glenn Hoye (pers. comm.)
Scotorepens orion (Orion Broad-nosed Bat)	Bender (2005); Ian Kitchen (pers. comm.)
Tadarida australis (White-striped Free-tailed Bat)	Bender (2005); Lisa Godinho (pers. comm.); Ian Kitchen (pers. comm.)
Vespedelus darlingtoni (Large Forest Bat)	Bender (2005); Lisa Godinho (pers. comm.); Ian Kitchen (pers. comm.)
Vespedelus regulus (Southern Forest Bat)	Bender (2005); Lisa Godinho (pers. comm.)
Europe	
Barbastella barbastellus (Western Barbastelle)	Laurent Arthur (pers. comm.); Steve Laurence (pers. comm.)
Eptesicus nilssonii (Northern Bat)	Baranauskas (2010); Michaelsen (2011)
Eptesicus serotinus (Common Serotine)	Juškaitis (1999); Hubert Krättli (pers. comm.); Steve Laurence (pers. comm.); Hugo Rebelo (pers. comm.)
Hypsugo savii (Savi's Pipistrelle)	Hubert Krättli (pers. comm.)
Myotis bechsteinii (Bechstein's Myotis)	Gianna Dondini (pers. comm.); Kerth and König (1999); Hubert Krättli (pers. comm.); Schlapp (1999); Simone Vergari (pers. comm.)
Myotis blythii (Lesser Mouse-eared Myotis)	Hubert Krättli (pers. comm.)
Myotis brandtii (Brandt' s Myotis)	Noel Home (pers. comm.); Stebbings and Walsh (1991)
Myotis dasycneme (Pond Myotis)	Baranauskas (2010)
Myotis daubentonii (Daubenton's Myotis)	Noel Home (pers. comm.); Hubert Krättli (pers. comm.); McAney (2006)
Myotis emarginatus (Geoffroy's Myotis)[1]	Dirk Swaenen (pers. comm.)
Myotis myotis (Mouse-eared Myotis)	Hubert Krättli (pers. comm.); Lesiński *et al.* (2009)
Myotis mystacinus (Whiskered Myotis)	Noel Home (pers. comm.); Hubert Krättli (pers. comm.); Stebbings and Walsh (1991)
Myotis nattereri (Natterer's Myotis)	Laurent Arthur (pers. comm.); Gianna Dondini (pers. comm.); Lena Godlevska (pers. comm.); Noel Home (pers. comm.); Hubert Krättli (pers. comm.); Steve Laurence (pers. comm.); Simone Vergari (pers. comm.); Stebbings and Walsh (1991)
Nyctalus lasiopterus (Giant Noctule)	Gianna Dondini (pers. comm.); Simone Vergari (pers. comm.)
Nyctalus leisleri (Leisler's Noctule)	Laurent Arthur (pers. comm.); Gianna Dondini (pers. comm.); Lena Godlevska (pers. comm.); Noel Home (pers. comm.); Hubert Krättli (pers. comm.); McAney (2006); Simone Vergari (pers.comm.); Stebbings and Walsh (1991)

Appendix 9-2 continued

Countries and Bat Species	References
Nyctalus noctula (Common Noctule)	Laurent Arthur (pers. comm.); Baranauskas (2010); Gianna Dondini (pers. comm.); Lena Godlevska (pers. comm.); Hubert Krättli (pers. comm.); Steve Laurence (pers. comm.); Lesiński *et al.* (2009); Eran Levin (pers. comm.); Simone Vergari (pers. comm.); Stebbings and Walsh (1991)
Pipistrellus kuhlii (Kuhl's Pipistrelle)	Hubert Krättli (pers. comm.); Eran Levin (pers. comm.)
Pipistrellus nathusii (Nathusius's Pipistrelle)	Baranauskas (2010); Lena Godlevska (pers. comm.); Hubert Krättli (pers. comm.); Lesiński *et al.* (2009); McAney (2006)
Pipistrellus pipistrellus (Common Pipistrelle)	Gianna Dondini (pers. comm.); Lena Godlevska (pers. comm.); Noel Home (pers. comm.); Hubert Krättli (pers. comm.); Steve Laurence (pers. comm.); Simone Vergari (pers. comm.); Stebbings and Walsh (1991)
Pipistrellus pygmaeus (Soprano Pipistrelle)	Baranauskas (2010); Bartonička and Řehák (2007); Noel Home (pers. comm.); Hubert Krättli (pers. comm.); Steve Laurence (pers. comm.); Michaelsen (2011); Hugo Rebelo (pers. comm.)
Plecotus auritus (Brown Long-eared Bat)	Baranauskas (2010); Gianna Dondini (pers. comm.); Lena Godlevska (pers. comm.); Noel Home (pers. comm.); Juškaitis (1999); Hubert Krättli (pers. comm.); Steve Laurence (pers. comm.); Lesiński *et al.* (2009); McAney (2006); Simone Vergari (pers. comm.); Stebbings and Walsh (1991)
Plecotus austriacus (Grey Long-eared Bat)	Hugo Rebelo (pers. comm.)
Vespertilio murinus (Particolored Bat)	Lena Godlevska (pers. comm.); Hubert Krättli (pers. comm.)
Cayman Islands	
Molossus molossus tropidorhynchus (Pallas's Mastiff Bat)	Lois Blumenthal (pers. comm.)
Costa Rica	
Carollia castanea (Chestnut Short-tailed Bat)	Kelm *et al.*, 2008
Carollia perspicillata (Seba's Short-tailed Bat)	Kelm, 2008; Kelm *et al.*, 2008
Carollia sowelli (Sowell's Short-tailed Bat)	Kelm *et al.*, 2008
Glossophaga commissarisi (Commissari's Long-tongued Bat	Kelm *et al.*, 2008
Glossophaga soricina (Pallas's Long-tongued Bat)	Kelm *et al.*, 2008
Micronycteris hirsuta (Hairy Big-eared Bat)	Kelm *et al.*, 2008
Micronycteris microtis (Common Big-eared Bat)	Kelm, 2008; Kelm *et al.*, 2008
Micronycteris schmidtorum (Schmidt's Big-eared Bat)	Kelm *et al.*, 2008
Mimon crenulatum (Striped Hairy-nosed Bat)	Reid *et al.*, 2009
Trachops cirrhosus (Fringe-lipped Bat)	Kelm, 2008; Kelm *et al.*, 2008
India	
Pipistrellus mimus (Indian Pygmy Bat)	Bat Conservation India (http://www.nichiyu.in/bats/inno.htm)
Japan	
Eptesicus japonensis (Japanese Short-tailed Bat)	Kuniko Kawai (pers. comm.)
Eptesicus nilssonii (Northern Bat)	Kuniko Kawai (pers. comm.)
Myotis brantii gracilis (Brandt's Myotis)	Kuniko Kawai (pers. comm.)
Myotis frater (Fraternal Myotis)	Kuniko Kawai (pers. comm.)
Myotis ikonnikovi (Ikonnikov's Myotis)	Kuniko Kawai (pers. comm.)
Myotis petax (Sakhalinien Myotis or Eastern Water Myotis)	Kuniko Kawai (pers. comm.)
Nyctalus aviator (Birdlike Noctule)	Kuniko Kawai (pers. comm.)
Plecotus auritis sacrimontis (Brown Long-eared Bat)	Kuniko Kawai (pers. comm.)
Vespertilio sinensis (Asian Particolored Bat)	Fukui *et al.* (2010); Kuniko Kawai (pers. comm.)

Appendix 9-2 continued

Countries and Bat Species	References
New Zealand	
Chalinolobus tuberculatus (Long-tailed Wattled Bat)	*The Timaru Herald* (2008)
South Africa	
Chaerephon pumilus (Little Free-tailed Bat)	Nigel Fernsby (pers. comm.); Marais (2009); Wendy White (pers. comm.)
Mops condylurus (Angolan Free-tailed Bat)	Nigel Fernsby (pers. comm.); Wendy White (pers. comm.)
Mops midas (Midas' Free-tailed Bat)	Nigel Fernsby (pers. comm.)
Neoromica capensis (Cape Serotine)	Nigel Fernsby (pers. comm.)
Nycteris thebaica (Egyptian Slit-faced Bat)[2]	Wendy White (pers. comm.)
Pipistrellus hesperidus (Dusky Pipistrelle)	Nigel Fernsby (pers. comm.)
Scotophilus dinganii (Yellow-bellied House Bat)	Nigel Fernsby (pers. comm.); Wendy White (pers. comm.)
Scotophilus nigrita (Giant House Bat)	Nigel Fernsby (pers. comm.); Wendy White (pers. comm.)
Scotophilus viridus (Green House Bat)	Wendy White (pers. comm.)
Tadarida aegyptiaca (Egyptian Free-tailed Bat)	Nigel Fernsby (pers. comm.); Wendy White (pers. comm.)
United States and Canada	
Antrozous pallidus (Pallid Bat)	Kiser (1999); Kiser and Kiser (2002); Tuttle *et al.* (2004)
Eptesicus fuscus (Big Brown Bat)	Horncastle and Bayless (2005); Kiser and Kiser (2002); Tuttle *et al.* (2004)
Corynorhinus rafinesquii (Rafinesque's Big-eared Bat)	Bayless (2006); Laura Finn (pers. comm.); Kiser and Kiser (2003)
Eumops floridanus (Florida Bonneted Bat)	Tuttle *et al.* (2004)
Molossus molossus (Pallas's Mastiff Bat)[3]	Kiser and Kiser (2002)
Myotis auriculus (Southwestern Myotis)	Horncastle and Bayless (2005); Katherine M. Thibault (pers. comm.)
Myotis austroriparius (Southeastern Myotis)[3]	Tuttle *et al.* (2004)
Myotis evotis (Western Long-eared Myotis)	Horncastle and Bayless (2005); Kiser (2001); Kiser and Kiser (2002)
Myotis keenii (Keen's Myotis)	Tuttle (1989)
Myotis lucifugus (Little Brown Mytois)	Kiser and Kiser (2002); Tuttle *et al.* (2004)
Myotis occultus (Arizona Myotis)	Horncastle and Bayless (2005)
Myotis septentrionalis (Northern Myotis)	Mooseman (2005); Tuttle *et al.* (2004)
Myotis sodalis (Indiana Myotis)	Kiser and Kiser (2002); Mooseman (2005); Ritzi *et al.* (2005)
Myotis thysanodes (Fringed Myotis)	Horncastle and Bayless (2005)
Myotis velifer (Cave Myotis)[3]	Kiser and Kiser (2002); Tuttle *et al.* (2004)
Myotis volans (Long-legged Myotis)	Horncastle and Bayless (2005)
Myotis yumanensis (Yuma Myotis)	Kiser and Kiser (2002); Tuttle *et al.* (2004)
Nycticeius humeralis (Evening Bat)	Kiser and Kiser (2002); Tuttle *et al.* (2004)
Perimyotis (Pipistrellus) subflavus (Tri-colored Bat or Eastern Pipistrelle)	Tuttle *et al.* (2004)
Tadarida brasiliensis (Mexican Free-tailed Bat)[3]	Kiser and Kiser (2002); Tuttle *et al.* (2004)
Virgin Islands	
Molossus molossus fortis (Pallas's Mastiff Bat)	Phyllis Parizek Benton (pers. comm.)

[1] Willing to occupy bat boxes during hibernation when they are at the appropriate temperatures.
[2] Willing to occupy special-purpose, artificial caves during maternity season.
[3] Not found in Canada.

Appendix 9-3. A Sampling of Successful Bat House Designs

NURSERY HOUSE

North American nursery-style houses popularized by Bat Conservation International evolved from experimental models designed in the 1980s and early 1990s. Researchers such as Tony Koch in Oregon and Cal Butchkoski and Lisa Williams-Whitmer in Pennsylvania (among many others) discovered that tall and wide (measured side to side) boxes proved attractive to species like little brown myotis (*Myotis lucifugus*) and big brown bats (*Eptesicus fuscus*) (Williams-Whitmer, 1994). Widths of nursery models vary anywhere from 14 to 48 in. (35.6 to 121.9 cm); heights vary from 20 to 36 in. (50.8 to 91.4 cm). Nursery houses as a rule are not deep from front to back; most are no more than 5 to 7 in. (12.7 to 17.8 cm) deep. The number of roost chambers dictates how deep the bat house actually is. The following plans are for building a four-chambered nursery house.

INSTRUCTIONS FOR BUILDING A FOUR-CHAMBERED NURSERY HOUSE (FIG. 9-11)

(adapted from The Bat House Builder's Handbook, 2004; © Bat Conservation International)

MATERIALS (MAKES TWO HOUSES; SEE FIGURES 9-12 AND 9-13 FOR ASSEMBLY AND SAWING DIAGRAMS)

- ½ sheet 4 ft. x 4 ft. x ½ in. AC, BC, or T1-11 exterior plywood
- ½ sheet 4 ft. x 4 ft. x ⅜ in. AC or BC exterior plywood
- 2 pieces 1 in. x 6 in. (¾ in. x 5½ in. finished) x 8 ft. pine or cedar
- 1 lb. coated deck or exterior-grade, 1 5/8-in. screws
- 20 to 25 coated deck or exterior-grade, 1¼-in. screws
- 20 to 25 exterior-grade, 1-in. screws
- 1 qt. dark, exterior water-based stain
- 1 qt. exterior, water-based primer
- 2 qts. flat, exterior water-based paint or stain
- 1 tube paintable latex caulk
- Black asphalt shingles or galvanized metal
- 12 to 20 7/8-in. roofing nails

Fig. 9-11. A typical North American-style nursery house, with multiple, tall chambers. Image © 2010 R. Ball, Habitat For Bats, McDonough, Georgia.

RECOMMENDED TOOLS

Table saw or circular saw
Variable-speed reversing drill
Screwdriver bit for drill
1½-in. hole saw or spade bit
Tape measure or yardstick
Caulking gun
Paint brushes
Hammer (optional)
Tin snips (optional)
Bar clamp (optional)
Sander (optional)

Appendix 9-3 continued

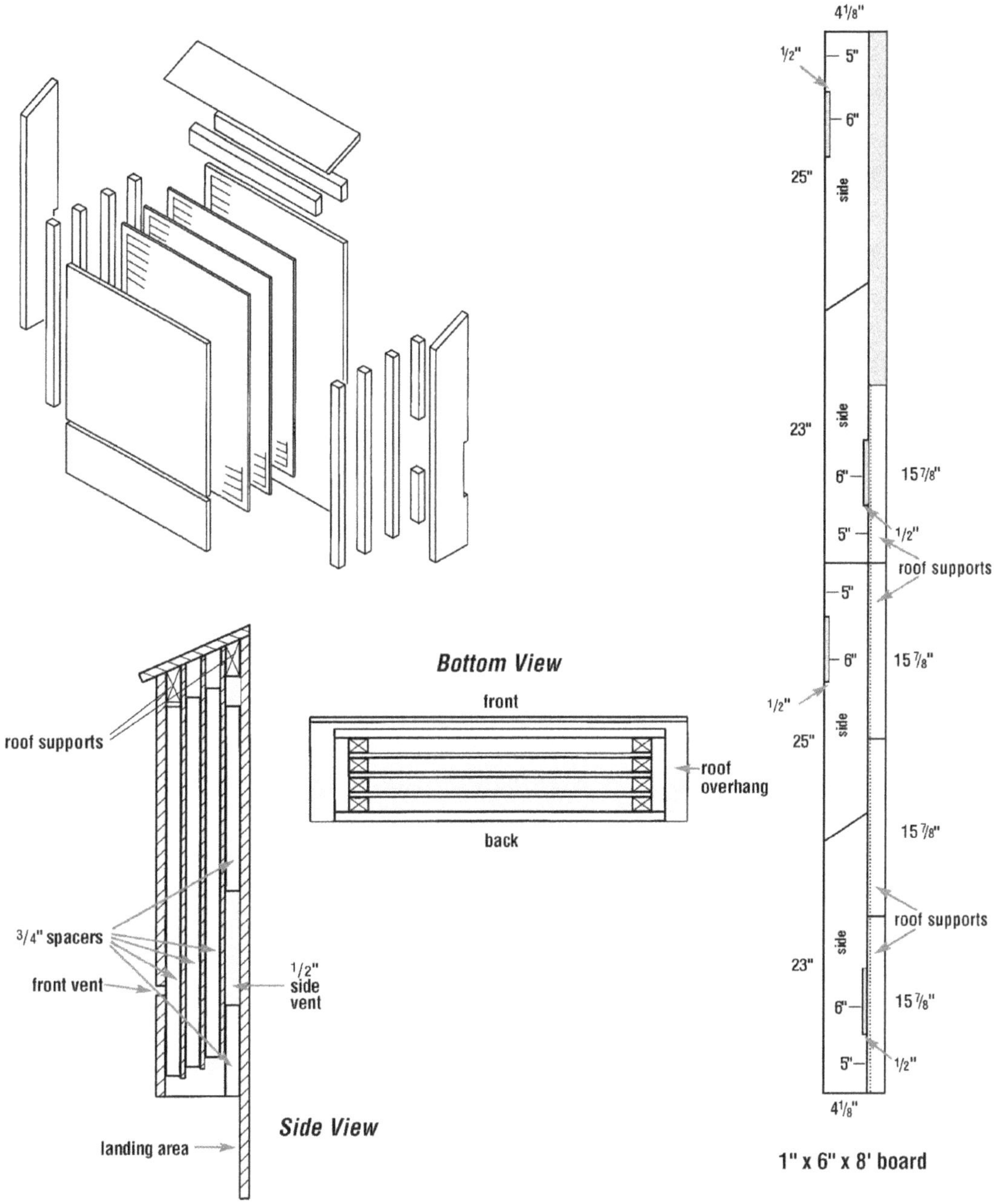

Fig. 9-12. Four-chambered nursery house assembly diagrams. Illustrations from the *Bat House Builder's Handbook* (2004).

Appendix 9-3 continued

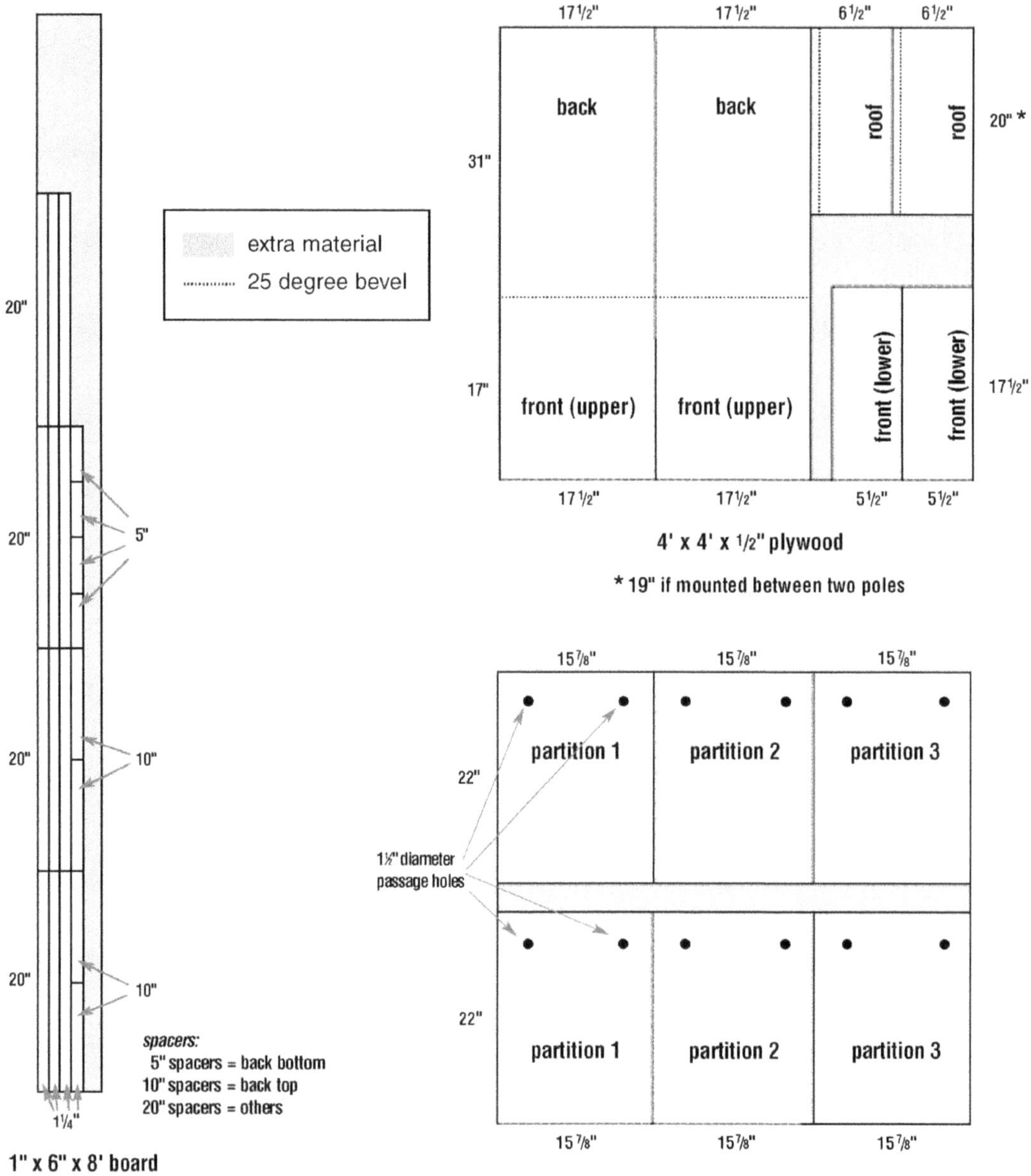

Fig. 9-13. Four-chambered nursery house sawing diagrams. Illustrations from the *Bat House Builder's Handbook* (2004).

Appendix 9-3 continued

Construction

1. Measure, mark and cut out all wood according to the diagrams in Figures 9-12 and 9-13.
2. Roughen interior and landing surfaces by cutting horizontal grooves with sharp object or saw. Space grooves ¼ in. to ½ in. apart, cutting 1/32 in. to 1/16 in. deep.
3. Apply two coats of dark, water-based stain to interior surfaces. Do not use paint, as it will fill grooves.
4. Attach side pieces to back, caulking first. Use 1 5/8-in. screws. Make sure top angles match.
5. Attach 5-in. and 10-in. spacers to inside corners per drawings in Figure 9-12 [assembly]. Use 1-in. screws. Roost chamber spacing will be ¾ in. (front to back). Do not block side vents.
6. Place first roosting partition on spacers even with bottom edge of roof. Place 20-in. spacers on partition and screw to first spacers (through partition), using 1 5/8-in. screws.
7. Repeat step 6 for remaining spacers and partitions.
8. Attach front to sides, top piece first (caulk seams). Be sure top angles match (sand if necessary). Leave ½-in. vent space between top and bottom front pieces. A bar clamp may be useful if sides have flared out during construction.
9. Attach roof supports to the top inside of front and back pieces with 1-in. screws. Do not let screws protrude into roosting chambers.
10. Caulk around all top surfaces, sanding first if necessary to ensure good fit with roof.
11. Attach roof to sides and roof supports with 1¼-in. screws. Caulk around roof and side joints to further guard against leaks and drafts. Do not let screws protrude into roosting chambers.
12. Paint or stain exterior three times (use primer for first coat).
13. Cover roof with shingles or galvanized metal.

Optional modifications

1. These nursery-house dimensions were chosen to permit construction of two bat houses per half-sheet of plywood. Increasing house width to 24 in. or more or adding partitions benefits bats and attracts larger colonies. Additional spacers are required to prevent warping of roost partitions for houses more than 24 in. wide.
2. Taller bat houses provide improved temperature gradients and may be especially useful in climates where daily temperatures fluctuate widely. Bat houses 3 ft. or taller should have the horizontal vent slot 12 in. from the bottom of the roosting chambers.
3. Two bat houses can be placed back-to-back mounted on poles. Before assembly, a horizontal ¾-in. slot should be cut in the back of each house about 10 in. from the bottom edge of the back piece to permit movement of bats between houses. Two pieces of wood, 1 in. x 4 in. x 1¾ in., screwed horizontally to each side, will join the two boxes. Leave a ¾-in. space between the two houses, and roughen the wood surfaces or cover the back of each with plastic mesh. One 2-in. x 4-in.

Appendix 9-3 continued

x 40-in. vertical piece, attached to each side, over the horizontal pieces, blocks light but allows bats and air to enter. Use a 2-in. x 6-in. vertical piece if securing houses with U-bolts to metal poles. A galvanized metal roof that covers both houses protects them and helps prevent overheating. Eaves should extend about 3 in. in front in southern areas and about 1½ in. in the north.

4. Ventilation may not be necessary in cold climates. In that case, the front of the bat house should be a single, 23-in.-long piece. Far-northern bat houses may also benefit from a partial bottom to help retain heat. Slope the sides and bottom at an angle of 45° or greater to reduce guano build-up. Leave a ¾-in. entry gap at the back, and be sure the bottom does not interfere with access to the front crevices. A hinged bottom is required to permit annual cleaning.
5. Durable plastic mesh can be substituted for roughening. Attach mesh to backboard, landing area, and one side of each partition after staining interior, but prior to assembly. Use ⅛-in. or ¼-in. plastic mesh [see *Products Mentioned in Text*) and attach every two inches with a 5/16-in. stainless steel staples.
6. Make partitions removable by attaching small cleats with thumbscrews to the bottom of side pieces for support. Spacer strips are unnecessary if grooves for partitions are cut in the side pieces with a router or dado saw blade.

ROCKET BOX

The "rocket box "or "post bat house" (Fig. 9-14) was created in the mid 1990s by Dan Dourson and John MacGregor, two U.S. Forest Service biologists in Kentucky. This design is essentially a four-sided wooden shell, 3 ft. (0.9 m) tall or taller, that slides over a rough-cut, wooden 4 x 4-in. (10.2 x 10.2-cm) post (Dourson and MacGregor, 1997). Untreated, oak or locust posts are recommended for longevity. A 360-degree, ¾-in. (19-mm) chamber between the post and shell allows bats to select warm or cool temperature zones. Later iterations of this design include two-chambered models (Tuttle *et al.*, 2004), a larger Dourson and MacGregor design that fits over an 8 x8-in. (20.3x20.3-mm) wooden post spliced to a steel pole (Carter, 2002; Carter *et al.*, 2001), and a variety of commercially-built, round, plastic-shelled boxes that attach to metal poles. Plans for building a rocket box can be accessed at http://www.batsnorthwest.org/rocketbox_plans.pdf.

Fig. 9-14. Rocket box erected at McNabney Marsh, Contra Costa County, California. Photo by the authors, © Bat Conservation International, Austin, Texas.

Appendix 9-3 continued

CJM BAT BOX/"SLOT HOUSE"

The CJM bat box or "slot house" (Fig. 9-15) is designed to imitate niches where crevice-dwelling bats roost, such as beneath loose bark or a split in a tree trunk. Three vertical chambers or "slots" of 0.6 in. (15 mm), 0.7 in. (18 mm) and 0.8 in. (20 mm) accommodate bats of different sizes. A small attic space permits bats to cluster, thereby conserving energy (Vincent Wildlife Trust, 2011). To date, at least 7 (possibly 8) species have been recorded using them, including western barbastelle (*Barbastella barbastellus*), Leisler's noctule (*Nyctalus leisleri*), common pipistrelle (*Pipistrellus pipistrellus*), soprano pipistrelle (*Pipistrellus pygmaeus*), brown long-eared bat (*Plecotus auritus*), Natterer's myotis (*Myotis nattereri*), whiskered myotis (*Myotis mystacinus*), and possibly Brandt's myotis (*Myotis brandtii*). The plans for constructing the CJM bat box are shown in Figure 9-16.

Fig. 9-15. CJM bat box or "slot house." Photo courtesy of Johnny Birks.

All timber is untreated, rough-sawn 180mm x 25mm

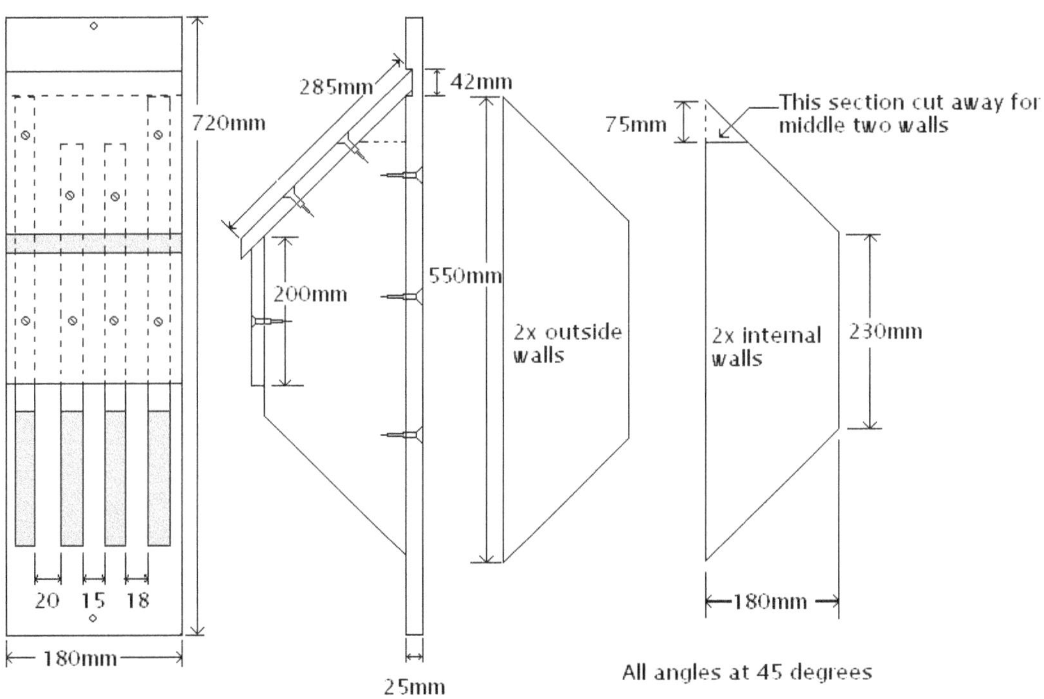

Fig. 9-16. CJM bat box construction plans. Illustrations courtesy of Colin J. Morris.

Appendix 9-3 continued

ARTIFICIAL HIBERNATION ROOST FOR GEOFFROY'S MYOTIS

This simple bat box was designed by Dirk Swaenen of the Bat Conservation Group in Flanders, Belgium specifically for hibernating Geoffroy's myotis (*Myotis emarginatus*) in an abandoned fortress in Schoten, Belgium (Fig. 9-17). The box is attached to the ceiling (Figs. 9-18 and 9-19) where bats typically roost, providing them additional protection against drafts. The box is made from 1-in. (2.5-cm) lumber and is 7.9 in. (20 cm) square x 5.9 in. (15 cm) tall. There are 4 compartments and no bottom. Up to 42 *Myotis emarginatus* and 1 *M. nattereri* have used four of these boxes inside the fortress.

Fig. 9-17. Abandoned fortress at Schoten, Belgium, where Geoffroy's myotis (*Myotis emarginatus*) hibernates throughout winter. Photo courtesy of Dirk Swaenen.

Appendix 9-3 continued

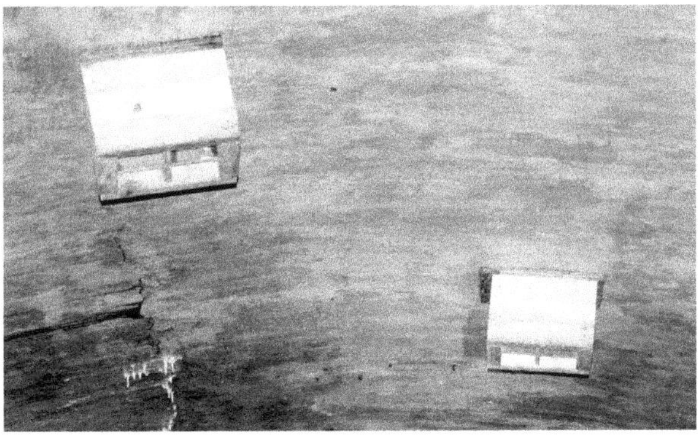

Fig. 9-18. Artificial hibernation roost boxes attached to a chamber ceiling in an abandoned fortress in Schoten, Belgium. Photo courtesy of Dirk Swaenen.

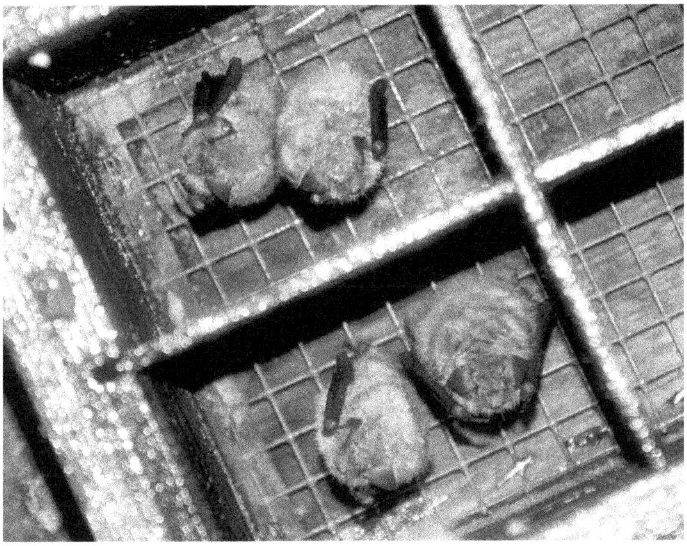

Fig. 9-19. *Myotis emarginatus* hibernating in an artificial roost. Photo courtesy of Dirk Swaenen.

Appendix 9-3 continued

INSTRUCTIONS FOR CONSTRUCTING AN ARTIFICIAL HIBERNATION ROOST FOR GEOFFROY'S MYOTIS (FIG. 9-20)

A) Drill a hole in each corner of a board measuring 7.87-in. (20-cm) x 11.81-in. (30-cm) x 1-in. (2.54-cm) thick.

B) Using appropriately sized screws, join together 4 boards measuring 7.87 in. (20 cm) high x 5.91 in. (15 cm) deep x 7.87 in. (20 cm) wide x 1 in. (2.54 cm) thick to form a double open-ended box.

C) Cut wood lath to a size that fits firmly inside the box. Attach it to the center of the board shown in A.

D) Staple ¼-in. (0.64 cm) polyethylene mesh or hardware cloth to the molding.

E) Affix the box shown in B to the board shown in A with screws.

F) Divide the box into 4 chambers with door skin or other very thin type of lumber (Fig. 9-21).

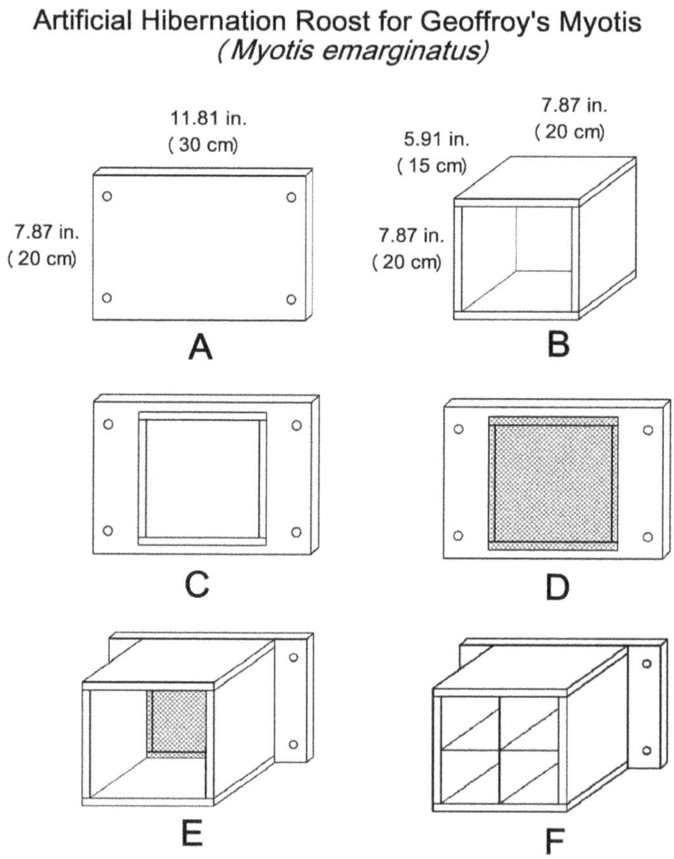

Fig. 9-20. Artificial hibernation roost construction plans. Illustrations courtesy of Dirk Swaenen.

Appendix 9-3 continued

Fig. 9-21. Newly completed artificial hibernation roost. Photo courtesy of Dirk Swaenen.

INDIAN BAT POT

Researchers with Bat Conservation India developed an artificial roost for bats made from a 6-in. (15.2-cm) diameter clay pot, although larger sizes can be used. At the base of the pot, a ¾ x 5-in. (19 x 127-mm) rectangular slot is cut. A mixture of cement and sand is poured into the pot, and a block of wood and brick tiles are inserted at a downward sloping angle to form a hollow chamber inside the pot. After the mixture hardens, the block of wood is removed. The entrance slot is reduced to ¾ x 2 in. (19 x 50 mm) by the cement mixture. Sturdy wire loops are used to attach the pot flush against a vertical substrate such as an exterior wall. First occupancy in Tirunelveli, South India was observed in 2005. Selected pots have been continuously used by Indian pygmy bats (*Pipistrellus minus*) throughout 2009 and 2010. Photos and instructions for constructing the bat pot are available on the Bat Conservation India website at http://www.nichiyu.in/bats/inno.htm.

ATTIC VENT BAT HOUSE

The concept of building bat houses like attic vents (Figs. 9-22 and 9-23) arose because of the thousands of urban bats dwelling in the attic vents of buildings. As with many bat houses, these are best constructed for bats that are being excluded from man-made structures. These bat houses are built around a standard, rectangular wooden

Appendix 9-3 continued

Fig. 9-22. Two attic vent bat houses placed back to back on telephone poles and erected at the Clayton County Water Authority in Morrow, Georgia. Photos courtesy of Susan M. Barnard.

Fig. 9-23. Four attic vent bat houses, each facing in a different direction (i.e., north, east, south, and west). The houses were occupied by a large group of *Tadarida brasiliensis cynocephala* and a few *Eptesicus fuscus* at Chehaw Wild Animal Park in East Dougherty, Georgia. Photo courtesy of John Fowler.

Appendix 9-3 continued

gable vent that can be purchased at many contractor's outlet stores (see http://www.architecturaldepot.com/mm5/merchant.mvc?Screen=CTGY&Category_Code=gable-vents&keyword=attic%20vents&&gclid=CNGYzPOzs6gCFcO8Kgodc0CKFA or http://www.cedaratticvents.com/square-rectangle-vents.php). The screen backing of the vents must be removed to allow bats to enter the boxes. The boxes can also be made by modifying the North American nursery-style bat house popularized by Bat Conservation International (see *Nursery House* above).

THE KENT BAT BOX

The Kent Bat Box (Fig. 9-24) is simple to construct, and because it has an open bottom, no guano accumulates inside, making it easy to maintain. Although the box can be constructed in a variety of sizes, the width of the crevices must not be changed. The crevices are 0.59 in. (15 mm) and 0.79 in. (20 mm), which are ideal for *Pipistrellus pipistrellus* (Fig. 9-25). Nevertheless, it is likely that many other small bat species will occupy this box. The Kent Bat Group requests feedback on other bat species found in boxes of their design. Their website is listed in the instructions for constructing the box (Fig. 9-26).

Appendix 9-3 continued

The Kent Bat Box

(Courtesy of Kent Bat Group - http://www.kentbatgroup.org.uk)

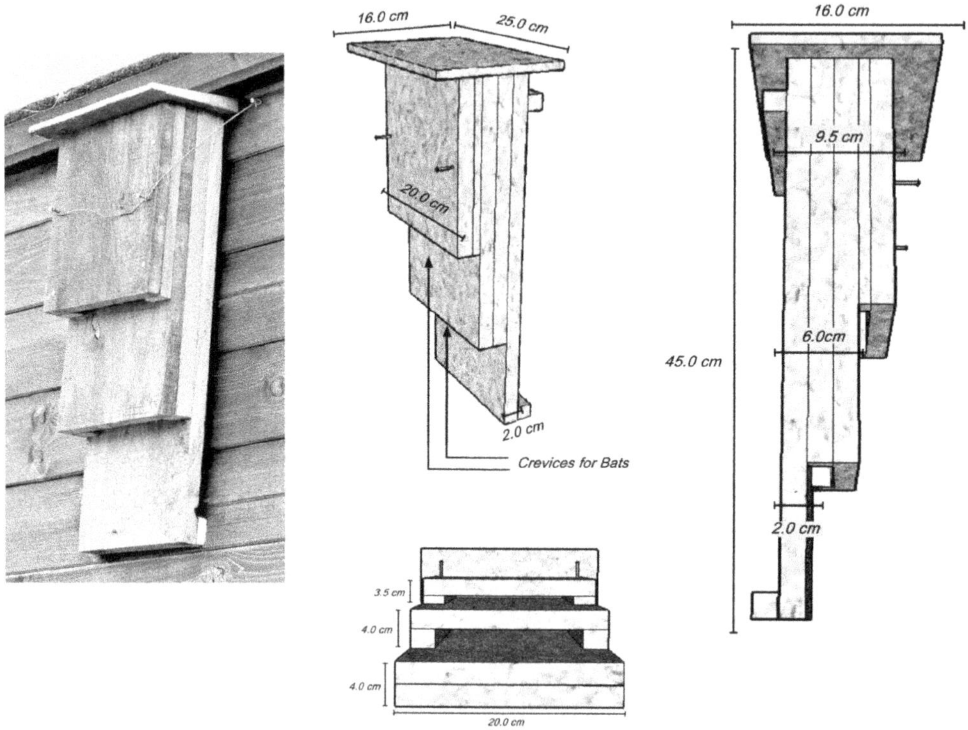

Part	Quantity	Size in cm
Roof	1	25.0 x 16.0 x 2.0
Back Panel	1	45.0 x 20.0 x 2.0
Center Panel	1	33.0 x 20.0 x 2.0
Front Panel	1	21.0 x 20.0 x 2.0
Back Rails	2	33.0 x 2.0 x 2.0
Front Rails	2	21.0 x 1.5 x 1.5
Stand Offs	2	20.0 x 2.0 x 2.0

Fig. 9-24. The Kent bat box. Photo and illustrations courtesy of the Kent Bat Group.

Fig. 9-25. Common pipistrelles (*Pipistrellus pipistrellus*) occupying a Kent bat box at Priority Country Park in Befordshire, England. Photo courtesy of Daniel Fellman.

Appendix 9-3 continued

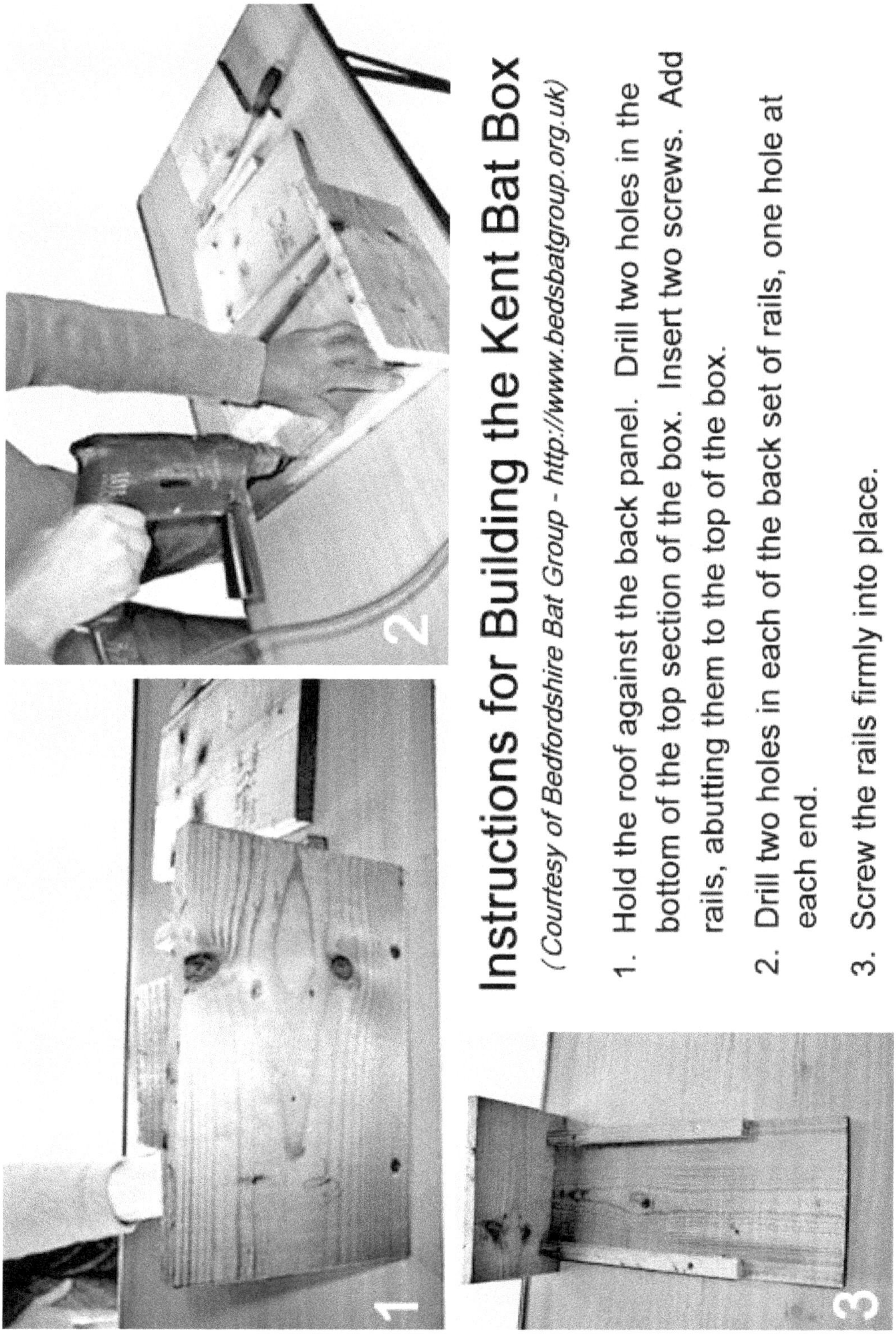

Fig. 9-26A. Instructions for building the Kent bat box (steps 1 thru 3). Photos courtesy of the Bedfordshire Bat Group.

Appendix 9-3 continued

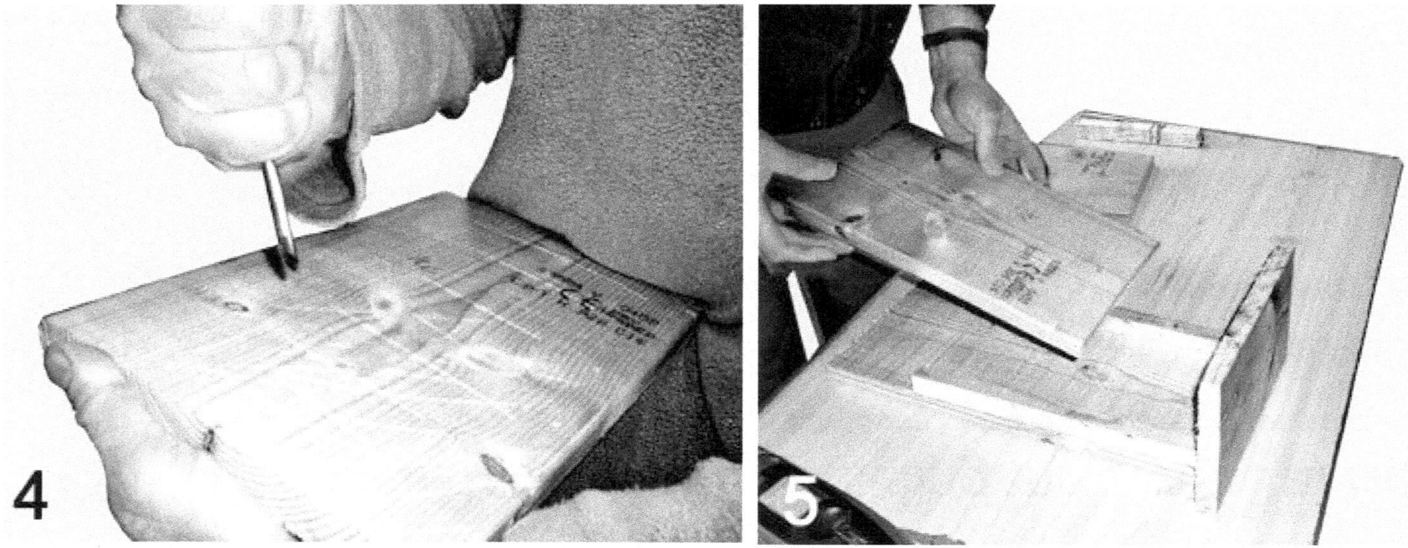

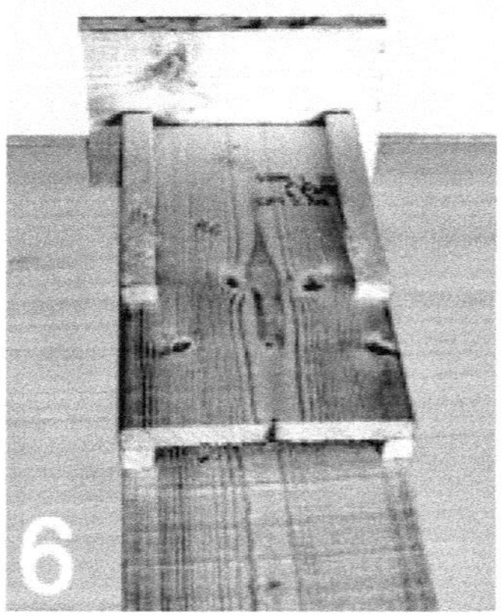

4. Before attaching the center panel to the rails, roughen the smooth side. This can be done with either a screw driver or a rasp.
5. Place the center panel firmly against the top of the box. Drill two sets of holes, offsetting them from the screws in the underlying rails. Screw the panel into place.
6. Add the front set of rails. Drill two holes in each, avoiding the screws in the underlying panel.

Fig. 9-26B. Instructions for building the Kent bat box (steps 4 thru 6). Photos courtesy of the Bedfordshire Bat Group.

Appendix 9-3 continued

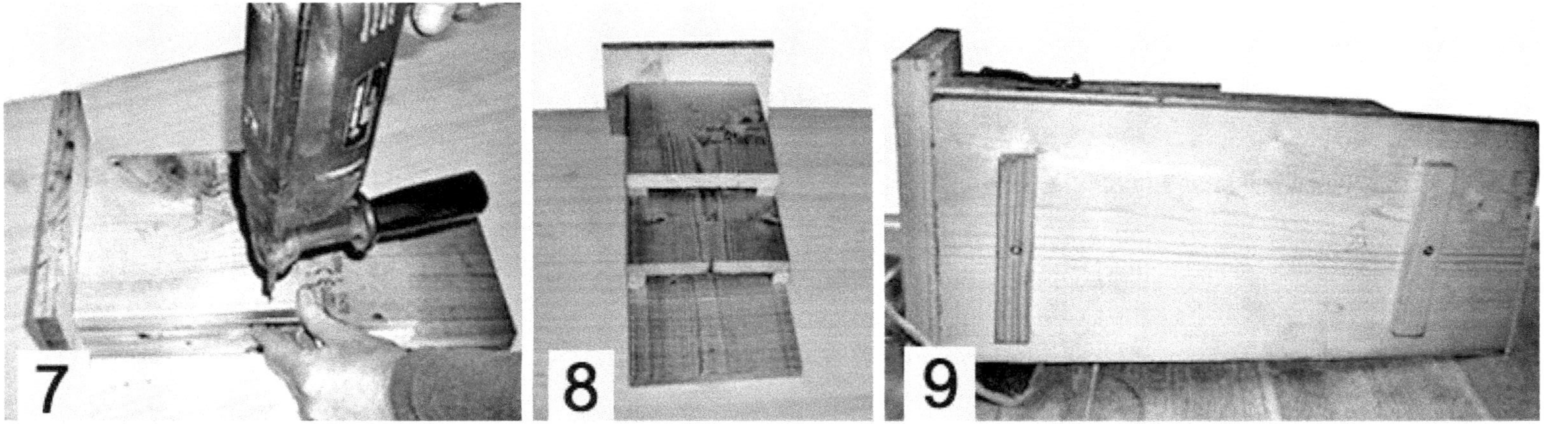

7. Place the front panel on the box, abutting it firmly to the roof. Drill four holes, one at each corner, and insert the screws.
8. The completed box.
9. If the box is to be placed on a building, add two stand offs to the back
10. The box can be hung with bailer twine (shown here), wire, or other material. The hanging material can be attached with washers and screws, but eye screws are preferable. Do not allow screws to penetrate the inside of the box.

Fig. 9-26C. Instructions for building the Kent bat box (steps 7 thru 10). Photos courtesy of the Bedfordshire Bat Group.

Appendix 9-3 continued

PLANTER/GUANO-CATCHING BAT BOX

The Planter/Guano-Catching Bat Box (Fig. 9-27) was designed by Susan M. Barnard in response to calls about bats roosting over areas where falling guano was a problem (e.g., over doorways and windows). When this box is mounted near roosts where evictions have occurred, bats sometimes adopt it as an alternate roost. There are no limitations on box size, and it can be made of any non-toxic exterior lumber. Roosting chambers should be sized for the species in question, for example, ¾ to 1 in. (1.9 to 2.54 cm) for *Eptesicus fuscus.* The number of roosting chambers is determined by the size of the box, which in turn may be determined by the number of bats expected to use it. Plans for constructing the Planter/Guano-Catching Bat Box are shown in Figure 9-28.

SCHWEGLER BAT BOXES

Schwegler bat boxes are commercially manufactured in Schorndorf, Germany. Because they are manufactured of concrete and wood, they are very heavy, making them very expensive to ship throughout the world. Consequently, these boxes are used primarily in Europe, with the majority of them being sold in the United Kingdom. In addition to the information presented below, Appendix 9-4 lists the European bats that have been observed using the many Schwegler boxes, and the potential use of these boxes by bats in other parts of the world. Anyone wishing to experiment with 'woodcrete' boxes should also review Appendix 9-4.

BAT BOX 1FD WITH TRIPLE FRONT PANEL (FIG. 9-29)

The 1FD bat box has been developed specifically for small bat species such as *Pipistrellus nathusii*, *P. pipistrellus*, *Myotis daubentonii*, *M. nattereri* (Fig. 9-29c) and *Plecotus auritus* (see also Appendix 9-4). The box measures 6.3 in. (16 cm) in diameter x 14.2 in. (36 cm) high. The interior, and the type and size of the box entrance meet the requirements of small species with varied roosting habits. It features a domed roof, increased interior height, and two grooved, internal wooden front panels for crevice-roosting species.

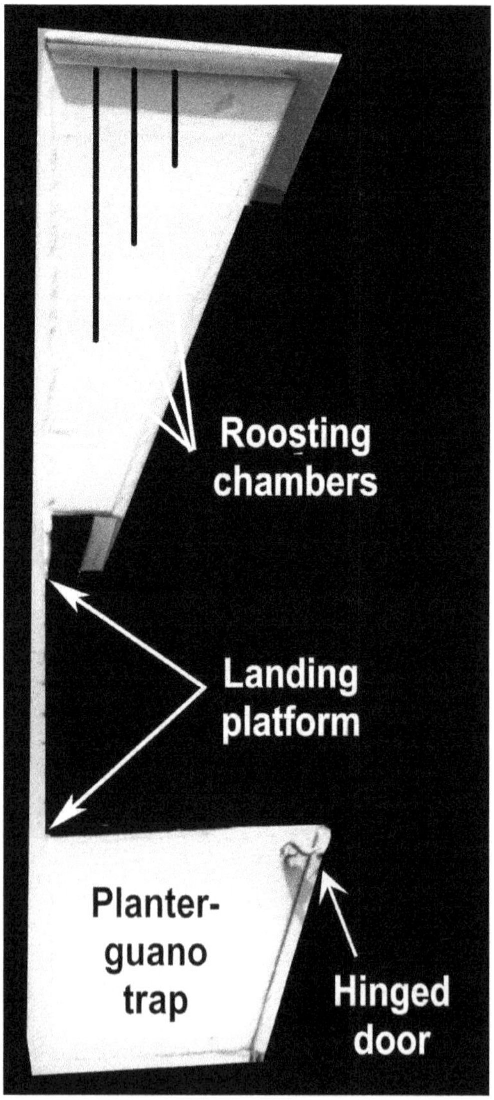

Fig. 9-27. Planter/guano-catching bat box. Photo courtesy of Susan M. Barnard.

Appendix 9-3 continued

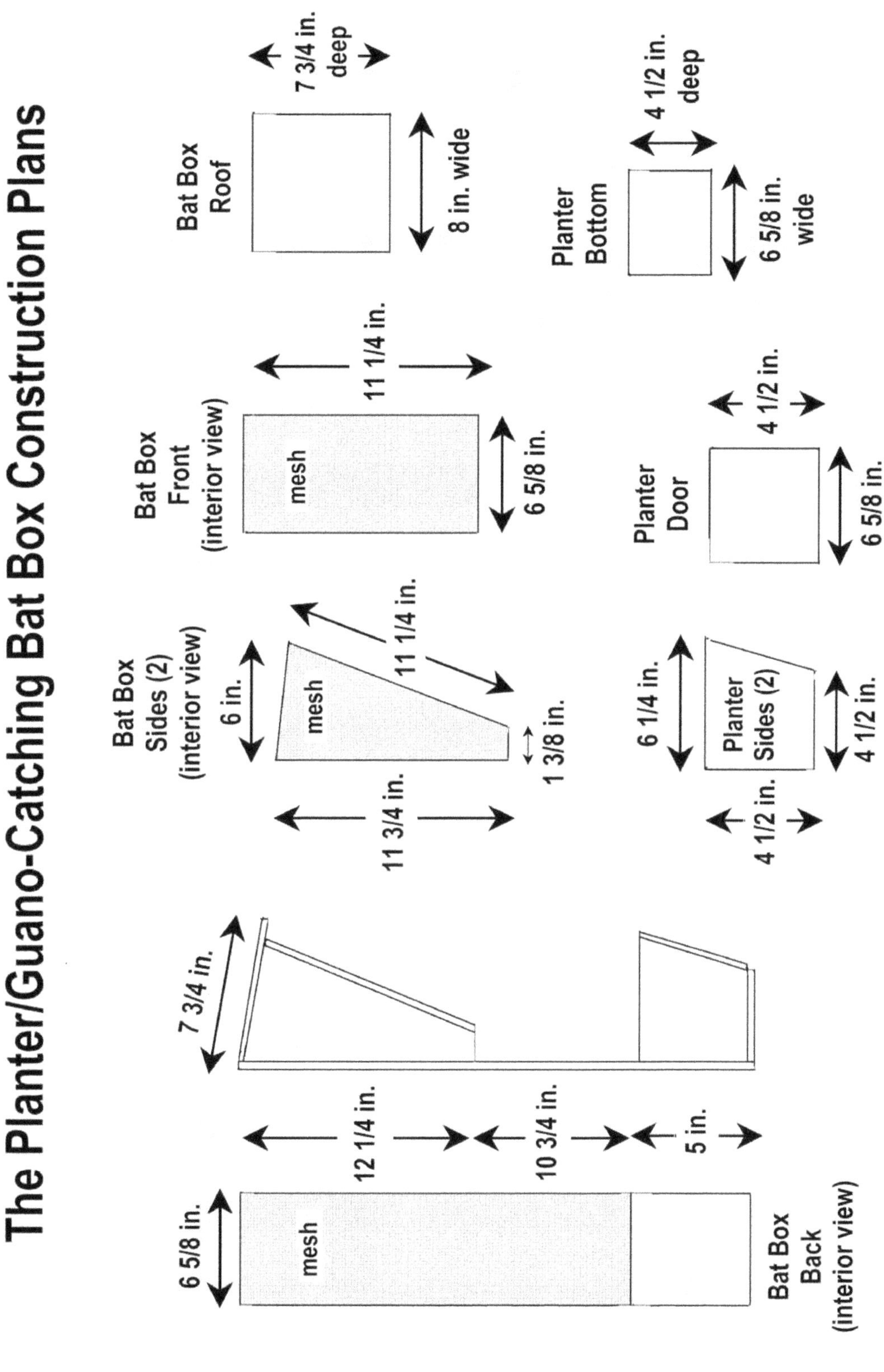

Fig. 9-28. Construction plans for the planter/guano-catching bat box. Illustrations courtesy of Susan M. Barnard.

Appendix 9-3 continued

Fig. 9-29. Schwegler bat box 1FD. A) front panel closed; B) front panel open showing two internal, grooved wooden panels; C) Natterer's myotis (*Myotis nattereri*) roosting in the domed top of the box. Photos courtesy of Schwegler Vogel- und Naturschutzprodukte GmbH.

The 1FD has also shown to be effective as a nursery roost. It is best used in areas inhabited by a variety of bat species, or when efforts are being made to encourage bats to colonize new locations. The front panel can be removed for cleaning and inspection of the box interior.

BAT BOX 1FF WITH BUILT-IN WOODEN REAR PANEL (FIG. 9-30)

The 1FF bat box is the typical flat design with an open bottom that allows guano to drop out. In addition to being attached to flat surfaces, the recess at the back allows the box to also be attached to poles or tree trunks. The 1FF box measures 5.5 in. (14 cm) deep (at the box top) x 10.6 in. (27 cm) wide x 16.9 in. (43 cm) high and is designed for crevice-roosting species (see Appendix 9-4). The 1FF bat box is popular for use in gardens and woodlands, the sides of buildings (see Fig. 9-30c), and in flight cages housing captive bats (see Fig. 9-14, Volume 3). The large front flap allows the box to be inspected easily for occupancy. At the interior back of the box is a roughened panel of durable wood. When ambient temperatures fluctuate during spring and autumn, this feature allows the bats to choose between the cooler wood-concrete surface and the warmer, wooden panel.

BAT BOX 1FFH OR UNIVERSAL BAT BOX (FIG. 9-31)

The universal bat box is designed as a summer roost for bats that inhabit forests and rear young in tree hollows (see Appendix 9-4). The universal box provides roosting choices with both hollows and crevices. The rear compartment has sufficient space for large bat species, or for females to rear young. The forward compartment can accommodate bats that prefer crevices in which to roost. The narrow entrance helps to reduce predation. Because the box is open-bottomed, guano can drop out, helping to maintain cleanliness. The clinging surfaces are grooved and made of durable, untreated wood. This feature allows bats to maintain footholds. To ensure stability of the box during windy weather, it contains four supports that are built into the semicircular, rear surface. This box can

Appendix 9-3 continued

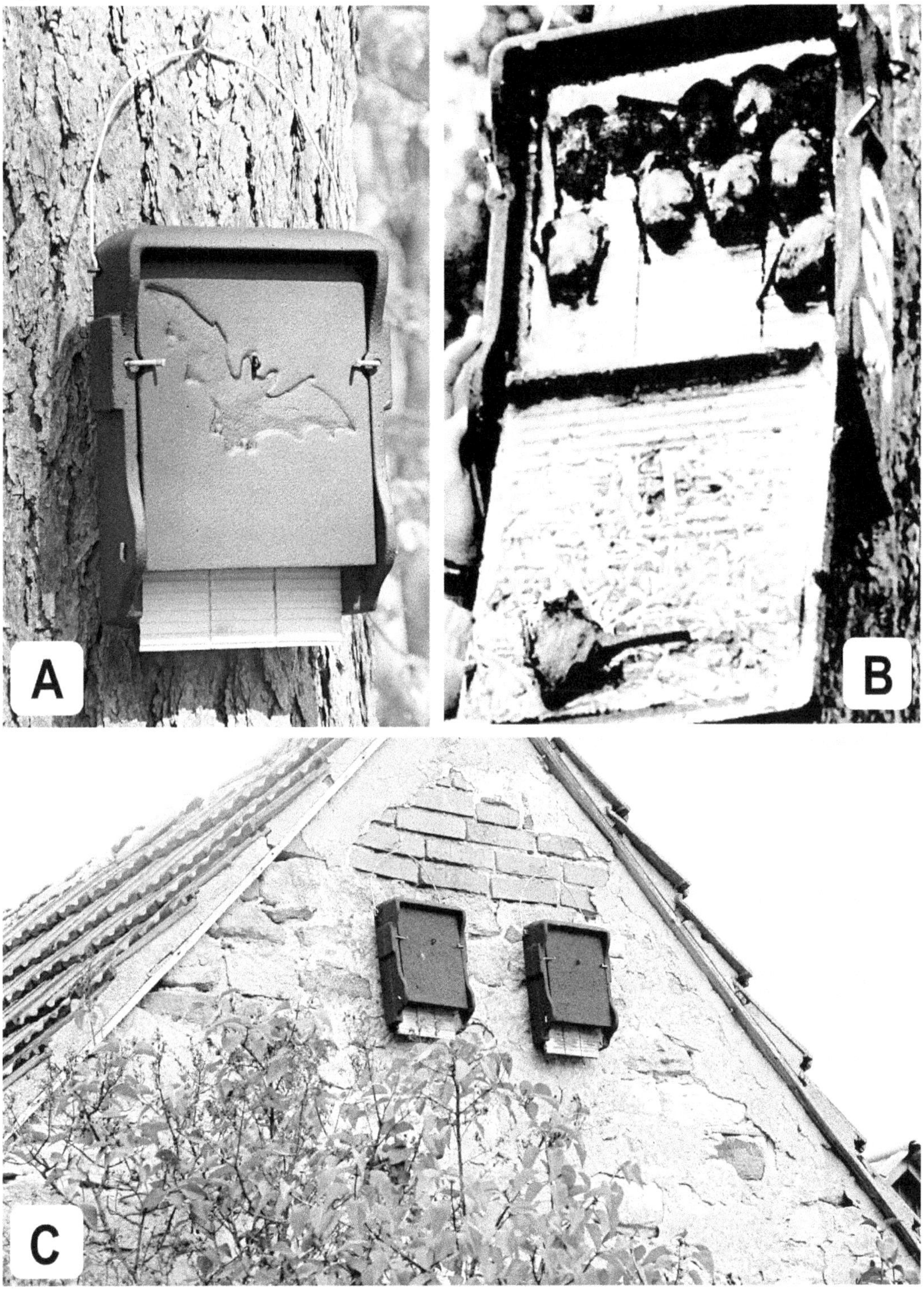

Fig. 9-30. Schwegler bat box 1FF. A) front panel closed; B) front panel open showing grooved, wooden rear panel and a group of roosting noctules (*Nyctalus noctula*); C) two 1FF bat boxes placed on the side of a building. Photos courtesy of Schwegler Vogel- und Naturschutzprodukte GmbH.

Appendix 9-3 continued

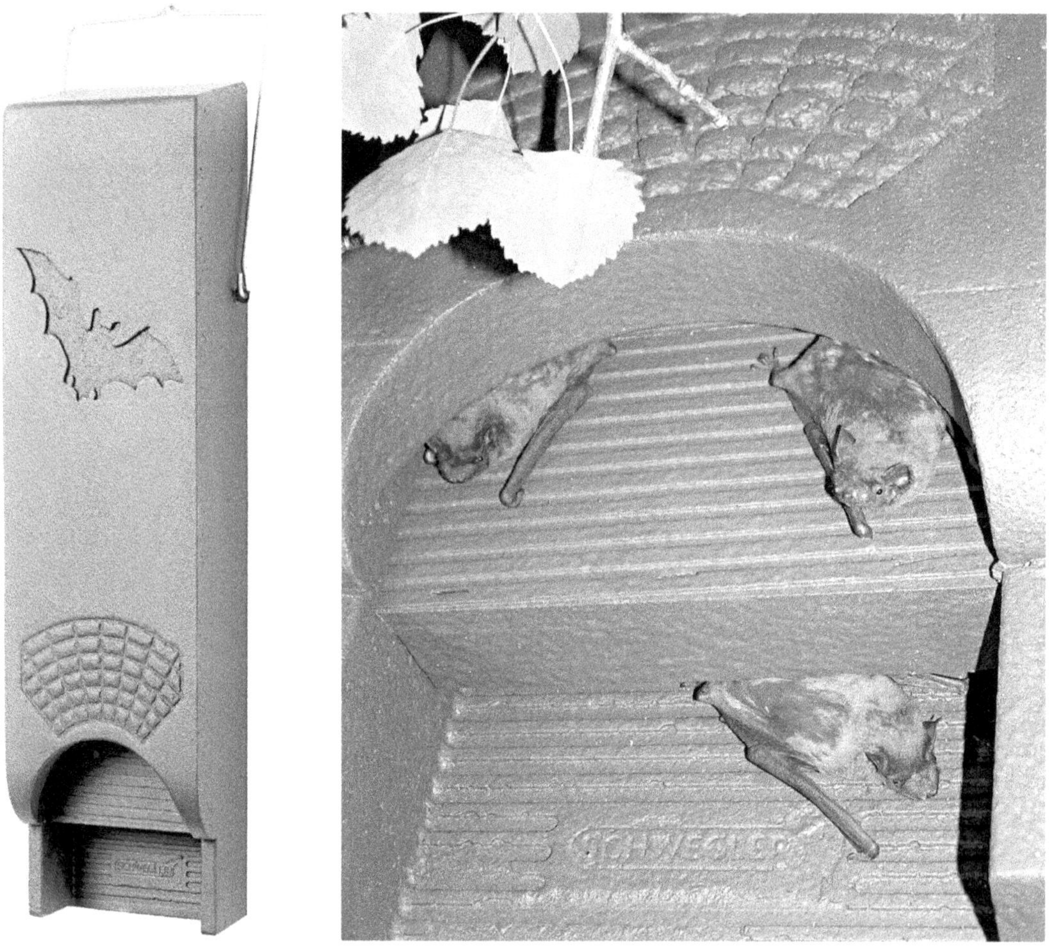

Fig. 9-31. Schwegler bat box 1FFH or universal bat box. Left) box exterior; Right) box interior showing roosting noctules (*Nyctalus noctula*). Photos courtesy of Schwegler Vogel- und Naturschutzprodukte GmbH.

not be opened; therefore, a mirror must be used on the underside to check for occupancy. The external dimensions of the universal box are approximately 7.5 in. (19 cm) deep x 9.6 in. (24.5 cm) wide x 34.5 in. (87.5 cm) high.

COLONIAL BAT BOX 1FS (FIG. 9-32)

The colonial bat box is designed to accommodate nursery groups of 70-100 individuals during summer (e.g., *Nyctalus noctula* [Fig. 9-32b], *Pipistrellus nathusii*, and *Plecotus auritus*; see also Appendix 9-4). Having a closed bottom, droppings must be removed periodically. The species and numbers of individuals can be checked at this time.

To provide roosting choices within the box, there are three upright, grooved wooden panels and an insulated, grill panel on the box roof (Fig. 9-32b). The front panel can be removed for cleaning (see Fig. 9-33b). The external dimensions of the colonial bat box are 11 in. (28 cm) in diameter x 17.3 in. (44 cm) high.

Appendix 9-3 continued

Fig. 9-32. Schwegler bat box 1FS or colony bat house. A) colonial box mounted in a tree; B) box interior with roosting noctules (*Nyctalus noctula*). Photos courtesy of Schwegler Vogel- und Naturschutzprodukte GmbH.

HIBERNATION BAT BOX 1FW (FIG. 9-33)

The hibernation box is the same internally as the colonial box, 1FS, discussed above. However, it is designed with several insulating layers to provide protection for hibernating bats during winter and nursery colonies from excessive heat during summer. The inside measures 7.9 in. (20 cm) in diameter x 15 in. (38 cm) high; the external dimensions are 15 in. (38 cm) in diameter x 19.7 in. (50 cm) high.

BAT BOX 2F (WITH OR WITHOUT DOUBLE FRONT PANEL) (FIG. 9-34)

Bat box 2F is a general purpose box, particularly attractive to small bat species (e.g., *Pipistrellus nathusii, P. pipistrellus* (see Fig. 9-34c), *P. pygmaeus, Myotis bechsteinii, M. daubentonii, M. nattereri,* and *Plecotus auritus*; see also Appendix 9-4). However, larger species such as *Nyctalus leisleri* (Fig. 9-35c) and *N. noctula* have also occupied this box. The domed roof provides an ideal roost for bats that hang freely from horizontal surfaces. When used with the double wooden panel, a crevice is created in the box front (see Fig. 9-34b) for species that roost in tight places. The front panel can be removed to check for occupancy and for cleaning. However, when using the box with the double wooden panel, extra care must be taken not to injure bats when the panel is being removed. The external dimensions are 6.3 in. (16 cm) in diameter x 13 in. (33 cm) high.

BAT BOX 2FN (FIGS. 9-4, 9-35A,B)

Bat box 2FN is attractive to bats of all sizes such as *Myotis myotis* (Fig. 9-35a), *Nyctalus lasiopterus* (Fig. 9-35b), *Myotis daubentonii, M. nattereri, and Plecotus auritus* (Fig. 9-4b) (see also Appendix 9-4). The intermediate floor gives ideal protection against small predators, draughts, and light. This box has two entrances, one in front and another at the rear. The rear entrance situates against a tree trunk, whereby bats often crawl into this entrance and leave by the front. The domed roof provides ideal roosting conditions for bats that hang from the ceiling of their roosts. Box 2FN is particularly suitable for use

Appendix 9-3 continued

in parks and forests. The front panel can be removed to check for occupancy and for cleaning. The external dimensions of the box are 6.3 in. (16 cm) in diameter x 14.2 in. (36 cm) high.

Fig. 9-33. Schwegler bat box 1FW or hibernation box. A) box closed, mounted in a tree; B) box open showing removable wooden panels, a featured shared with box 1FS; C) front panel open showing multi-layered cavity wall that protects bats from extreme temperatures, not only in winter, but also in summer; D) roosting noctules (*Nyctalus noctula*). Photos courtesy of Schwegler Vogel- und Naturschutzprodukte GmbH.

Appendix 9-3 continued

Fig. 9-34. Schwegler bat box 2F or general purpose box. A) box mounted on a tree with the front panel closed; B) front panel open showing grooved, wooden panel attached to the front panel of the box; C) roosting common pipistrelles (*Pipistrellus pipistrellus*). Photos courtesy of Schwegler Vogel- und Naturschutzprodukte GmbH.

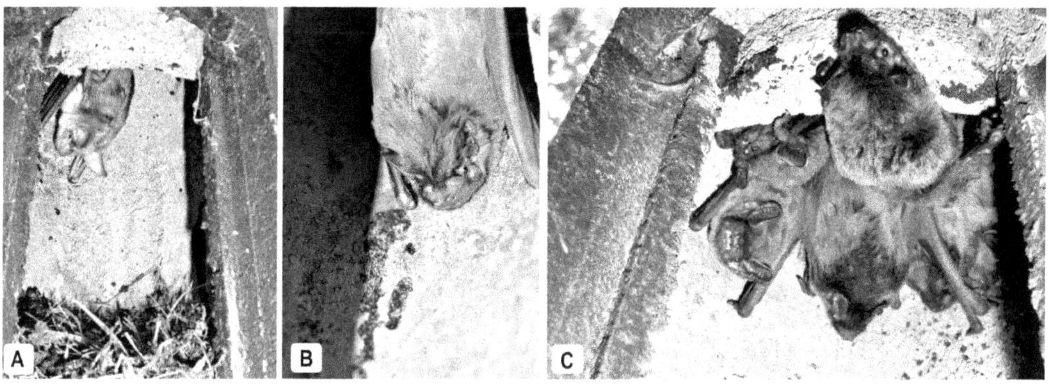

Fig. 9-35. A) Schwegler bat box 2FN with a roosting mouse-eared bat (*Myotis myotis*); B) Schwegler bat box 2FN with a roosting giant noctule (*Nyctalus lasiopterus*); C) Schwegler bat box 2F with a group of roosting Leiser's noctules (*Nyctalus leisleri*). Photos courtesy of Gianna Dondini and Simon Vergari.

Appendix 9-4. Types of Schwegler bat boxes*, European bat species that have occupied them, and the potential for occupancy by other bat species

Schwegler Bat Box	Occupied by European Bat Species	Potential for Occupancy by Other Species?[5]
1FD	*Myotis daubentonii*[1] *Myotis nattereri*[1] (see Fig. 9-29c) *Pipistrellus nathusii*[1] *Pipistrellus pipistrellus*[1] *Plecotus auritus*[1]	*Chalinolobus morio* *Chalinolobus tuberculatus* *Mormopterus norfolkensis* *Myotis auriculus* *Myotis austroriparius* *Myotis brandtii* *Myotis evotis* *Myotis keenii* *Myotis mystacinus* *Myotis septentrionalis* *Myotis sodalis* *Myotis thysanodes* *Myotis volans* *Myotis yumanenisis* *Neoromicia capensis* *Perimyotis subflavus* *Pipistrellus hesperidus* *Pipistrellus kuhlii* *Pipistrellus pygmaeus* *Scotorepens balstoni* *Vespedelus darlingtoni* *Vespedelus regulus*
1FF	*Barbastella barbastellus*[11] *Myotis bechsteinii*[2,11] *Myotis daubentonii*[2] *Myotis nattereri*[2,11] *Nyctalus leisleri*[1] *Nyctalus noctula*[1,2] (see Fig. 9-30b) *Pipistrellus pipistrellus*[11] *Pipistrellus pygmaeus*[11] *Pipistrellus* spp.[2] *Plecotus auritus*[2,11]	*Antrozous pallidus* *Chaerephon pumilus* *Chalinolobus gouldii* *Chalinolobus morio* *Eptesicus fuscus* *Eptesicus nilssonii* *Eptesicus serotinus* *Mops condylurus* *Mops midas* *Myotis auriculus*[6] *Myotis austroriparius* *Myotis brandtii* *Myotis evotis* *Myotis keenii* *Myotis lucifugus* *Myotis occultus* *Myotis septentrionalis* *Myotis sodalis* *Myotis thysanodes* *Myotis volans* *Myotis yumanensis* *Neoromica capensis* *Nycticeius humeralis* *Nyctophilus bifax* *Nyctophilus geoffroyi* *Nyctophilus gouldi* *Pipistrellus kuhlii* *Pipistrellus nathusii* *Pipistrellus pipistrellus* *Pipistrellus pygmaeus* *Plecotus austriacus* *Scotophilus dinganii* *Scotorepens orion* *Tadarida aegyptiaca* *Vespertilio murinus*

Appendix 9-4 continued

Schwegler Bat Box	Occupied by European Bat Species	Potential for Occupancy by Other Species?[5]
1FFH	*Myotis nattereri*[1] *Nyctalus noctula*[1] (see Fig. 9-31)	*Chalinolobus gouldii* *Chalinolobus morio* *Mops condylurus* *Mops midas* *Mormopterus norfolkensis* *Myotis bechsteinii* *Myotis daubentonii* *Myotis keenii* *Myotis evotis* *Myotis volans* *Nyctalus lasiopterus* *Nyctalus leisleri* *Nycteris thebaica* *Nycticeius humeralis* *Nyctophilus bifax* *Nyctophilus geoffroyi* *Nyctophilus gouldi* *Pipistrellus nathusii* *Scoteanax rueppellii* *Scotophilus viridus* *Scotorepens balstoni* *Scotorepens orion* *Tadarida australis* *Vespedelus darlingtoni* *Vespedelus regulus*
1FS	*Myotis bechsteinii*[2] *Myotis daubentonii*[2] *Myotis nattereri*[2] *Nyctalus noctula*[1,2] (see Fig. 9-32b) *Pipistrellus nathusii*[1] *Pipistrellus* spp.[2] *Plecotus auritus*[1,2]	*Antrozous pallidus* *Chaerephon pumilus* *Chalinolobus gouldii* *Chalinolobus morio* *Eptesicus fuscus* *Eptesicus serotinus* *Molossus molossus* *Mops condylurus* *Mops midas* *Mormopterus norfolkensis* *Myotis brandtii* *Myotis evotis* *Myotis leisleri* *Myotis lucifugus* *Myotis mystacinus* *Myotis septentrionalis* *Myotis sodalis* *Myotis yumanensis* *Nyctalus leisleri* *Nycteris thebaica* *Nycticeius humeralis* *Nyctophilus bifax*[7] *Nyctophilus geoffroyi* *Nyctophilus gouldii* *Plecotus austriacus* *Scotophilus viridus* *Scotorepens balstoni* *Scotorepens orion* *Tadarida brasiliensis* *Vespedelus darlingtoni* *Vespedelus regulus*

Appendix 9-4 continued

Schwegler Bat Box	Occupied by European Bat Species	Potential for Occupancy by Other Species?[5]
1FW[8]	*Myotis bechsteinii*[2] *Myotis daubentonii*[2] *Myotis nattereri*[2,11] *Nyctalus noctula*[2,11,12] (see Fig. 9-33d) *Pipistrellus nathusii*[12] *Pipistrellus* spp.[2] *Plecotus auritus*[2,11]	For summer use, perhaps those listed above for box 1FS, and for winter use, *Eptesicus fuscus*[9]
2F	*Myotis bechsteinii*[2,3] *Myotis daubentonii*[1,2] *Myotis nattereri*[2,3,10] *Nyctalus leisleri*[3] (see Fig. 9-35c) *Nyctalus noctula*[2,3] *Pipistrellus nathusii*[1] *Pipistrellus pipistrellus*[1,3] (see Fig. 9-34c) *Pipistrellus pygmaeus*[4] *Pipistrellus* spp.[2] *Plecotus auritus*[1,2]	Same as bat species that might roost in bat box 1FD. When used with the double front panel, same species that might roost in box 1FF.
2FN	*Eptesicus serotinus*[11] *Myotis bechsteinii*[2,3,4,11] *Myotis daubentonii*[1,2,4] *Myotis myotis*[3] (see Fig. 9-35a) *Myotis nattereri*[2,4,11] *Nyctalus lasiopterus*[1,3] (see Fig. 9-35b) *Nyctalus noctula*[1,2,3,4] *Pipistrellus pipistrellus*[11] *Pipistrellus pygmaeus*[11] *Pipistrellus* spp.[2] *Plecotus auritus*[1,2,3,4,11] (see Fig. 9-4b)	*Antrozous pallidus* *Chalinolobus gouldii* *Chalinolobus morio* *Eptesicus fuscus* *Mops condylurus* *Mops midas* *Myotis austroriparius* *Myotis brandtii* *Myotis leisleri* *Myotis lucifugus* *Myotis septentrionalis* *Myotis sodalis* *Myotis thysanodes* *Myotis velifer* *Myotis yumanensis* *Nycteris thebaica* *Nycticeius humeralis* *Nyctophilus bigax* *Nyctophilus geoffroyi* *Nyctophilus gouldi* *Perimyotis subflavus* *Plecotus austriacus* *Pipistrellus nathusii* *Scoteanax rueppellii* *Scotophilus viridus* *Scotorepens balstoni* *Scotorepens orion* *Tadarida australis* *Tadarida brasiliensis* *Vespedelus darlingtoni* *Vespedelus regulus*

* For a description of Schwegler bat boxes and their intended use, see Appendix 9-3.
[1] Schwegler Vogel- & Naturschutzprodukte GmbH Catalog No. 68.
[2] Poulton (2006).
[3] Gianna Dondini and Simone Vergari (pers comm.), Florence, Italy.
[4] Sally Humphreys (pers. comm.), Dorset, England (also, *Nyctalus nyctalus* and *Plecotus auritus* were observed hibernating in bat box 2FN).
[5] Based on published body sizes and roosting habits of species observed occupying bat boxes.
[6] Found roosting in bat box 1FF at Kirtland Airforce Base, Albuquerque, New Mexico USA (Katherine M. Thibault, pers. comm.).

Appendix 9-4 continued

[7] Has been roosting in a homemade bat box year-round at Tolga Bat Hospital, Atherton Tableland, Queensland, Australia (Jennefer Maclean, pers. comm.).

[8] Poulton (2006) did not report if bats were hibernating in bat box 1FW. However, Henry W. Schofield (pers. comm.) reported that it is not uncommon to find individuals or small numbers of all UK species (except *Rhinolophus ferrumequinum, R. hipposideros, Eptesicus serotinus, Myotis bechsteinii,* and *Plecotus austriacus*) in a range of boxes over the winter.

[9] Tuttle and Hensley, 1993.

[10] Bat box 2F, with the double front panel, served as a hibernaculum for this species when the box was placed in a man-made tunnel in Hertfordshire, England (Patty Briggs, pers. comm.).

[11] Steve Laurence (pers. comm.) reported the following roosting events occurring in Wiltshire, England:

- A solitary *Eptesicus serotinus* hibernating in bat box 2FN that was mounted on the wall of a disused railway tunnel.
- Groups of hibernating *Nyctalus noctula* in bat box 1FW.
- Maternity roosts of *Myotis bechsteinii* in bat boxes 1FF and 2FN.
- Maternity roosts of *Myotis nattereri* and *Plecotus auritus* in bat boxes 1FF, 1FW, and 2FN.
- Maternity and mating roosts of *Barbastella barbastellus* in bat box 1FF. It should be noted that the bats had both 1FF and 2FN available to them, but 1FF was preferred.
- Mating roosts of *Pipistrellus pipistrellus* and *P. pygmaeus* in bat boxes 1FF and 2FN.

[12] Hubert Krättli (pers. comm.) reported *Nyctalus noctula* and *Pipistrellus nathusii* hibernating in bat box 1FW in Switzerland.

Appendix I

Measurement Conversion Factors

LENGTH CONVERSIONS

When you know ...	Multiply by ...	To find ...
Centimeters (cm)	0.328	Feet (ft)
Centimeters (cm)	0.3937	Inches (in)
Centimeters (cm)	0.01	Meters (m)
Centimeters (cm)	10.0	Millimeters (mm)
Centimeters (cm)	0.0109	Yards (yds)
Feet (ft)	30.48	Centimeters (cm)
Feet (ft)	12.0	Inches (in)
Feet (ft)	0.3048	Meters (m)
Feet (ft)	304.8	Millimeters (mm)
Feet (ft)	0.333	Yards (yd)
Inches (in)	2.54	Centimeters (cm)
Inches (in)	0.08333	Feet (ft)
Inches (in)	0.0254	Meters (m)
Inches (in)	25.4	Millimeters (mm)
Inches (in)	0.0278	Yards (yd)
Meters (m)	3.2808	Feet (ft)
Meters (m)	39.369	Inches (in)
Meters (m)	1.0936	Yards (yd)
Micrometers (μm)	0.0001	Centimeters (c)
Micrometers (μm)	1000	Nanometers (nm)
Millimeters (mm)	0.1	Centimeters (c)
Millimeters (mm)	0.00328	Feet (ft)
Millimeters (mm)	0.03937	Inches (in)
Millimeters (mm)	0.001	Meters (m)
Nanometers (nm)	0.001	Micrometers (μm)

TEMPERATURE CONVERSIONS

When you know ...	Multiply by ...	To find ...
Degrees Fahrenheit (F)	(°F – 32) x (5/9) = °C	Degrees Celsius (C)
Degrees Celsius (C)	°C (5/9) + 32 = °F	Degrees Fahrenheit (F)

ENERGY CONVERSIONS

When you know ...	Multiply by ...	To find ...
Calories (cal)	4.2	Joules (J)
Joules (J)	0.2388	Calories (cal)

VOLUME CONVERSIONS

When you know ...	Multiply by ...	To find ...
Cups (c)	8.0	Fluid Ounces (fl oz)
Cups (c)	0.2366	Liters (L)
Cups (c)	225.0	Milliliters (dry)
Cups (c)	236.588	Milliliters (liquid)
Deciliters (dL)	0.1	Liters (L)
Drops (gtt)	0.065	Milliliters (ml)
Drops (gtt)	0.01316	Teaspoons (tsp)
Fluid Ounces (fl oz)	0.125	Cups (c)
Fluid Ounces (fl oz)	0.02957	Liters (L)
Fluid Ounces (fl oz)	29.57	Milliliters (ml)
Fluid Ounces (fl oz)	0.03125	Quarts (qt)
Fluid Ounces (fl oz)	2.0	Tablespoons (tbsp)
Fluid Ounces (fl oz)	6.0	Teaspoons (tsp)
Gallons (gal)	128.0	Fluid Ounces (fl oz)
Gallons (gal)	3.785	Liters (L)
Liters (L)	4.227	Cups (c)
Liters (L)	10.0	Deciliters (dL)
Liters (L)	33.814	Fluid Ounces (fl oz)
Liters (L)	0.264	Gallons (gal)
Liters (L)	100	Milliliters (ml)
Liters (L)	1.057	Quarts (qt)
Milliliters (ml)	0.004227	Cups (c)
Milliliters (ml)	0.0338	Fluid Ounces (fl oz)
Milliliters (ml)	0.01	Liters (L)
Milliliters (ml)	0.676	Tablespoons (tbsp)
Milliliters (ml)	0.20	Teaspoons (tsp)
Quarts (qt)	4.0	Cups (c)
Quarts (qt)	0.25	Gallons (gal)
Quarts (qt)	0.9464	Liters (L)
Quarts (qt)	32.0	Ounces (oz)
Quarts (qt)	64.0	Tablespoons (tbsp)
Tablespoons (tbsp)	0.0625	Cups (c)
Tablespoons (tbsp)	14.79	Milliliters (ml)
Tablespoons (tbsp)	0.5	Ounces (oz)
Tablespoons (tbsp)	3.0	Teaspoons (tsp)
Teaspoons (tsp)	0.020833	Cups (c)
Teaspoons (tsp)	4.93	Milliliters (ml)
Teaspoons (tsp)	0.1666	Ounces (oz)
Teaspoons (tsp)	0.3333	Tablespoons (tbsp)

MASS (WEIGHT) CONVERSIONS

When you know ...	Multiply by ...	To find ...
Grains (gr)	64.799	Milligrams ((mg)
Grains (gr)	0.0022857	Ounces (oz)
Grams (g)	15.432	Grains (gr)
Grams (g)	0.001	Kilograms (kg)
Grams (g)	1000000	Micrograms (µg)
Grams (g)	1000	Milligrams (mg)
Grams (g)	0.0353	Ounces (oz)
Kilograms (kg)	2.2046	Pounds (lb)
Kilograms (kg)	1000	Grams (g)
Kilograms (kg)	1000000	Milligrams (mg)
Kilograms (kg)	35.2739	Ounces (oz)
Micrograms (µg)	0.001	Milligrams (mg)
Milligrams (mg)	0.0154	Grains (gr)
Milligrams (mg)	0.001	Grams (g)
Milligrams (mg)	1000	Micrograms (µg)
Ounces (oz)	437.5	Grains (gr)
Ounces (oz)	28.35	Grams (g)
Ounces (oz)	0.02835	Kilograms (kg)
Pounds (lb)	453.59	Grams (g)
Pounds (lb)	0.4536	Kilograms (kg)
Pounds (lb)	16.0	Ounces (oz)

CONCENTRATION CONVERSIONS

When you know ...	Multiply by ...	To find ...
Micromoles (µmol)	1000	Nanomoles (nmol)
Micromoles (µmol)	0.001	Millimoles (mmol)
Millimoles (mmol)	1000	Micromoles (µmol)
Millimoles (mmol)	0.001	Moles (mol)
Moles (mol)	1000	Millimoles (mmol)
Nanomoles (nmol)	0.001	Micromoles (µmol)

Index

www.ingramcontent.com/pod-product-compliance
Ingram Content Group UK Ltd.
Pitfield, Milton Keynes, MK11 3LW, UK
UKHW050418051026
15470UKWH00022B/1681

9 781934 899083